Lecture Notes in Mathematics

A collection of informal reports and seminars
Edited by A. Dold, Heidelberg and B. Eckmann, Zürich

35

N. P. Bhatia
Western Reserve University · Cleveland, Ohio, USA

G. P. Szegö
Università degli Studi di Milano · Milano, Italy

Dynamical Systems: Stability Theory and Applications

1967

Springer-Verlag · Berlin · Heidelberg · New York

Library of Congress Catalog Card Number 67-25757 . Title No. 7355.

PREFACE

This book began as a series of lecture notes of the course given by N.P. Bhatia at the Western Reserve University during the Spring of 1965 and the lecture notes of the courses given by G.P. Szegö at the University of Milan during the year 1964 - 65 and at Case Institute of Technology during the summer of 1965. These courses were meant for different audiences, on one side graduate students in mathematics, and on the other graduate students in systems theory and physics.

However in the process of developing these notes we have found a number of other results of interest which we decided to include (See 1.9, 2.7, 2.8, 2.11, 2.14, 3.3, 3.4, 3.5, 3.7, 3.8, 3.9). Therefore, this monograph is of a dual nature involving both a systematic compilation of known results in dynamical systems and differential equations and a presentation of new Theorems and points of view. As a result, a certain lack of organizational unity and overlapping are evident.

The reader should consider this monograph not as a polished, finished product, but rather as a complete survey of the present state of the art including many new open areas and new problems. Thus, we feel that these notes fit the special aims of this Springer-Verlag series. We do hope that this monograph will be appropriate for a one year graduate course in Dynamical Systems.

This monograph is still devoted to a mixed audience so we have tried to make the presentation of Chapter I (Dynamical Systems in Euclidean Space) as simple as possible, using the most simple mathematical techniques and proving in detail all statements, even those which may be obvious to more mature readers. Chapter 2 (Dynamical Systems in Metric Spaces) is more advanced. Chapter 3 has a mixed composition : Sections 3.1, 3.2, 3.6, 3.7 and 3.8 are quite elementary, while the remaining part of the chapter is advanced. In this latter part we mention many problems which are still in an early developmental stage. A sizeable number of the results contained in this monograph have never been published in book form before.

We would like to thak Prof. Walter Leighton of Western Reserve University, Prof. Mihailo Mesarović of Case Institute of Technology, and Prof. Monroe Martin, Director of Institute for Fluid Dynamics and Applied Mathematics of the University of Maryland, under whose sponsorship the authors had the chance of writing this monograph. We wish to thank several students at our universities, in particular, A. Cellina, P. Fallone, C. Sutti and G. Kramerich for checking parts of the manuscript. We are also indebted to Prof. A. Strauss and Prof. O. Hajek for many helpful suggestions and inspiring discussions and to Prof. J. Yorke for allowing to present his new results in Sec. 3.4. We wish also to express our appreciation to Mrs. Carol Smith of TECH - TYPE Corp., who typed most of the manuscript.

The work of the first author has been supported by the National Science Foundation under Grants NSF-GP-4921 and NSF - GP-7057, while the work of the second author has been sponsored by the CNR, Comitato per la Matematica, Gruppo N° 11, and by the National Foundation under Grant NSF-GP-6114.

The authors

March 1967

TABLE OF CONTENTS

CHAPTER 0

Notation , Terminology and Preliminary Lemmas

0.1 *Notation*

T: topological space

X: metric space with metric ρ

E: real euclidean space of n-dimensions

E^2: the real euclidean plane

G: group

R: set of real numbers.

R^+: non-negative real numbers

R^-: non-positive real numbers

I: set of integers

I^+: set of non-negative integers

I^-: set of non-positive integers

In the sequel, when not otherwise stated, capital letters will denote matrices and sets, small latin letters vector (notable exceptions t,s,k,v and w which have been used to denote real numbers), small greek letters real numbers (notable exception π, which denotes a mapping).

If $x = (x_1, \ldots, x_n) \in E$, $||x||$ will denote the euclidean norm of x i.e.,

0.1.1 $$||x|| = (\Sigma_1^n x_i^2)^{1/2}$$

while

0.1.2 $$|x| = \max (|x_i|, i = 1,\ldots,n)$$

Given two points $x,y \in E$ $\rho(x,y)$ will denote the euclidean distance between x and y, i.e.,

0.1.3 $$\rho(x,y) = [\sum_{1}^{n} {}_i (x_i - y_i)^2]^{1/2}$$

If M is a non-empty subset of X, $x \in X$, and $\alpha > 0$, then we write

0.1.4 $$\rho(x,M) = \inf\{\rho(x,y): y \in M\},$$

0.1.5 $$S(M,\alpha) = \{x \in X: \rho(x,M) < \alpha\},$$

0.1.6 $$S[M,\alpha] = \{x \in X: \rho(x,M) \leq \alpha\},$$

0.1.7 $$H(M,\alpha) = \{x \in X: \rho(x,M) = \alpha\}.$$

$S(M,\alpha)$, $S[M,\alpha]$, and $H(M,\alpha)$ will sometimes be referred to as the open sphere, the closed sphere, and the spherical hypersurface (of radius α about M).

The closure, boundary, complement, and interior of any set $M \subset X$ is denoted respectively by $\bar{M}$, ∂M, $C(M)$, and $I(M)$.

If $\{x_n\}$ is any sequence such that $\lim_{n \to \infty} x_n = x$, then this fact is simply denoted by $x_n \to x$.

We shall frequently be concerned with transformations Q from X to 2^X (the set of all subsets of X). Given $Q: X \to 2^X$, and $M \subset X$, we write

0.1.8 $$Q(M) = \cup\{Q(x): x \in M\}.$$

where

0.1.9 $$\cup\{Q(x): x \in M\} = \bigcup_{x \in M} \{Q(x)\}$$

If $\{Q_i\}$, $i \in I$, is a family of transformations from X to 2^X with I as an index set, then

0.1.10 $$Q = \bigcup\{Q_i : i \in I\}$$

denotes the transformation from X to 2^X defined by

0.1.11 $$Q(x) = \bigcup\{Q_i(x) : i \in I\}.$$

Given two sets $M, N \subset X$, their difference is denoted by $M \setminus N$. Given two maps π_1 and π_2 with $\pi_1 \circ \pi_2$ we will denote the composition map.
ometimes we will use the logic symbols $\exists$, $\in$, $\ni$, $\forall$ and $\Rightarrow$ meaning "there exists", belonging to", "such that", "for all" and "implies".
ometimes the following simplified symbols will be used :

$$\sum_1^n{}_i \triangleq \sum_{i=1}^n$$

nd

$$\bigcup(\gamma(x) : x \in M) \triangleq \bigcup_{x \in M} \gamma(x) .$$

0.2 *Terminology*

0.2.1 *DEFINITION*

Given a compact set $M \subset E$, *a continuous scalar function* $v = \phi(x)$, *defined in an open neighborhood* $N(M)$ *of* M *is said to be positive (negative) semidefinite for the set* M *in the open neighborhood* $N(M)$ *if*

$$\phi(x) = 0 \qquad \textit{for all} \quad x \in M$$

$$\phi(x) \geq 0 \quad (\phi(x) \leq 0) \qquad \textit{for all} \quad x \in N(M) \setminus M$$

If $N(M) = E$, *then the scalar function* $v = \phi(x)$ *is said to be positive (negative) semidefinite for the set* M. *If* $M = \{0\}$ *and* $N(M) = E$, *then the scalar function* $v = \phi(x)$ *is called positive (negative) semidefinite. If for the set* M, *a function* $v = \phi(x)$ *defined in a neighborhood* $N(M)$ *with* $\phi(x) = 0$ *for* $x \in M$ *is not semidefinite, we shall call it indefinite.*

0.2.2 *Remark*

The definition (0.2.1) as well as the following definitions (0.2.4) applies to a slightly larger class of sets than the compact sets, namely for the class of closed sets with a compact vicinity; viz closed sets M, such that for some $\beta > 0$ the set $\overline{S(M,\beta) \setminus M}$ is compact.

0.2.3 *Example*

If X is locally compact, then for sufficiently small $\delta > 0$, the set $C(S[x,\delta])$ is a set with a compact vicinity.

0.2.4 *DEFINITION*

Given a compact set $M \subset E$, *a continuous scalar function* $v = \phi(x)$, *defined in an open neighborhood* $N(M)$ *of* M *is said to be positive (negative) definite for the set* M *in the neighborhood* $N(M)$ *if it is*

$$\phi(x) = 0 \qquad x \in M$$

$$\phi(x) > 0 \quad (\phi(x) < 0) \quad \textit{for all} \quad x \in N(M) \setminus M.$$

If $N(M) = E$, *then the real-valued function* $v = \phi(x)$ *is said to be positive (negative) definite for the set* M. *If* $M = \{0\}$ *and* $N(M) = E$, *then the scalar function* $\phi(x)$ *is called positive (negative) definite.*

0.2.5 *DEFINITION*

A scalar function $\alpha = \alpha(\mu)$ *is called strictly increasing if* $\alpha(\mu_1) > \alpha(\mu_2)$ *whenever* $\mu_1 > \mu_2$, *and it is called increasing if* $\alpha(\mu_1) \geq \alpha(\mu_2)$ *whenever* $\mu_1 > \mu_2$.

0.2.6 *DEFINITION*

Given a scalar function $v = \phi(x)$, *if there exists an increasing function* $\alpha = \alpha(\mu)$ *such that*

0.2.7 $$\alpha(\mu) \to +\infty \quad \textit{as} \quad \mu \to +\infty$$

and such that $\phi(x)$ *satisfies in* E *the inequality*

0.2.8 $$\alpha(\rho(M,x)) \leq |\phi(x)| \quad , \qquad M \textit{ a compact set}$$

then the real-valued function $v = \phi(x)$ *is called radially unbounded for the set* M.

0.2.9 *DEFINITION*

If M *is closed set (not necessarily compact) and the function* $v = \phi(x)$ *satisfies the requirements of definition (0.2.1) (or 0.2.4) then* $\phi(x)$ *is called weakly semidefinite (or weakly definite) for the set* M *in the open set* $N(M)$. *If further* $\phi(x)$ *is defined in* $S(M,\delta)$ *for some* $\delta > 0$, *and if there is a strictly increasing function* $\alpha(\varepsilon), \alpha(0) = 0$, *such that*

0.2.10 $$\alpha(\rho(x,M)) \leq \phi(x), \qquad x \in S(M,\delta),$$

holds, then $\phi(x)$ *is called* *(positive) definite for the set* M *in the neighborhood* $S(M,\delta)$.

0.2.11 *DEFINITION*

If M *is a* closed *set and in the neighborhood* $N(M)(\supset S(M,\alpha))$ *the real-valued function* $v = \phi(x)$ *satisfies the condition*

0.2.12 $$|\phi(x)| \leq \beta(\rho(M,x))$$

where $\beta = \beta(\mu)$ *is an increasing function, then the function* $v = \phi(x)$ *is called* *uniformly bounded for the set* M *in* $N(M)$.

0.2.13 *DEFINITION*

If $M \subset E$ *is a* closed *set and there does not exist an* $\eta > 0$ *such that the real-valued function* $v = \phi(x)$ *is at least weakly semidefinite for the set* M *in the set* $S(M,\eta)$*, then* $\phi(x)$ *will be called* *indefinite for the set* M.

If $M \subset E$ *is a* closed *set, a continuous real-valued function* $\phi(x)$ *which is not at least weakly semidefinite for the set* M *in an open neighborhood* $N(M)$*, will be called* *indefinite for the set* M *in* $N(M)$.

The properties of the scalar function $v = \phi(x)$ can be investigated in two different spaces: the $(n + 1)$ dimensional Euclidean space (v,x) and the n-dimensional Euclidean space (x). In this latter case one actually considers the properties of the sets $\phi(x) = k$ $(-\infty < k < +\infty)$.

0.2.14 *DEFINITION.*

A set D *of real numbers is called relatively dense if there is a* $T > 0$ *such that*

$$D \cap (t - T, t + T) \neq \phi \quad \text{for all} \quad t \in R.$$

0.3 *Preliminary Lemmas*

We shall now state a few obvious properties of definite (or semidefinite) functions both in the space (v,x) and in the space (x). We shall define in the following corollaries properties of real-valued functions with respect to a compact set. The statements are identical in the case of sets with a compact vicinity and weaker when, instead of considering compact sets, one considers closed, non compact sets. In particular, the statements concerning definite functions become statements on weakly definite functions, as it must be obvious to the reader by comparing definitions (0.2.1) and (0.2.4) with the definition (0.2.9).

0.3.1 *LEMMA*

A continuous scalar function $v = \phi(x)$ *is positive (negative) definite for a compact set* M *if* M *is the absolute minimum (maximum) of the function.*

0.3.2 *LEMMA*

A continuous scalar function $v = \phi(x), \phi(x) = 0$ *for* $x \in M$, *is at least semidefinite for the compact set* M *if and only if there does not exist in* E *any hypersurface on which* $\phi(x)$ *changes its sign and it is definite if there does not exist any point* $y \notin M$ *such that* $\phi(y) = 0$.

0.3.3 *LEMMA*

Necessary and sufficient condition for the continuous real-valued function $v = \phi(x)$ *to be positive definite for the compact set* M *in some open neighborhood* N(M) *is that there exists two strictly increasing, continuous functions* $\alpha = \alpha(\mu)$ *and* $\beta = \beta(\mu)$ *such that*

0.3.4 $$\alpha(\rho(M,x)) \leq \phi(x) \leq \beta(\rho(M,x)), \quad \alpha(0) = \beta(0) = 0$$

Proof: The condition (0.3.4) is clearly sufficient. To see the necessity, define

$$\alpha^*(\gamma) = \inf\{\phi(x):\gamma \leq \rho(x,M) \leq \delta\},$$

and

$$\beta^*(\gamma) = \sup\{\phi(x):\rho(x,M) \leq \gamma\},$$

where $\delta > 0$ is such that $N(M) \supset S(M,\delta)$. Then indeed

$$\alpha^*(\rho(x,M)) \leq \phi(x) \leq \beta^*(\rho(x,M)) \qquad ,$$

and $\alpha^*(\gamma)$ and $\beta^*(\gamma)$ are continuous.

Notice that $\alpha^*(\gamma) > 0$, $\beta^*(\gamma) > 0$ for $\gamma \neq 0$ and $\alpha^*(0) = \beta^*(0) = 0$, and the functions $\alpha^*(\gamma)$, $\beta^*(\gamma)$ are increasing. Now, there exists strictly increasing functions $\alpha(\gamma)$ and $\beta(\gamma)$ defined over an interval $0 \leq \gamma \leq \delta' < \delta$, such that

$$\alpha(\gamma) \leq \alpha^*(\gamma) \leq \beta^*(\gamma) \leq \beta(\gamma)$$

and $\alpha(0) = \beta(0) = 0$. For example, $\alpha(\gamma)$ may be chosen as follows. Let $\alpha^*(\delta') = \eta$, $\eta > 0$. Then there is a sequence of points $\gamma_1 > \gamma_2 > \gamma_3 > \dots > 0$, $\gamma_n \to 0$ as $n \to \infty$, such that $\alpha^*(\gamma_n) \geq \frac{\eta}{n}$, and $\gamma_1 = \delta'$. Now define

$$\alpha(\gamma) = \frac{\eta}{n+1} - \frac{\eta(\gamma_n - \gamma)}{(n+1)(n+2)(\gamma_n - \gamma_{n-1})}$$

$$\text{for } \gamma_n \geq \gamma \geq \gamma_{n+1},\ n = 1,2, \quad \dots$$

The existence of $\beta(\gamma)$ may be demonstrated in the same way and (0.3.4) holds with these $\alpha(\gamma)$ and $\beta(\gamma)$. The theorem is proved.

CHAPTER 1

DYNAMICAL SYSTEMS IN A EUCLIDEAN SPACE

1.1 *Definition of a continuous dynamical system.*

1.1.1 *DEFINITION*

A transformation $\pi : E \times R \to E$ *is said to define a dynamical system* (E,R,π) *(or continuous flow* $\mathcal{F}$*) on* E *if it has the following properties:*

i) $\pi(x,0) = x$ *for all* $x \in E$

1.1.2 ii) $\pi(\pi(x,t),s) = \pi(x,t+s)$ *for all* $x \in E$ *and all* $t,s \in R$.

iii) π *is continuous*

For every $x \in E$ the mapping π induces a continuous map $\pi_x : R \to E$ of R into E such that $\pi_x(t) = \pi(x,t)$. This mapping π_x is called the motion through x.

For every $t \in R$ the mapping π induces a continuous map $\pi^t : E \to E$ such that $\pi^t(x) = \pi(x,t)$. The map π^t is called transition (or action).

1.1.3 *THEOREM*

The mapping π^{-t} *defined by*

$$\pi^{-t}(x) = \pi(x,-t)$$

is the inverse of the mapping π^t.

Proof. It must be proved that $(\pi^t)^{-1} = \pi^{-t}$. This can be easily shown by applying to the point $x \in E$ the mapping π^t, then to the image point of $x : y = \pi(x,t)$ the mapping π^{-t}. The image point of y under this mapping:

$z = \pi^{-t}(y)$ must coincide with x. In fact, using axioms (i) and (ii) we have

$$z = \pi^{-t}(\pi(x,t)) = \pi(\pi(x,t),-t) = \pi(x,t-t) = \pi(x,0) = x,$$

which proves the theorem.

1.1.4 *THEOREM*

The mapping π^t *is a topological transformation of* E *onto itself.*

Proof. The map π^t is an onto mapping. In fact, all points $x \in E$ are image points of points $\pi(x,-t) \in E$. For the same reasons the map π^t is one to one. In fact the statement

$$\pi(x,t) = \pi(y,t) = z \qquad x,y,z \in E \qquad t \in R \text{ fixed}$$

implies, by application of the inverse map π^{-t}, that

$$x = y = \pi(z,-t)$$

which shows that π^t is one to one.

Since, by the definition 1.1.1, π^{-t} is obviously continuous the theorem is proved.

As a consequence of this fact, it follows that the dynamical system $\mathcal{F}$ is a one-parameter group of topological transformations, meaning by this that for each value of $t \in R$ a topological transformation is defined and, furthermore, the transformation π^t forms a group. We claim that the set $\{\pi^t\}$, $t \in R$ is a group with the group operation defined by

1.1.5 $$\pi^t \pi^s = \pi^{t+s}$$

Hereby π^o is the identity element and for any π^t, π^{-t} is the inverse. In fact

i) $\pi^t\pi^o = \pi^{t+o} = \pi^t$

ii) $\pi^t\,\pi^{-t} = \pi^{t-t} = \pi^o$ and furthermore

iii) $\pi^t(\pi^s\pi^q) = \pi^t\,\pi^{s+q} = \pi^{t+(s+q)} = \pi^{(t+s)+q} = \pi^{t+s}\,\pi^q = (\pi^t\pi^s)\pi^q$

so that all axioms of a group are satisfied. Notice also that we have in fact a commutative group as:

iv) $\pi^t\,\pi^s = \pi^{t+s} = \pi^{s+t} = \pi^s\,\pi^t$

1.1.6 *A simplified notation*

In most of the following work it will be inessential to distinguish a particular mapping π. When its use will not be misleading, we shall, therefore, introduce the notation xt instead of $\pi(x,t)$. For a fixed t, xt is, therefore, the image point of a point $x \in E$ under the mapping π^t induced by $\mathcal{F}$.

In this simplified notation the first two axioms of (1.1.2) take the following very simple form: $x0 = x$ and $(xt)s = x(t+s)$.

In line with the above notation if $M \subset E$ and $S \subset R$ we define

1.1.7 $$MS = \{xt : x \in M,\ t \in S\} \quad .$$

Whenever M or S is a singleton, namely, $M = \{x\}$ or $S = \{t\}$, we write xS and Mt for $\{x\}S$ and $M\{t\}$ respectively.

1.1.8 *Remark*

One can define dynamical systems in a more general framework as the triplet (T,G,π), where T is a topological space, G a topological group and π the map which satisfies axioms similar to 1.1.2. In this chapter besides (E,R,π) we shall once in a while discuss properties of (E,I,π), where I is the group of integers. The dynamical system (E,I,π) is called a discrete dynamical system or continuous cascade. In the advanced Chapter 2 we shall discuss the more general case of the dynamical system (X,R,π), where X is a metric space and mention more general problems related to the dynamical system (T,G,π), where T is a topological space and G is any topological group.

1.1.9 *Notes and references.*

The introduction of the definition of a dynamical system cannot be attributed to any one person. Some historical remarks on the generation of such concepts can be found in a paper by V. V. Nemytskii [10] and in a paper by G. D. Birkhoff [1, Vol. 2 pg. 710].

The first abstract definitions of a dynamical system can be found in the works of A. A. Markov [1] and of H. Whitney [1,II]. Most concepts have been introduced by Poincaré and his successor, G. D. Birkhoff, in the framework of the theory of dynamical systems defined by ordinary differential equations.

The theory of dynamical systems received new impetus by the publication of the books by Nemytskii and Stepanov, G. T. Whyburn, Gottschalk and Hedlund [4] and Montgomery and Zippin.

1.2 *Elementary concepts.*

1.2.1 *DEFINITION*

For any fixed $x \in E$ *and* $a \leq b \in R$, *the trajectory segment is the set*

1.2.2 $$x[a,b] = \{xt : t \in [a,b]\}$$

For every fixed $x \in E$ *the trajectory or orbit*[1] *through* x *is the set*

1.2.3 $$xR = \{xt : t \in R\}$$

The sets xR^+ *and* xR^- *are respectively called positive and negative semi-trajectory through* x.

By the axioms defining a dynamical system, it follows that:

1.2.4 For all $t \in R$ $\quad xR = (xt)R$

From the properties of π_x it follows that the trajectory segment is a closed and bounded set.

1.2.5 *Remarks on trajectories and motions.*

The trajectory xR is a set, a curve through the point x. Therefore, a trajectory is a purely geometrical concept in which the dependence upon the time does not show. On the whole trajectory xR not even the direction of the motion appears. By direction of motion we mean the direction in which the point $yt : y \in xR$ moves with increasing t on xR. In some cases it may be possible to recognize on a trajectory a positive and a negative direction of motion, that is, the case if one maps xR^+ and xR^- separately

---

(1) Throughout this book the word trajectory will be preferentially used.

for any trajectory which is not closed and bounded. It can be seen in many drawings showing various flows that the trajectories are represented as lines or sets of points with arrows. These arrows show the direction of the motion on xR. In the case of discrete dynamical systems, the trajectory xI is in many cases a disjoint set of points. For this reason in the literature the set xI is very often called a punctual trajectory.

In some parts of these notes, in particular in Chapter 2, the following notation for trajectories and semi-trajectories will be adopted.

1.2.6 $xR \triangleq \gamma(x), \quad xR^+ \triangleq \gamma^+(x), \quad xR^- \triangleq \gamma^-(x).$

The symbols γ, γ^+ and γ^- denote the maps from E to 2^E defined by 1.2.6. Thus the notation $\gamma(x)$ etc. will be adopted when it is desired to emphasize that the trajectory is an element of the maps $E \to 2^E$, while the notation xR will be used when the simple geometrical concept of trajectory is predominant.

The motion π_x through the point $x \in E$ is a mapping which maps R into E or to be more exact maps R onto xR. One can also say that the motion π_x through a point $x \in E$ is the locus of xt for all $t \in R$, parametrized by t. A motion can be visualized as the law with which the point xt moves on xR. In order to be absolutely clear in this basic distinction between the concept of trajectory xR, the concept of motion π_x, we may think of xR as the rail on which a material point moves according to the law π_x.

1.2.7 *DEFINITION*

A point $x \in E$ *having the property that*

$$xR = \{x\}$$

is called *critical* *or* *stationary* *or* *equilibrium* *or* *rest point.*

1.2.8 *REMARK*

Critical points are the fixed points of the mapping $\pi^t: E \to E$. The definition 1.2.7 has defined rest points as a particular type of trajectories. It must be remarked that a critical point can be defined also from the properties of the corresponding motion π_x.

1.2.8 *DEFINITION*

A point $x \in E$ *to which there corresponds a motion* π_x, *having the property that*

$$\pi_x(t) = x(t) = x \qquad \text{for all} \quad t \in R$$

is called a critical point.

Some basic properties of critical points shall now be proved.

1.2.9 *THEOREM*

If for $a < b, a, b \in R, x \in E$

1.2.10 $$x[a,b] = \{x\}$$

x *is a critical point.*

Proof. We shall give the proof for the case of the discrete system (E, I, π). For the case of the theorem a very simple proof shall be given as Corollary 1.2.24. For the case of discrete systems the statement of the theorem could be rephrased as follows: If for an $h_1 \in I$,

1.2.11 $$x(h_1 + 1) = xh_1 \quad ,$$

then for all $h \in I$

$$x(h + 1) = xh$$

and x is a critical point.

In fact, by the axiom 1.1.2

$$x(h + 1) = x(h_1 + 1 + h-h_1) = (x(h_1 + 1))(h-h_1)$$

and because of 1.2.11 it follows that

$$x(h + 1) = (x(h_1))(h-h_1) = x(h_1 + h-h_1) = x(h).$$

1.2.12 *Exercise.*

Prove the analogue of theorem 1.2.9 for (E,R,π).

1.2.13 *THEOREM*

The set of critical points is closed.

Proof. It must be shown that the limit of a sequence $\{x_n\}$ of critical points is a critical point.

From the definition 1.2.7, for all $t \in R$, it follows that $x_n t = x_n$. On the other hand, from the continuity of the mapping defining $\mathcal{F}$, we have that if $x_n \to x$, then $x_n t \to xt$. Thus, $xt = x$ for all $t \in R$.

1.2.14 *THEOREM*

If for every $\varepsilon > 0$ *there exists at least one* $y \in S(x,\varepsilon)$ *such that either* $yR^+ \subset S(x, \varepsilon)$ *or* $yR^- \subset S(x, \varepsilon)$, *then* x *is a rest point.*

Proof. If x is not a critical point, then there is a $\tau > 0$ such that $x \neq xt$ for $0 < t \leq \tau$. Let $\varepsilon > 0$ be such that $x \notin S(x\tau,\varepsilon)$. By continuity of the map π there is a $\delta > 0$ such that $y \in S(x,\delta)$ implies $y\tau \in S(x\tau,\varepsilon)$ and we may indeed assume that $S(x,\delta) \cap S(x\tau,\varepsilon) = \emptyset$. Thus we have proved that if x is not a critical point, then x has a neighborhood which contains no positive semi-trajectory. Similarly one can show that in this case there is a neighborhood of x which contains no negative semi-trajectory. This proves the theorem.

The following theorems can be regarded as the same as theorem 1.2.14 from the point of view of the motions near a critical point. The proof is analogous to 1.2.14.

1.2.16 *COROLLARY* of 1.2.14

If

1.2.17 $$\lim_{t \to \infty} yt = x \qquad y, x \in E$$

then x *is an equilibrium point.*

As a consequence of the properties of motions defining a critical point, it is easy to prove that:

1.2.18 *THEOREM*

No motion reaches a critical point within a finite value of t.

Proof. Assume for contradiction that $y\tau = x$ where $y \neq x$ and x is a critical point. By applying the inverse map π^{-1} we obtain $y = x(-\tau)$, which contradicts the hypothesis that $xt = x$ for all $t \in R$.

The last theorem even if it is just a restatement of properties already emphasized in proving that π^t is a topological transformation gives a very intuitive interpretation of the basic property of equilibrium point from the point of view of the motions.

The next concept characterizes certain properties of some motions.

1.2.19 *DEFINITION*

A motion which for all $t \in R$ *and some* $\tau \neq 0$ *satisfies the condition*

1.2.20 $$\pi_x(t + \tau) = \pi_x(t)$$

is called <u>*periodic*</u>.

By axiom (ii) of the definition of dynamical system it follows that (1.2.20) implies that, for all integers n,

1.2.21 $$\pi_x(t + n\tau) = \pi_x(t)$$

The smallest positive number τ satisfying the property 1.2.20 is called the <u>period</u> of the periodic motion π_x. If a periodic motion <u>does not have such a least period</u> τ, then π_x defines a <u>critical point</u>.

It is immediate to prove that

1.2.22 *THEOREM*

If there exist at least one $s \in R$ *and one* $\tau \neq 0, \tau \in R$ *such that*

1.2.23 $$\pi_x(s + \tau) = \pi_x(s),$$

then π_x *is periodic.*

Proof. We need only to prove that 1.2.23 implies 1.2.20. For any $t' \in R$ we have, using axiom 1.1.2 and the equality 1.2.23,

$$\pi(\pi(x,s),t') = \pi(x,s+t') = \pi(\pi(x,s+\tau),t') = \pi(x,s+t'+\tau),$$

so that for arbitrary $t' \in R$, 1.2.20 holds with $t = s+t'$.

1.2.24 *COROLLARY* of Theorem (1.2.22)

If $x[a,b] = \{x\}$, $a < b$, x *is a critical point.*

Proof. Since $xa = x$, it follows from 1.2.22 that π_x is periodic. For all τ with $0 \leq \tau \leq b-a$ it is also $x(a + \tau) = x$. Thus the periodic motion π_x does not have a least period and, therefore, it defines a critical point. Q.E.D.

1.2.25 *THEOREM*

The set of all periodic orbits with period $\tau \in [0,T], T \in R^+$ *is a closed set.*

The proof of this theorem is quite simple and it is left as an exercise to the reader. Theorem 1.2.25 is essentially a generalization of Theorem 1.2.13. Notice that the set above is not necessarily compact as can be shown by considering the linear second order oscillation, represented by the equations

$$\begin{cases} \dot{x}_1 = x_2 \\ \dot{x}_2 = -x_1 \end{cases}$$

whose solutions define periodic orbits with the same period completely filling the plane E^2.

Since the property 1.2.20 is a property which holds for all $t \in R$, it is possible to characterize the trajectory associated with a periodic motion as the set of all $x \in E$, such that for all $t \in R$

1.2.26 $$x(t + \tau) = x(t)$$

We can, in addition, say that the trajectory associated with a periodic motion is closed and bounded. This statement will be stressed in following theorems 1.3.21, 1.3.25 and 2.5.12.

1.2.27 *Remark on trajectory segments.*

If x is not a rest point and xR is not a periodic orbit, then ,for all $a,b \in R$ $a \neq b$, the trajectory segments $x[a,b]$ are arcs (homeomorphic images of segments). In the case that xR is a periodic orbit $x[a,b]$ is still an arc for all $a,b \in R$ with $a < b$ and $(b-a) < \tau$, where τ is the period.

1.2.28 *Remark on discrete systems.*

The concept of periodic solutions is very particular for the case of a discrete system. Assume in fact that a continuous dynamical system $\mathcal{F}$ is given, and suppose that this dynamical system presents some periodic motions. Consider now the same system as a discrete system, then the trajectories instead of curves will be disjoint sets of points. Now, besides the very pathological case that all the periods of all periodic motions have the form $z_i = m_i \xi$ where m_i is any integer and ξ is a real number, the motions which are periodic in the continuous case will not remain periodic in the discrete case. In this case, of course, we allow ourselves the freedom to choose ξ as a scaling factor in the transformation of the system from R to I.

The concept of critical point can be imbedded in the concept of invariant set which will be defined next.

1.2.29 *DEFINITION*

A set $M \subset E$ *is called* *invariant* *if under all transformations of the group* $\{\pi^t\}$ *it is transformed into itself. That is, if for all* $t \in R$

1.2.30 $$\pi^t(M) = M .$$

With a more compact notation it will be said that $M \subset E$ *is invariant if*

$$MR = M .$$

Similarly, one can define *positively* *and* *negatively* *invariant sets* $M \subseteq E$ *as sets having the properties* $MR^+ = M$ *and* $MR^- = M$ *respectively. (Obviously, since if* $x \in M$, *then* $xR \subseteq MR$). *Sets which are either positively or negatively invariant will sometimes be called* *semi-invariant.*

1.2.31 *THEOREM*

The set $M \subset E$ *is invariant if and only if* $x \in M$ *implies*

1.2.32 $$xR \subseteq M.$$

Thus, if a point x belongs to an invariant set M, then the whole trajectory xR belongs to M. This is equivalent to saying that invariant sets consist of whole trajectories.

The property of invariance of a set $M \subset E$ within the flow $\mathcal{F}$ has been presented as a property of a set of trajectories. Obviously, such an invariant set can also be defined by the property of the motions associated with its points. In fact,

1.2.33 *THEOREM*

A set M *is invariant if and only if all the motions* π_x *defined by points* $x \in M$ *have the property that for all* $t \in R$

1.2.34 $$\pi_x(t) \in M$$

1.2.35 *Examples of invariant sets.*

We shall discuss their properties later. All trajectories, and therefore, critical points, defined by periodic motions are invariant sets. Invariant sets may have the same dimension as the space in which the dynamical system is defined such as for instance the disc defined by the interior of a closed and bounded trajectory for the case of a second order system. At the limit the whole space is an invariant set.

Invariant sets are extremely important in the theory of dynamical systems as it will be shown in Section 1.5. In the following sections, we shall be mostly concerned with the study of their properties. Given an invariant set

M of dimension less than that of the space E in which a dynamical system $\mathcal{F}$ is defined, it is possible to define only on M a new dynamical system $\mathcal{F}'$ whose trajectories coincide with those defined by $\mathcal{F}$ on M. This technique will be extensively used and better clarified while dealing with the analysis of dynamical systems with a given structure, for instance, for the case of dynamical systems defined by differential equations.

The following basic theorem will clarify the structure of invariant sets:

1.2.35 *THEOREM*

If a set $M \subset E$ *is invariant, then both* ∂M *and* $I(M)$ *are invariant.*

The proof (1.2.38) of this theorem is based upon the two following lemmas.

1.2.36 *LEMMA*

If the sets $M_i \subset E(i = 1, \cdots, n)$ *are invariant then the sets* $\cup_i M_i, \cap_i M_i$, *and* $M_j \setminus M_i (j = 1, \cdots, n)$ *are invariant.*

Proof. That $\cup_i M_i$ are invariant sets is obvious. Consider now the set $P = \cap_i M_i$ which is a subset of all sets M_i. Thus for all $x \in P$ $x \in M_i$ for all M_i. From the hypothesis it then follows that $xR \subset M_i$ for all i, thus $xR \subset \cup_i M_i = P$ for all $x \in P$. Thus because of Theorem 1.2.31 P is invariant. Finally, if $M_j \setminus M_i$ were not invariant, then there would exist a $t \in R$ and an $x \in M_j \setminus M_i$ with $xt = y \in M_i$. Since M_i is invariant $x = y(-t) \in M_i$, which is a contradiction. This proves the lemma.

1.2.37 *LEMMA*

If the set $M \subset E$ *is invariant, also* $\bar{M}$ *is invariant.*

Proof. It must be shown that for all $x \in \partial M$, $xR \subset M$. In fact, consider the sequence $\{x_n\} : x_n \in M; x_n \to x$, because of the continuity axiom it follows that for all $t \in R: x_n t \to xt$. Since the invariance of M implies that $x_n t \in M$, it follows that $xt \in \bar{M}$. From the definition 1.2.29 the lemma follows.

1.2.38 *Proof of the Theorem* 1.2.35

Since M is invariant also $C(M)$ is invariant and so are $\bar{M}$ and $\overline{C(M)}$. Consequently $\partial M = \bar{M} \cap \overline{C(M)}$ and $IM = \bar{M} \setminus \partial M$ are invariant.

Since, for $M \subset E$ if ∂M is invariant M is obviously invariant from Theorem 1.2.35 it follows that:

1.2.39 *COROLLARY* of Theorem 1.2.35

Let $M \subset E$. *If* ∂M *is invariant, then* $\bar{M}$ *is invariant. If* M *is either closed or open and* ∂M *is invariant, then* M *is invariant.*

It's easy to show that if M is neither closed nor open, then from the fact that ∂M is invariant it does not necessarily follow that M is invariant.

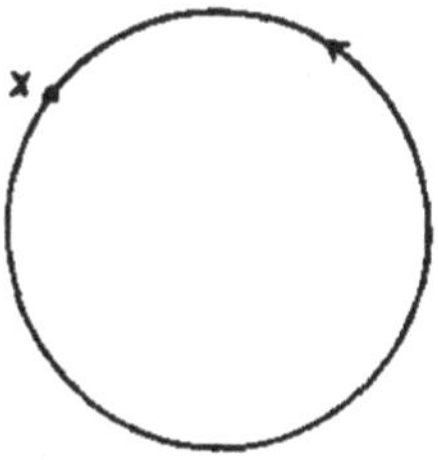

Consider, for instance the flow in E^2 shown aside where the orbit xR is periodic. Let M be the interior of the disc. Consider the set $M \cup \{x\}$. The boundary of which is xR and is invariant. Clearly, however, $M \cup \{x\}$ is not invariant.

A weaker version of Theorem 1.2.35 will be proved next for the case of semi-invariant sets.

1.2.40 *THEOREM*

If $M \subset E$ *is a positively (negatively) invariant set, then both* IM *and* $\bar{M}$ *are positively (negatively) invariant sets.*

Proof. If IM is not positively inv·iant then there is an $x \in IM$ and a $t > 0$ such that $xt \in \partial M$. There is then a sequence $\{x_n\}, x_n \notin M$, $x_n \to xt$. Consider the sequence $\{x_n(-t)\}$. Clearly $x_n(-t) \to xt(-t) = x(t-t) = x0 = x$. Since $x \in IM$ we have $x_n(-t) = y_n \in IM$ for sufficiently large n, but $x_n = y_n t \notin M$, which contradicts the positive invariance of M, and proves the theorem.

The proof of the second assertion is left as an exercise.

Obviously if M is positively invariant, but not invariant, ∂M does not necessarily have the same invariance properties as IM.

We shall now see what properties of equilibrium point (Theorems 1.2.13,14 and 18) are extendable to the more general case of invariant sets. Theorem 1.2.13 obviously does not have any meaning for the case of invariant sets since all trajectories are invariant sets. Theorem 1.2.14 does not hold for the case of invariant sets and it is incorrect also in the case of compact invariant sets. It is, furthermore, easy to show that the conditions of this theorem for the case of compact invariant sets do not even imply that the set M is either positvely or negatively invariant. On the contrary it is easy to extend theorem 1.2.18 to the case of invariant sets. In fact, since an invariant set $M \subset E$ consists of complete trajectories, the statement that a point $x \notin M$ reaches M in a finite time is equivalent to the existence of a $t \in R$ such that $xt \in M$. As M is invariant, this implies that $x \in M$ which is a contradiction. We have then proved that:

1.2.41 *THEOREM*

No invariant set $M \subset E$ *is reached by a point* $x \notin M$ *in a finite time.*

Next we shall introduce an important subclass of invariant sets: minimal sets:

1.2.42 *DEFINITION*

A set $Q \subseteq E$ *is called* <u>*minimal*</u> *if it is non-empty, closed and invariant and does not have any proper subset with these three properties.*

1.2.43 *Examples of minimal sets*

Equilibrium points are (compact) minimal sets. Compact trajectories not containing equilibrium points are minimal sets. One can also construct dynamical systems which have the surface of a torus as a minimal set.

The class of noncompact minimal sets in E^2 contains only one element as shown by the following:

1.2.44 *THEOREM*

A minimal set $M \subset E^2$ *consists of a single trajectory.*

This theorem will be proved in (1.3.26).

The interest for compact minimal sets is, on the other hand, justified by the following:

1.2.45 *THEOREM*

Every non empty compact invariant set $M \subset E$ *contains some minimal set .*

Proof. If M itself is minimal, the theorem is proved. If M is not minimal then there exists a closed set $M_1 : M_1 \subset M$ which is invariant. If M_1 is not minimal, then there exists a closed set $M_2 \subset M_1$ which is invariant. The set M of all closed and invariant subsets of M is thus clearly a partially ordered set by the relation $\subset$. Since E is complete the intersection of any chain is non-empty, closed and invariant and thus is an upperbound in that ordering. Therefore, by Zorn's Lemma it has a maximal element which is a

1.2.46 *THEOREM*

A set $M \subset E$ *is minimal if and only if for each* $x \in M$, $\overline{xR} = M$.

Proof. Let M be minimal, then for each $x \in E$, it is $\overline{xR} = M$. Since $\overline{xR}$ is nonempty, closed and invariant and it cannot be a subset of M, $\overline{xR} = M$.

Now suppose, for each $x \in M$, $\overline{xR} = M$. If M were not minimal, then it would contain a nonempty, closed, invariant proper subset N. Then for $x \in N$, $\overline{xR} \subset N \subset M$ since N is closed and invariant. Thus $\overline{xR} \neq M$ which contradicts the assumption and proves the theorem.

1.2.47 *THEOREM*

If a minimal set $M \subset E$ *has an interior point, then all its points are interior points.*

Proof. Let $x \in M$ be an interior point of M. Then there exists a $\delta > 0$ such that $S(x,\delta) \subset M$. For each $t \in R$ $S(x,\delta)t \subset M$ and $S(x,\delta)t$ is a neighborhood of xt. Thus if one point of a trajectory in M is an interior point of M every point of that trajectory is an interior point of M. Now $\overline{xR} = M$ (Theorem 1.2.46) and let $y \in \overline{xR} \setminus xR$. Then indeed $\overline{yR} = \overline{xR} = M$ and there exists a point $z \in yR$ such that $z \in S(x,\delta)$. But then z is an interior point of M and so is y as $y \in zR = yR$. Q.E.D.

Additional properties of minimal sets will be given in sections (1.3.23, 1.3.26)

1.2.48 *Notes and references*

The definition of minimal sets is due to G. D. Birkhoff [1, Vol. 1 pp. 654-672]. Notice that the definition given there (M is minimal if $x \in M$ implies $\Lambda^+(x) = \Lambda^-(x) = M$) applies only for compact sets.

The proof given here of Theorem 1.2.45 is different from the one given by Nemytskii and Stepanov. The proof of Zorn's Lemma can be found, for instance, in the book by Dugundji[1, pg 31].

Theorem 1.2.47 is attributed by Nemytskii and Stepanov [1] to G.T. Tumarkin.

1.3 *Limit sets of trajectories*

The concept of limit sets is one of the most useful concepts in the theory of dynamical systems. The existence or absence of limit sets, their location and their properties will characterize the asymptotic properties of trajectories and motions and will provide us of one of the basic tools for our analysis of dynamical systems. In fact, limit sets and their properties will allow us to give a complete qualitative description of the behavior of dynamical systems.

1.3.1 *DEFINITION*

A point $y \in E$ *is called a positive (or omega) limit point of a point* $x \in E$ *if there exists a sequence* $\{t_n\}: t_n \in R^+; t_n \to +\infty$ *such that*

1.3.2 $$xt_n \to y$$

The set of all positive limit points of a point $x \in E$ *will be called the positive limit set of* x *and denoted by* $\Lambda^+(x)$. *Thus*

1.3.3 $$\Lambda^+(x) = \{y \in E: \exists \{t_n\} \subset R^+ \text{ such that } t_n \to +\infty \text{ and } xt_n \to y\}$$

The set of all positive limit points of all points $x \in B \subset E$ *will be called the positive limit set of the set* B *and denoted* $\Lambda^+(B)$. *Thus*

1.3.4 $$\Lambda^+(B) = \bigcup \{\Lambda^+(x): x \in B\}$$

Similarly one can define negative (or alpha) limit points y *of a point* $x \in E$ *and the negative limit set* $\Lambda^-(x)$ *as follows:*

1.3.5 $$\Lambda^-(x) = \{y \in E: \exists \{t_n\} \subset R^- \text{ such that } t_n \to -\infty \text{ and } xt_n \to y\}$$

Similarly to 1.3.4 *the negative limit set of a set* $B \subset E$ *can be defined as*

1.3.6 $$\Lambda^-(B) = \bigcup\{\Lambda^-(x) : x \in B\}.$$

If a point $x \in E$ has a limit point y, this limit point is common to all points of xR. In fact, if there exists a sequence $\{t_n\}; t_n \in R^+$; $t_n \to +\infty$ such that $xt_n \to y$ then for all $\tau \in R^+$ it is also true that $x\tau(t_n - \tau) = xt_n \to y$ with $t_n - \tau \to +\infty$.

That of having a limit point can be regarded both as a property of the motion π_x defined by x as well as a property of the corresponding trajectory.

1.3.7 *REMARK*

By the preceding considerations, the definition 1.3.3 of a positive limit set can be shown to be equivalent to

1.3.8 $$\Lambda^+(x) = \bigcap_{t \in R} \overline{(xt)R^+}$$

and similarly for the negative limit sets.

1.3.9 *Examples of limit sets.*

i) Consider the planar system in polar coordinates

$$\dot{r} = r(1-r) \qquad \left(\cdot = \frac{d}{dt}\right),$$
$$\dot{\theta} = 1.$$

The phase portrait is easily seen to consist of a closed trajectory γ_1 coinciding with the unit circle $r = 1$, a point trajectory given by $r = 0$, and trajectories which approach spirally to the unit circle as $t \to +\infty$. Further, trajectories in the interior of the unit circle all approach the point $r = 0$ as $t \to -\infty$ (see *Figure* 1.3.10). The closed trajectory is periodic and is the positive limit set of all trajectories (including itself) except the point trajectory $r = 0$. The point trajectory $r = 0$

is the negative limit set of all trajectories in the interior of the unit circle (including itself), and the positive limit set of the point trajectory $r = 0$ is the point $r = 0$, and the negative limit set of the periodic trajectory γ is the unit circle.

1.3.10 *Figure*

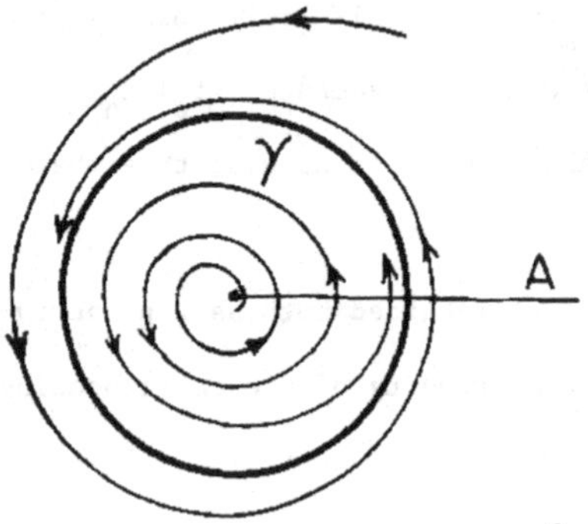

ii) Consider in the space E^2 the differential equations

$$\dot{\rho} = \frac{\rho}{1+\rho}$$

$$\dot{\theta} = \frac{1}{1+\rho}$$

which define solutions of the form $\rho = ce^{\theta}$, where ρ and θ are polar coordinates, in the whole space E. We shall now map the plane E^2 on the strip $-1 < x < 1$ by the transformation

$$\chi = \frac{x}{1-x^2} \qquad Y = y \quad .$$

In these new variables, the given differential equations take the form

1.3.11 *Figure*

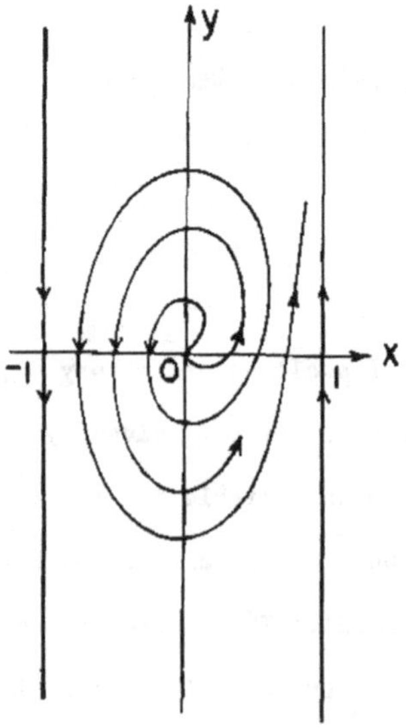

$$\dot{x} = \frac{x(1 - x^2) - y(1 - x^2)^2}{(1 + x^2)(1 + \rho)}$$

$$\dot{y} = \frac{y}{1 + \rho} + \frac{x}{1 - x^2} \cdot \frac{1}{1 + \rho}$$

We complete now the space with the straight lines $x = \pm 1$ (*Figure* 1.3.11) and the corresponding limiting equations.

$$\dot{x} = 0 \qquad \dot{y} = \pm 1.$$

The dynamical system so defined has the two straight lines $x = \pm 1$ as positive limit sets of all points in the strip $-1 < x < 1$.

1.3.12 *REMARK*

All the following theorems on the positive limit set $\Lambda^+(x)$, hold with obvious variations also for the case of the negative limit set $\Lambda^-(x)$.

1.3.13 *THEOREM*

For every $x \in E$

i) $\Lambda^+(x)$ *is closed and invariant*

ii) $\overline{xR^+} = xR^+ \cup \Lambda^+(x)$

iii) *If* $\Lambda^+(x)$ *is bounded it is connected, hence it is a continuum.*

iv) *If* $\Lambda^+(x)$ *is not bounded, none of its components is bounded.*

Proof. i) $\Lambda^+(x)$ is closed. Consider the sequence $\{y_k\}: y_k \in \Lambda^+(x); y_k \to y$. It must be shown that $y \in \Lambda^+(x)$. For each k there exists a sequence $\{t_n^k\}; t_n^k \to +\infty$ with $xt_n^k \to y_k$. We may assume without loss of generality that

$\rho(xt_k^k, y_k) < 1/k$, holds for each k, and $t_k^k > k$. Then considering the sequence $\{t_n\}$, where $t_n = t_n^n$, we see that $t_n \to +\infty$, and $xt_n \to y$, since

$$\rho(xt_n, y) \leq \rho(xt_n, y_n) + \rho(y, y_n) \leq \frac{1}{n} + \rho(y, y_n)$$

so that $y \in \Lambda^+(x)$.

Λ^+ is invariant. Consider the sequence $\{t_n\}: t_n \in R^+; t_n \to +\infty;$ $xt_n \to y \in \Lambda^+(x)$, it must be shown that $yR \subset \Lambda^+(x)$. Consider the point $y\tau$ where $\tau \in R$ is arbitrary and fixed. From the continuity axiom

$$x(t_n + \tau) \to y\tau \in \Lambda^+(x)$$

which holds from all τ. Thus $yR \subset \Lambda^+(x)$.

ii) Is obvious.

iii) Assume that $\Lambda^+(x)$ is compact, but not connected. Thus $\Lambda^+(x) = P \cup Q$ where $P, Q \subset E$ are compact and disjoint. Then there exists $\varepsilon, \eta > 0$ such that $S[P, \varepsilon] \cap S[Q, \eta] = \emptyset$. Let $y \varepsilon P$ and $z \varepsilon Q$.

From the definition 1.3.1 there are sequences $\{t_n\}: t_n \to +\infty$ and $\{\tau_n\}: \tau_n \to +\infty$ such that $xt_n \to y$ and $x\tau_n \to z$.

Assume, (if necessary by choosing suitable subsequences) that $\tau_n - t_n > 0$ for each n. Then for each n there exists a sequence $\{t_n'\}: t_n' \to +\infty$ such that

$$t_n < t_n' < \tau_n$$

such that because of the continuity of π^t

$$xt_n' \in \partial S[P, \varepsilon] \quad .$$

Since $\partial S[P, \varepsilon]$ is compact we may assume $xt_n' \to w \in \partial S[P,\varepsilon]$. Thus $w \in \Lambda^+(x)$, which is a contradiction since $\varepsilon > 0$.

iv) The proof of this will be given in 2.2.11. The following is an elementary indication of how the proof will be set up. For that we shall map the Euclidean space E on a spherical hypersurface H in the n + 1 dimensional Euclidean space. This mapping is an obvious generalization of the well-known idea of mapping a plane of the spherical hypersurface $\tilde{E}^2$ in the 3-space, by means of a family of straight lines through the point ω, which is the point of the sphere with the maximum distance from the plane E^2 (*Figure* 1.3.14). Each point on the surface $\tilde{E}^2$ of the sphere will correspond to one point of the plane E^2, with exception

1.3.14 *Figure*

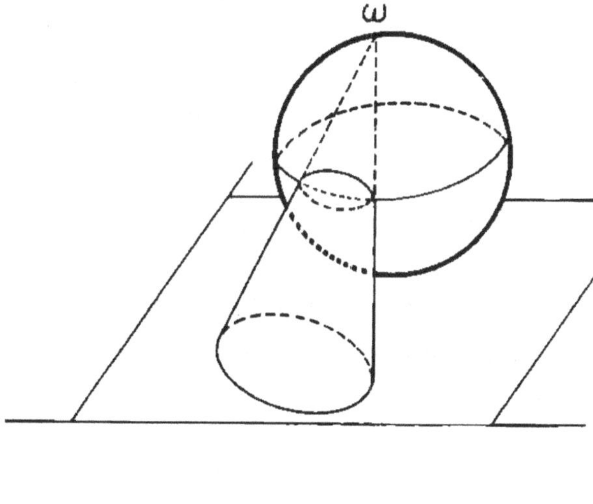

of the point ω which is the image point of all points at infinity of the plane E^2. We can then write that $\tilde{E}^2 = E^2 \cup \{\omega\}$ and for the n-dimensional case $\tilde{E} = E \cup \{\omega\}$. It must be noticed that $\tilde{E}$ is compact and that if

$\{\omega\} \in \tilde{A} \subset \tilde{E}$, the corresponding set $A \subset E$ is not compact in E. If $\pi: E \times R \to E$ is the mapping which defines the dynamical system on E, let $\tilde{\pi}: \tilde{E} \times R \to \tilde{E}$ be defined by the rules $\tilde{\pi}(x,t) = \pi(x,t)$ for all $x \in E, t \in R$ and $\tilde{\pi}(\omega,t) = \omega$. Assume that $x \in E$, $\Lambda^+(x)$ is not compact, $\tilde{\Lambda}^+(x) = \Lambda^+(x) \cup \{\omega\}$. Let $\Lambda^+(x)$ be nonconnected. It must be proved that no component of $\Lambda^+(x)$ is compact in E.

Since $\tilde{E}$ is compact $\tilde{\Lambda}^+(x) \subset \tilde{E}$ is compact and because of (iii) it is connected. On the other hand

$$\Lambda^+(x) = \tilde{\Lambda}^+(x) \setminus \{\omega\}$$

being a complement of a closed set it is open. Since $\tilde{\Lambda}^+(x)$ is connected and $\Lambda^+(x)$ is nonconnected $\{\omega\}$ is the (connecting) limit point of all the nonconnected components of $\Lambda^+(x)$. Thus no component of $\Lambda^+(x)$ is compact in E.

From the result (ii) of Theorem 1.3.13 it follows

1.3.15 *COROLLARY*

If xR^+ *is compact* $\Lambda^+(x)$ *is a continuum.*

The inverse of this statement presented in the next theorem requires a complete proof and it is not true for general metric space (see Theorem 2.2.13).

1.3.16 *THEOREM*

If $x \in E$ *and* $\Lambda^+(x)$ *is compact* $\overline{xR^+}$ *is compact.*

Proof. For every $\varepsilon > 0$, the set $S[\Lambda^+(x),\varepsilon]$ is compact, and there exists a $\tau = \tau(\varepsilon) > 0$, such that $(x\tau)R^+ \subset S[\Lambda^+(x), \varepsilon]$. For otherwise there must exist a sequence $\{t_n\}, t_n \to +\infty$ with $xt_n \in \partial S(\Lambda^+(x),\varepsilon)$. Since $\partial S(\Lambda^+(x),\varepsilon)$ is compact we may assume that $xt_n \to y \in \partial S(\Lambda^+(x),\varepsilon)$. Hence $y \in \Lambda^+(x)$ and also $y \in \partial S(\Lambda^+(x),\varepsilon)$, which is absurd. Then $xR^+ = x[0,\tau] \cup (x\tau)R^+$. Thus $\overline{xR^+} = x[0,\tau] \cup \overline{(x\tau)R^+}$. But $x[0,\tau]$ is compact and so is $\overline{(x\tau)R^+}$ because it is a closed subset of $S[\Lambda^+(x),\varepsilon]$. Hence $\overline{xR^+}$ is compact. Q.E.D.

1.3.17 *THEOREM*

If $x \in E$ *and* $\Lambda^+(x)$ *is compact and non empty, then*

1.3.18 $$\lim_{t \to +\infty} \rho[xt,\Lambda^+(x)] = 0 \quad .$$

Proof. If 1.3.18 were false, there could be found a sequence $\{t_n\}$; $t_n \in R^+; t_n \to +\infty$ and a $\gamma > 0$ such that

1.3.19 $$\rho(xt_n,\Lambda^+(x)) \geq \gamma > 0$$

The sequence $\{xt_n\}:xt_n \in \overline{xR^+}$, is such that $xt_n \to y \in \Lambda^+(x)$, since $\overline{xR^+}$ is compact (1.3.16). On the other hand from 1.3.19 it follows that

$$\rho[y,\Lambda^+(x)] \geq \gamma > 0.$$

This contradiction proves the theorem.

1.3.20 *REMARK*

If the set $\Lambda^+(x)$ is not compact the statement of Theorem 1.3.17 is incorrect. This can be seen, for instance, in the example 1.3.9 (ii) where $\Lambda^+(x)$ is noncompact and the limit 1.3.18 does not exist. The next theorems will relate the properties of the limit sets to those of periodic orbits and in general of minimal sets.

1.3.21 *THEOREM*

If for $x \in E, \pi_x$ *defines a periodic motion, then*

1.3.22 $$\Lambda^+(x) = \Lambda^-(x) = xR = \overline{xR}.$$

Proof. Let $y = xt_0 = x(t_0 \pm n\tau)$. Since $\lim_{n \to \infty} t_0 \pm n\tau = \pm \infty$, $y = \lim_{n \to \infty} x(t_0 \pm n\tau)$. Thus $y \in \Lambda^+(x)$ and $y \in \Lambda^-(x)$, i.e., $xR \subset \Lambda^+(x)$ and $xR \subset \Lambda^-(x)$.

To show for example that $\Lambda^+(x) \subset xR$, let $z \in \Lambda^+(x)$, i.e., $xt_n \to z$. Represent each t_n as $t_n = k\tau + t_n'$ where k is an integer and $0 \leq t_n' < \tau$. The bounded sequence $\{t_n'\}$ can be assumed convergent without loss of generality. Let $t_n' \to t_0$. Note now that

$$xt_n = x(k\tau + t_n') = xt_n' \to xt_0 \in xR$$

Thus $\Lambda^+(x) \subset xR$. The fact that $\Lambda^-(x) \subset xR$ is proved similarly Q.E.D.

From Definition 1.2.42 and Theorem 1.3.13 (ii), it follows that

1.3.23 *THEOREM*

If $x \in E$ *and* xR *is minimal, then either* $\Lambda^+(x) \cup \Lambda^-(x) = \emptyset$ *or* $(\Lambda^+(x) \cup \Lambda^-(x)) \subseteq xR$.

Proof.

$$xR = \overline{xR} = xR \cup (\Lambda^+(x) \cup \Lambda^-(x)).$$

1.3.24 *THEOREM*

Let $x \in E$ *and such that* $\Lambda^+(x) = \Lambda^-(x) = \emptyset$, *then* xR *is minimal.*

Proof. Clearly xR cannot contain any proper invariant subset. It remains to be proved that $xR = \overline{xR}$. Consider for that any sequence $\{x^n\}: x^n \in xR$. There exists then a sequence $\{t_n\}: t_n \in R$ such that $x^n = xt_n$. Let $x^n \to y$. The sequence $\{t_n\}$ must be bounded for otherwise $y \in \Lambda^+(x) \cup \Lambda^-(x)$ which is absurd. Since $\{t_n\}$ is bounded, we may assume that $t_n \to \tau$, then $x^n \to x\tau \in xR$, hence $y = x\tau$ and $xR = \overline{xR}$, which proves the theorem.

1.3.25 *THEOREM*

If $x \in E$ *is not an equilibrium point, then necessary and sufficient conditions for* xR *to be periodic is that* $xR = \overline{xR} = \Lambda^+(x) = \Lambda^-(x)$.

The proof of this theorem is given as 2.5.12.

The class of noncompact minimal sets in E^2 contains only one element as shown by the following:

1.3.26 *THEOREM*

If the minimal set $M \subset E^2$ *is not compact, then it consists of a single trajectory with empty positive and negative limit sets.*

Proof. If M is minimal and $\Lambda^+(x) \neq \emptyset$, then $x \in M$ implies $\overline{\gamma(x)} = M = \Lambda^+(x)$. On the other hand, $x \in \Lambda^+(x)$ implies that xR is periodic. In fact if $\gamma(x)$ is not periodic (and, in particular, not a rest point) consider any transversal ℓ through x (*Figure* 1.3.27 i). Then the trajectory $\gamma(x)$ must intersect ℓ

at a point $y \neq x$, since $x \notin \Lambda^+(x); \gamma^+(y)$ lies either inside or outside of the Jordan curve consisting of the arc of the trajectory $\gamma(x)$ between x and y and the transversal between x and y. In this case $x \notin \Lambda^+(x)$, since yR^+ cannot meet ℓ between x and y. Similarly we can dispose of the other case (*Figure* 1.3.27 ii). Then $\gamma(x)$ is periodic and M is compact which contradicts the assumption and proves the theorem.

1.3.27 *Figure*

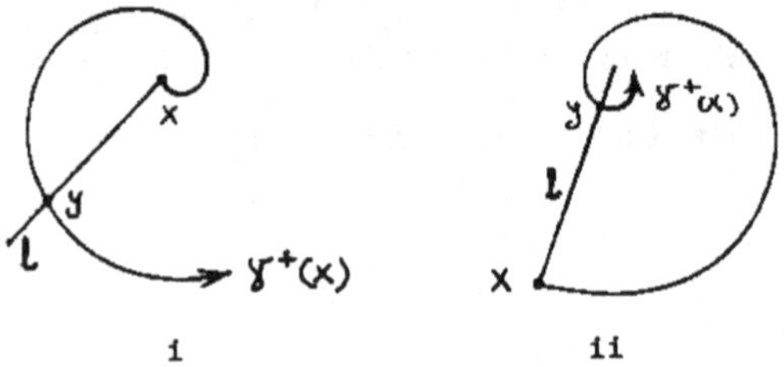

1.3.28 *REMARK*

Theorem 1.3.26 is false for $n > 2$. This will be shown in 2.9.13.

In some cases it may be important to distinguish the way with which trajectories tend to the limit set. This can be done by introducing the concept of asymptotical trajectory.

1.3.29 *DEFINITION*

A trajectory xR *is called* *__positively asymptotic__* *if*

1.3.30 $$xR^+ \cap \Lambda^+(x) = \emptyset \quad .$$

From the result (ii) of Theorem 1.3.13 it follows that in this particular case

1.3.31 $$\Lambda^+(x) = \overline{xR^+} \setminus xR^+.$$

That is, the positive limit set consists only of points on the boundary of xR^+.

For instance, a non-periodic trajectory which has as positive limit set a periodic motion is positively asymptotic, while the trajectory defined by a periodic motion is not, as shown by Theorem 1.3.21. If $\Lambda^+(x)$ is nonempty and compact, then as Theorem 1.3.17 shows, we can say that a positively asymptotic trajectory tends asymptotically to its positive limit set.

If $M \subset E$ is not a singleton, the set $\Lambda^+(M)$ has rather weak properties. For instance, while it is easy to prove that:

1.3.32 *THEOREM*

If M *is a continuum and* $\Lambda^+(M)$ *is compact, then also* $\Lambda^+(M)$ *is a continuum.*

1.3.33 *Remark*

1.3.34 *Figure*

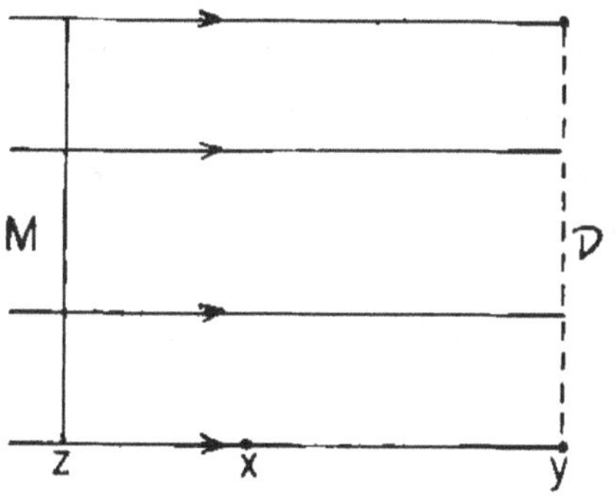

Without the assumption of compactness of $\Lambda^+(M)$, the set $\Lambda^+(M)$ could be disconnected as shown by the following example. Consider the flow shown in Figure 1.3.34, which consists of parallel lines through the segment M tending to the segment $\mathcal{D}$ which consists of equilibrium points.

The trajectory $\gamma(z)$, however, has the limit point $\Lambda^+(z) = \{x\}$, which does not belong to $\bar{\mathcal{D}}$. Thus $\Lambda^+(M) = (\bar{\mathcal{D}} \setminus \{y\}) \cup \{x\}$. This set is neither closed nor connected.

1.3. 35 Notes and References

The definition of limit sets is due to G.D. Birkhoff [I, Vol. I, pp. 654-672]. This concept has been used by H. Poincaré [I , Vol.I] without a formal definition. Alternative definitions were given by S.Lefschetz [2] and by T.Ura [4] (see also 2.2.17)

Theorem 1.3.13 (iv) is due to N.P. Bhatia [3].

Theorem 1.3.17 is due to G.D. Birkhoff (reference above).

The concept of positively asymptotic trajectory is due to V.V.Nemytskii.

The proof of theorem 1.3.26 uses lemmas on transversals on the plane which can be found, for instance, in Coddington and Levinson [2, Ch. 16].

Notice that theorem 1.3.26 can be proved, with almost no variation also for the case of compact sets after having assumed that the minimal set is not a rest point, since then $\Lambda^+(x) = \emptyset$. The proof given here holds only for the case of flows defined by the solutions of ordinary differential equations which define dynamical systems. The theorem is however true for the case of general dynamical systems. Its proof requires the general theory of dynamical system on the plane developed by O. Hajek [5].

1.4 *Prolongations.*

The concept of trajectory has been described in detail in the previous sections. Given any point $x \in E$ the set $\overline{xR}$ associated with it may be very small, in particular, in the case when xR has compact limit sets. Furthermore, the form and the properties of xR do not contain any information about the properties of neighboring trajectories. To overcome these limitations, the concept of prolongation has been introduced.

1.4.1 *DEFINITION*

If $x \in E$ *the (first positive) prolongation* $D^+(x)$ *of* x *is the set of all* $y \in E$ *such that there are sequences* $\{x_n\}: x_n \in E$ *and* $\{t_n\}$ $t_n \in R^+$ *with* $x_n \to x$ *and* $x_n t_n \to y$. *Thus*

1.4.2 $D^+(x) = \{y \in E: \exists \{x_n\} \subset E$ *and* $\{t_n\} \subset R^+$ *such that* $x_n \to x$ *and* $x_n t_n \to y\}$.

Similarly we can define the (first) negative prolongation $D^-(x)$ *of* x *as*

1.4.3 $D^-(x) = \{y \in E: \exists \{x_n\} \subset E$ *and* $\{t_n\} \subset R^-$ *such that* $x_n \to x$ *and* $x_n t_n \to y\}$

and the prolongation $D^+(M)$ *of a set* $M \subset E$ *as:*

1.4.4 $$D^+(M) = \cup\{D^+(x); x \in M\}.$$

1.4.5 *THEOREM*

For any $x \in E$

1.4.6 $$D^+(x) = \cap\{\overline{S(x,\delta)R^+}; \delta > 0\}.$$

Proof. $D^+(x) \subset \bigcap \{\overline{S(x,\delta)R^+} : \delta > 0\}$ is clear. To prove that $\bigcap\{\overline{S(x,\delta)R^+} : \delta > 0\} \subset D^+(x)$, we let $y \in \bigcap\{\overline{S(x,\delta)R^+} : \delta > 0\}$. That is, $y \in \overline{S(x,\delta)R^+}$ for every $\delta > 0$. Thus for any $\varepsilon > 0$ and $\delta > 0$ there is a $z \in S(x,\delta)R^+$ such that $\rho(z,y) < \varepsilon$. Now $z \in S(x,\delta)R^+$ means that there is a $w \in S(x,\delta)$ and a $t \geq 0$ such that $z = wt$. We thus see that for any $\varepsilon > 0$ and $\delta > 0$ there is a $w \in E$ and a $t \geq 0$ such that $\rho(x,w) < \delta$ and $\rho(y,wt) < \varepsilon$. Thus for any sequences $\{\varepsilon_n\},\{\delta_n\}$ of positive numbers with $\varepsilon_n \to 0$ and $\delta_n \to 0$ we can find sequences $\{x_n\}$ in E and $\{t_n\}$ in R^+ such that $\rho(x_n,x) < \delta_n$ and $\rho(x_nt_n,y) < \varepsilon_n$, i.e., also $x_n \to x$ and $x_nt_n \to y$. Hence $y \in D^+(x)$. This proves the theorem.

1.4.7 *Examples of* $D^+(x)$.

i) The simplest non-trivial example of a prolongation is found in a dynamical system defined in the plane and having a saddle point. The simplest system with a saddle point is given by the differential system

$$\dot{x}_1 = -x_1, \qquad \dot{x}_2 = + x_2.$$

1.4.8 *Figure*

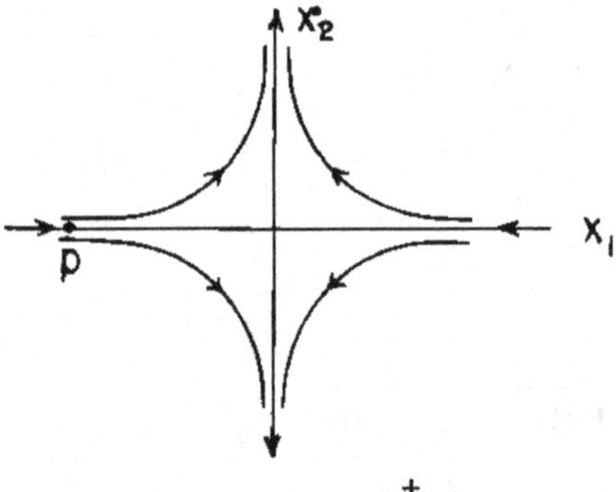

If p is any point on the x_1-axis, then $D^+(p)$ consists of all the points on $\gamma^+(p)$ as well as all the points on the x_2-axis. For a point p not on the x_1-axis, $D^+(p) = \gamma^+(p)$. Similarly, if p is a point on the x_2-axis, then

$D^-(p)$ consists of points $\gamma^-(p)$ and points on the x_1-axis . Notice that we have in this example $\overline{\gamma^+(p)} \subset D^+(p)$, and $\overline{\gamma^-(p)} \subset D^-(p)$ for all p. In fact, these relations hold always. Significant, however, is the fact that for p on the x_1-axis $D^+(p) \neq \overline{\gamma^+(p)}$, and for p on the x_2-axis, $D^-(p) \neq \overline{\gamma^-(p)}$.

ii) In Example (i) $D^+(x)$ is always connected. We now give an example to show that this need not be the case in general. Consider in the plane a dynamical system given by

$$\dot{x}_1 = \sin x_2, \qquad \dot{x}_2 = \cos^2 x_2.$$

The phase portrait as shown in the figure, consists, in particular, of trajectories γ_k given by $\gamma_k = \{(x_1,x_2): x_2 = k\pi + \frac{\pi}{2}\}, k = 0, \pm 1, \pm 2, \ldots$. These are lines parallel to the x_1-axis. Between any two consecutive γ_k's the trajectories are given by $\gamma = \{(x_1,x_2): x_1 + c = \sec x_2\}$, where c is some constant depending on the trajectory. The phase portrait between the lines $x_2 = -\pi/2$ and $x_2 = +\pi/2$ is shown in Figure 1.4.9.

1.4.9 *Figure*

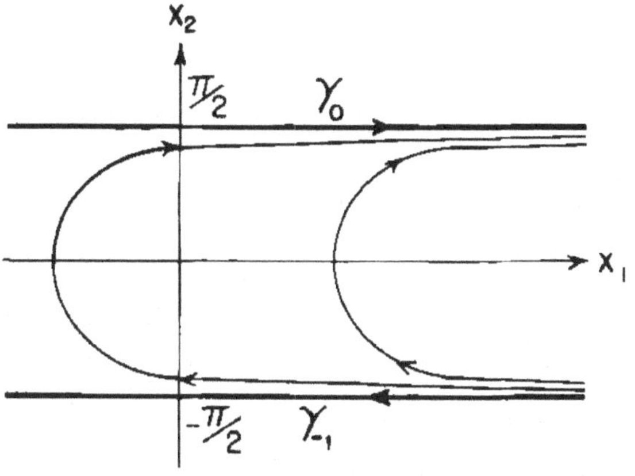

Notice that for any point $p \in \gamma_{-1}$, $D^+(p) = \gamma^+(p) \cup \gamma_0 \cup \gamma_{-2}$. Here $D^+(p)$ is not connected. Notice also that $\Lambda^+(p) = \emptyset$, for every p in the plane. We shall refer to this example later in other connections.

iii) The first prolongation $D^+(x)$ is not always a "curve" as shown by the following flow (*Figure* 1.4.10). The point y is an equilibrium point which has the property that for all $x \in E$, $\Lambda^+(x) = y$. The behavior of the trajectories is different, however, from M to $C(M)$. In fact, if $x \in C(M)$, $\Lambda^-(x) = \emptyset$, while if $x \in M$, $\Lambda^-(x) = y$. One can see that in this case $D^+(y) = \overline{M}$

1.4.10 *Figure*

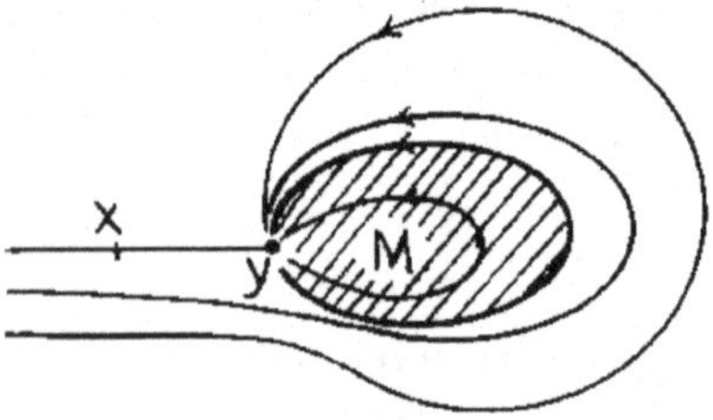

1.4.11 *THEOREM*

If $x \in E$, $\overline{xR^+} \subset D^+(x)$

Proof. If $y \in \overline{xR^+}$, then $y \in xR^+$ or $y \in \Lambda^+(x)$. If $y \in xR^+$, then indeed $y \in D^+(x)$. If $y \in \Lambda^+(x)$, then there is a sequence $\{t_n\}, t_n \to +\infty$ with $xt_n \to y$. The choice of sequences $\{x_n\}, \{t_n\}$, with $x_n = x$ for each n shows that $y \in D^+(x)$, which proves the theorem.

1.4.12 *THEOREM*

If $x \in E, D^+(x)$ *is closed and positively invariant.*

Proof. Let $y \in D^+(x)$, and $t \geq 0$. There are sequences $\{x_n\}, \{t_n\}, x_n \to x$, $t_n \geq 0$, such that $x_n t_n \to y$. Consider the sequences $\{x_n\}, \{\tau_n\}$, with $t_n = \tau_n + t$. Clearly $\tau_n \geq 0$, $x_n \to x$, and $x_n(\tau_n + t) = x_n\tau_n(t) \to yt$. Here $yt \in D^+(x)$ and so $D^+(x)$ is positively invariant. To see that $D^+(x)$ is closed, consider a sequence $\{y_n\}$, with $y_n \to y$, $y_n \in D^+(x)$. It is to be shown th $y \in D^+(x)$. Now, there are sequences $\{t_n^k\}$, $t_n^k \geq 0$, $k = 1,2,\ldots$, and sequences

$\{x_n^k\}$, $k = 1,2,\ldots$, with $x_n^k \to x$ for each fixed k, and $x_n^k t_n^k \to y_k$ for each fixed k. We may assume without loss of generality that $(x_n^k, x) \leq 1/k$ and $\rho(x_n^k t_n^k, y_n) \leq 1/k$ for $n \geq k$. Now consider the sequences $\{x_n\},\{t_n\}$, with $x_n = x_n^n$, and $t_n = t_n^n$. We have $\rho(x,x_n) \leq \frac{1}{n}$, so that $x_n \to x$, and $\rho(y,x_n t_n) \leq \rho(y,y_n) + \rho(y_n,x_n t_n) \leq \rho(y,y_n) + \frac{1}{n}$, so that $x_n t_n \to y$. Hence $y \in D^+(x)$, and so $D^+(x)$ is closed. This completes the proof.

The set $D^+(M)$ has stronger properties than $\Lambda^+(M)$ in fact.

1.4.13 *THEOREM*

If $M \subset E$ *is compact,* $D^+(M)$ *is closed and positively invariant.*

Proof. $D^+(M)$ being the union of positively invariant sets is positively invariant. To see that $D^+(M)$ is closed, let $\{y_n\}$ be a sequence with $y_n \to y$, $y_n \in D^+(M)$. Then there is a sequence $\{x_n\}, x_n \in M$ and $y_n \in D^+(x_n)$. As M is compact, we may assume that $x_n \to x \in M$. We shall show that $y \in D^+(x)$. This is so, because there are sequences $\{x_n^k\},\{t_n^k\}$ $k = 1,2,\ldots$, with $t_n^k \geq 0$, $x_n^k \to x_k$, and $x_n^k t_n^k \to y_k$ for each fixed k. We may assume, without loss of generality that $\rho(x_k,x_n^k) \leq \frac{1}{k}$, and $\rho(y_k,x_n^k t_n^k) \leq \frac{1}{k}$ for $n \geq k$. Then considering the sequences $\{x_n^n\},\{t_n^n\}$ we notice that $\rho(x,x_n^n) \leq \rho(x,x_n) + \rho(x_n,x_n^n) \leq \rho(x,x_n) + \frac{1}{n}$, and $\rho(y,x_n^n t_n^n) \leq \rho(y,y_n) + \rho(y_n,x_n^n t_n^n) \leq \rho(y,y_n) + \frac{1}{n}$, which shows that $x_n^n \to x$ and $x_n^n t_n^n \to y$, i.e., also $y \in D^+(x)$. Hence also $y \in D^+(M)$ and the same is closed. Q.E.D.

Additional properties of prolongations will be presented in Sections 1.5 and 2.4 to which the reader is referred.

1.4.14 *Notes and References*

The concept of prolongation is due to T. Ura [2]. The example (1.4.7) in E^2 can be found in the work by H. Poincaré [1, Vol. 1, pp. 44] and in the work of I. Bendixson, but without a formal definition of prolongation.

1.5 *Lagrange and Liapunov Stability for Compact Sets*

In the last sections we have performed what can be called the anatomy of dynamical systems. In fact, we have been concerned with the definitions and the essential properties of the elements which constitute a dynamical system: trajectories, motions, invariant and minimal sets, prolongations, etc. The limit sets of trajectories and prolongations have been defined. We have proved (1.3 and 1.4) that these limit sets are closed and invariant sets. In the next sections we shall be concerned with what can be called the "physiology" of dynamical systems, i.e., the study of the behavior, the relations, and the relative properties of its elements. Our analysis will start from the most simple properties. Consider a point $x \in E$; the first properties of such a point within a dynamical system $\mathcal{F}$ can have is that the associated trajectory xR has a limit set. In fact, if xR does not have such a limit set, xR tends to infinity both for $t \to +\infty$ and $t \to -\infty$ in a certain way. If a trajectory xR does not have any limit set this fact classifies this trajectory in the dynamical system in a certain way which will be clarified in Chapter 2. Consider then the case of a trajectory which has a limit set (either positive or negative). If the limit set is not compact this would mean that the trajectory xR will cover a non compact region of the space (Example: an infinite strip), but not the whole space. The next, and more interesting case, is when a trajectory xR has a compact limit set.

1.5.1 *DEFINITION*

A point $x \in E$ *is said to be positively Lagrange stable (or* L^+*-stable) if* $\overline{xR^+}$ *is compact, negatively Lagrange stable (or* L^-*-stable) if* $\overline{xR^-}$ *is compact and Lagrange stable if* $\overline{xR}$ *is compact.*

In the space E the property of Lagrange stability is equivalent to the fact that $\overline{xR^+}$ is positively bounded. The property of a point L^+-stable can be generalized to a whole set $B \subset E$.

1.5.2 *DEFINITION*

A set $B \subset E$ *is called* <u>L^+*-stable*</u> *(*L^-*-stable,* L*-stable) if all points* $x \in B$ *are* L^+*-stable (*L^-*-stable,* L*-stable) . A dynamical system* $\mathcal{F}$ *is called* <u>L^+*-stable*</u> *(*L^-*-stable,* L*-stable) if all points* $x \in E$ *are* L^+*-stable (*L^-*-stable,* L*-stable) .*

1.5.3 *DEFINITION*

If a point $x \in E$ *is neither* L^+ *, nor* L^-*-stable it will be called* <u>L*-unstable*</u>*. A dynamical system* $\mathcal{F}$ *is called* <u>*unstable*</u> *if all points* $x \in E$ *are* L*-unstable.*

Lagrange stability is both a property of the trajectory and the motion associated with a given point $x \in E$ within a dynamical system $\mathcal{F}$. In the space E the statement that $x \in E$ is L^+-stable is equivalent to the concept that the motion π_x through $x \in E$ is positively bounded.

The properties of Lagrange stable points are essentially characterizable by the properties of their compact limit sets. These properties have been extensively investigated in Section 1.3. Thus for Lagrange stable points in the space E , from Theorem 1.2.35 and 1.3.13 it trivially follows that:

1.5.4 *COROLLARY*

If a point $x \in E$ *is* L^+*-stable, then* $\Lambda^+(x)$ *is compact, connected and contains a minimal set. If a* ***closed*** *set* $B \subset E$ *is* L^+*-stable also* ∂B *is* L^+*-stable.*

1.5.5 *Figure*

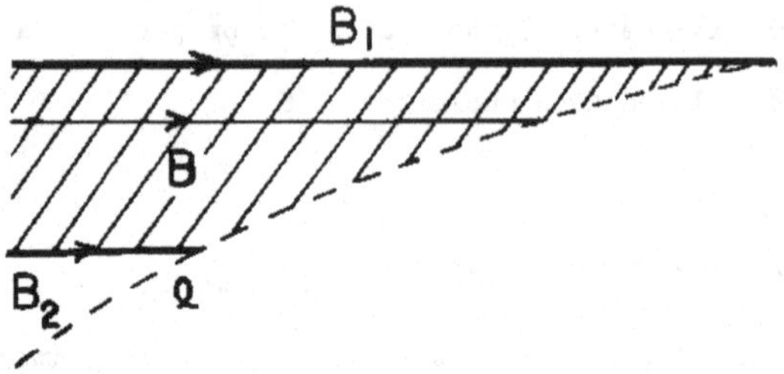

It must be pointed out that the second part of the Theorem 1.5.4 holds only if B is closed. It is, in fact, easy to produce an example of a non-closed set, which is L^+-stable, but ∂B is not. Consider, for example, the flow represented in Figure 1.5.5 whose trajectories are a family of parallel straight lines having their positive limit set of the curve Q each element of which is a critical point. The curve Q tends asympototically to the straight line B_1. Consider the open set B bounded by the two straight lines B_1 and B_2 such that B_1 and B_2 do not belong to B. This set is non-closed and it is L^+-stable since for all $x \in B$, $\overline{xR^+}$ is compact. The set $\partial B = B_1 \cup B_2$ is on the other hand not L^+-stable.

The concept of Lagrange stability, if applied to certain sets, fails completely to provide us with any additional information about the properties of such sets. For example, the expression "a compact positively invariant L^+-stable set B" is clearly redundant since, if the set B is compact and positively invariant, this implies that for all points $x \in B$, $\overline{xR^+}$ is compact and therefore B is L^+-stable. The same is true for ∂B. Clearly the property of being L-stable has a non trivial meaning if applied to non compact sets or, in particular, to the whole dynamical system $\mathcal{F}$. Also, the case of a L-stable dynamical system may be a rather pathological one like the case of a system in which all trajectories are compact (for example when all motions are periodic). From these remarks it must be obvious that Lagrange stability is a rather weak concept which does not provide us with much information regarding the qualitative behavior of dynamical systems. In particular, it is strictly a property of the trajectories within a set, not related with the properties of the system outside the set. The concept which will be introduced next will provide us with a much more precise characterization of the qualitative behavior of the dynamical systems. This will essentially be a property of the points in a neighborhood A of a set $M \subset E$ with respect to the set M. In the case of sets with compact vicinity, these properties will be characterizable in terms of the properties of the limit sets of the points in a set $N \supset M$, and it is therefore closely related with the idea of Lagrange stability.

1.5.6 *DEFINITION*

Let $M \subset E$ *be a compact set. Let* $B \subseteq E$ *be such that for all* $x \in B$, $\Lambda^+(x) \neq \emptyset$ *and* $\Lambda^+(x) \subseteq M$. *Then* M *is called an* *__attractor relative to__* B.

The largest set B *will be denoted* $A(M)$ *and called the* *__region of attraction of__* M. *Thus*

1.5.7 $$A(M) = \{x \in E: \quad \Lambda^+(x) \neq \emptyset, \text{ and } \Lambda^+(x) \subseteq M\}$$

If there exists a $\delta > 0$ *such that* $S(M,\delta) \subseteq A(M)$, *then* M *is called an* *__attractor__* (*). *If in addition* $A(M) = E$, *then* M *is called a* *__global attractor__*.

* Instead of (positive) attractor, a set M, which satisfies the conditions of definition 1.5.6 is sometimes called quasi asymptotically stable set.

If there exists a $\delta > 0$, *such that* $x \in S(M,\delta)$ *implies* $\Lambda^+(x) \cap M \neq \emptyset$, *then* M *is called* *weak attractor*.

The set

1.5.8 $$A_\omega(M) = \{x \in E: \Lambda^+(x) \neq \emptyset \text{ and } \Lambda^+(x) \cap M \neq \emptyset\}$$

is then called *region of weak attraction*.

If M *is an attractor and it is such that for all* $\delta > 0$ *and for all compact sets* $K \subset A(M)$ *there exists a* $T = T(K,\delta) \geq 0$ *such that* $Kt \subset S(M,\delta)$ *for all* $t > T$, *then* M *is called a* *uniform attractor*.

If for all $\varepsilon > 0$, *there exists a* $\delta(\varepsilon) > 0$, *such that* $x \in S(M,\delta)$ *implies that* $xR^+ \subset S(M,\varepsilon)$, M *is called* *stable*. *If a compact set is not stable, then it is called* *unstable*.

From the definition it obviously follows that if $M \subset E$ is an attractor, then

1.5.9 $$\partial A(M) \cap \partial M = \emptyset$$

Notice that $A_\omega(M) \subseteq A(M)$.

1.5.10 *DEFINITION*

Let $M \subset E$ *be a compact set. Then* M *is a* *negative attractor* *if there is a* $\delta >$ *such that* $x \in S(M,\delta)$ *implies that* $\Lambda^-(x) \neq \emptyset$ *and* $\Lambda^-(x) \subseteq M$. *The region of negative attraction or* *region of repulsion* $A^-(M)$ *is defined similarly to what was done in* (1.5.7).

We shall now study the basic properties of the set A(M); the set $A^-(M)$ has similar properties.

1.5.11 *THEOREM*

If $M \subset E$ *is a compact attractor, the set* $A(M) \setminus M$ *is open.*

Proof. We have to show that for all $x \in A(M)\setminus M$ there exists an $\varepsilon > 0$ such that $S(x,\varepsilon) \subset A(M)\setminus M$. The set $S(M,\delta)\setminus M$ is open, where δ is as in the definition of attractor. Now let $x \in A(M)\setminus M$. By the same argument as in Theorem 1.3.13 it can be shown that there exists a $\tau > 0$ such that $x\tau \in S(M,\delta)\setminus M$. Since $S(M,\delta)\setminus M$ is open we can find an $\varepsilon > 0$ such that $S(x\tau, \varepsilon) \subset S(M,\delta)\setminus M$. Because of continuity of the map π, $S(x\tau, \varepsilon)(-\tau)$ is open, it further contains x and is thus a neighborhood of x. By the definition of $A(M)$ it also follows that $S(x\tau, \varepsilon)(-\tau) \subset A(M)\setminus M$, hence $A(M)\setminus M$ is open. Q.E.D.

1.5.12 *Remark.* Theorem 1.5.11 is false if M is not an attractor. Consider for that the flow shown in Figure 1.5.13 which has the following properties:

1.5.13 *Figure.*

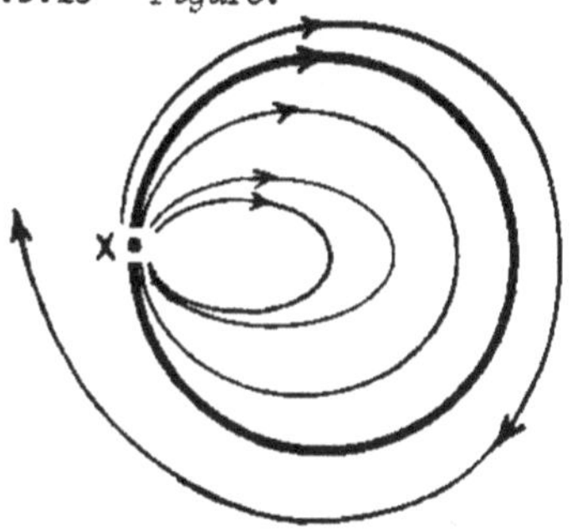

$x \in C(A\{x\}) \Rightarrow \Lambda^-(x) \subset A(\{x\})$

$x \in A(\{x\}) \Rightarrow \Lambda^+(x) = \Lambda^-(x) = \{x\}$.

Clearly, $\{x\}$ is not an attractor and its region of attraction $A(\{x\})$ is a compact set.

1.5.14 *THEOREM*

Let $M \subset E$ *be a compact set. Then if* M *is an attractor, the sets* $A(M)$ *and* $\partial(A(M))$ *are invariant.*

Proof. Notice that $x \in A(M)$ if and only if $\Lambda^+(x) \neq \emptyset$ and $\Lambda^+(x) \subseteq M$. Since, however for any $t \in R$ and $x \in E$, we have $\Lambda^+(x) = \Lambda^+(x\,t)$, the result follows, from 1.2.36. Q.E.D.

In what follows additional properties of attractors will be defined.

1.5.15 *DEFINITION*

Let $M \subset E$ *be a compact attractor. If there exists a point* $x \notin M$ *such that* $\Lambda^{-}(x) \cap M \neq 0$, *then the set* M *will be called an* *unstable attractor.*

1.5.16 *DEFINITION*

Let $M \subset E$ *be a positively invariant compact attractor. The set* M *will be called a* *stable attractor* *or an* *asymptotically stable set* *if there is a* $\eta > 0$ *such that* $x \in S(M,\eta) \setminus M$ *implies* $\Lambda^{-}(x) \cap M = \emptyset$. *The set* $A(M)$ *defined in* 1.5.7 *will in this case be called* *region of asymptotic stability of* M . *If* $A(M) = E$, *the set* M *will be called* *globally asymptotically stable**.

1.5.17 *DEFINITION*

Let $M \subset E$ *be a negative attractor as in definition* 1.5.10 , *if there exists a* $\eta > 0$ *such that* $x \in S(M,\eta) \setminus M$ *implies that* $\Lambda^{+}(x) \cap M = \emptyset$, M *will be said to be* *completely unstable* *(or negatively asymptotically stable).*

1.5.18 *Remark.* By reversing the direction of motion along the trajectories, sets which are completely unstable will become asymptotically stable and vice versa.

1.5.19 *THEOREM.*

If $M \subset E$ *is a compact attractor and it is positively invariant the condition that* $x \in S(M,\eta) \setminus M$ *implies that* $\Lambda^{-}(x) \cap M = \emptyset$ *is equivalent to the condition that* M *is stable.*

Proof. Let $\eta > 0$ be such that $x \in S(M,\eta) \setminus M$ implies $\Lambda^{-}(x) \cap M = \emptyset$. Now assume if possible, that there is an $\varepsilon > 0$ such that for every

---

* or asymptotically stable in the large.

$\delta > 0$ there is an $x \in S(M,\delta)$ such that $xR^+ \not\subset S(M,\varepsilon)$. We may assume $\varepsilon < \eta$. Clearly there is a sequence $\{x_n\}$, $x_n \to M$ and a sequence $\{t_n\}$, $t_n > 0$, and $\rho(x_n t_n, M) = \varepsilon$. As M is compact, we can assume that $x_n \to x \in M$. We will show that $\{t_n\}$ is not bounded. For otherwise we can find a convergent subsequence, and so assume that $t_n \to t \geq 0$. Since now $x_n \to x$, $t_n \to t$, we have $x_n t_n \to xt$. Since $xt \notin M$, $x \in M$ and $t \geq 0$, this contradicts positive invariance of M. Therefore, the sequence $\{t_n\}$ is not bounded. We may assume therefore that $t_n \to +\infty$. Setting now $x_n t_n = y_n$, we notice that $x_n = y_n(-t_n)$. Thus we have a sequence with $\rho(y_n, M) = \varepsilon$. Since the set $\{y: \rho(y,M) = \varepsilon\}$ is compact, we can assume $y_n \to y$, with $\rho(y,M) = \varepsilon$. Then $y \in S(M,\eta)$. Then, however, as $x_n = y_n(-t_n) \to x \in M$, whereas $-t_n \to -\infty$, we see that $x \in \Lambda^-(y)$. Thus $\Lambda^-(y) \cap M \neq \emptyset$, which is a contradiction. The last part of the argument shows that the converse is also true, and the theorem is proved.

1.5.20 *Remark.* The condition that M be positively invariant is essential in Theorem 1.5.19. It is in fact easy to produce the example of a compact attractor, which is not positively invariant and which does not satisfy Theorem 1.5.19 .

1.5.21 *Figure.*

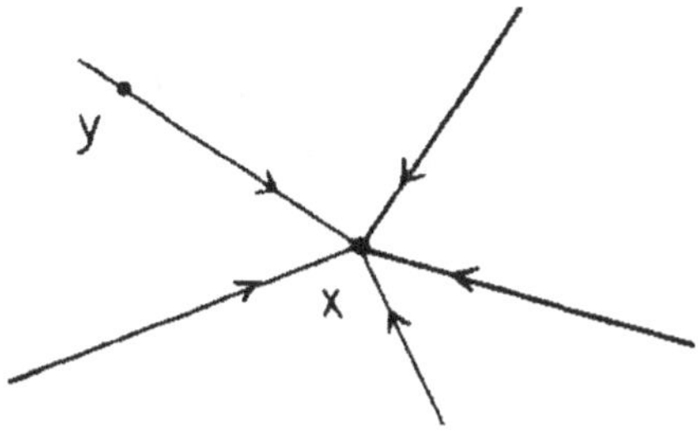

Consider in fact the flow represented in Figure 1.5.21 . The trajectories are a family of straight lines through the critical point x . On each trajectory the motion moves the point toward x . Thus all $p \in E$ $\Lambda^-(p) = \emptyset$ and $\Lambda^+(p) = \{x\}$. Consider a point $y \in E$, $y \neq x$ and the compact set $M = \{x\} \cup \{y\}$. M obviously has the property that $\Lambda^-(p) \cap M = \emptyset$ for all $p \in E$, and it is a compact attractor, however, Theorem 1.5.19 is obviously not satisfied.

1.5.22 *Remark*. Theorem 1.5.19 is also incorrect without the assumption that M be an attractor. In fact then it is not true that if M is positively invariant and compact and $\Lambda^-(x) \cap M = \emptyset$ for $x \notin M$, then M is stable. This can be shown by the following counterexample. Consider

1.5.23 *Figure*

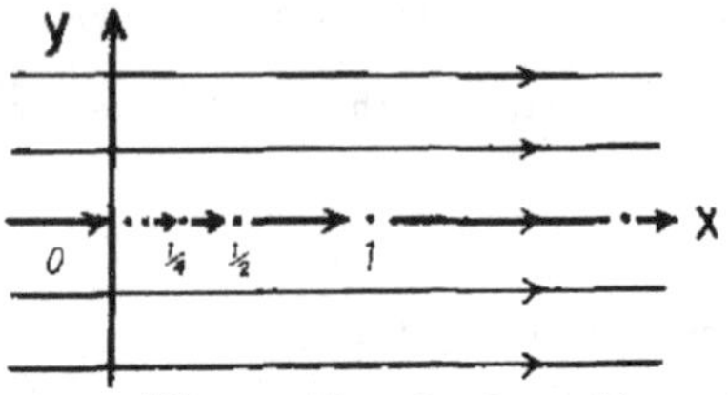

the sequence of equilibrium points on the axis X with abscisses 0, 1, ½, ¼, 1/8, ... and fill in the rest of the flow as shown in Figure 1.5.23. Consider the equilibrium point $\{0\}$. Clearly for all $x \in C(\{0\})$, $\Lambda^-(x) \cap \{0\} = \emptyset$, but $\{0\}$ is not a stable set.

1.5.24 *THEOREM.*

If $M \subset E$ *is closed and stable, then it is positively invariant.*

Proof. Notice that $x \in M$ implies $xR^+ \subset S(M,\varepsilon)$ for every $\varepsilon > 0$. Here $xR^+ \subset \bigcap_{\varepsilon>0} S(M,\varepsilon) = M$, as M is closed. Hence M

is positively invariant.

1.5.25 *Remark.* Theorem 1.5.24 shows that the fact that both in Definition 1.5.16 and in Theorem 1.5.19 it has been assumed that M is positively invariant is not a restriction. In fact, positive invariance is a necessary condition for stability. Thence

1.5.26 *COROLLARY*

If a compact set $M \subset E$ *is stable, it is positively invariant and if in addition* M *is an attractor it is asymptotically stable.*

We now investigate the relationship between uniform attraction and asymptotic stability.

1.5.27 *THEOREM*

Let M *be a compact asymptotically stable set. Then* M *is uniformly attracting.*

Proof. Notice first that the definition of a uniform attractor is equivalent to the following: a compact attractor M is a uniform attractor if given $\delta > 0$ and a compact set $K \subset A(M)$ there is a $T = T(K,\delta)$ such that $Kt \subset S(M,\delta)$ for $t > T$. Let now M be asymptotically stable. Let $K \subset A(M)$ be compact. And let $\varepsilon > 0$ be given. Since M is stable, there is a $\delta > 0$ such that $\gamma^{+}(S[M,\delta]) \subset S(M,\varepsilon)$. For any $x \in K$, define $\tau_x = \inf\{t > 0: xt \in S(M,\delta)\}$. τ_x is defined as M is an attractor, and $K \subset A(M)$. Set $T = \sup\{\tau_x : x \in K\}$. We claim that T is finite. For otherwise, there will be a sequence $\{x_n\}$ in K such that $\tau_{x_n} \to +\infty$. However, since K is compact we may assume that $x_n \to y \in K$. Then there is a $\tau > 0$ such that $y\tau \in S(M,\delta)$. As $S(M,\delta)$ is open, there is an open neighborhood N of $y\tau$ such that $N \subset S(M,\delta)$. The inverse image

$N^* = N(-\tau)$ of N by the transition π^τ is open and a neighborhood of y. Further, $N^*(\tau) = N \subset S(M,\delta)$, so that $N^*t \subset S(M,\varepsilon)$ for $t \geqq \tau$. Since $x_n \in N^*$ for large n, we have $\tau_{x_n} \leqq \tau$ for large n. This contradicts $\tau_{x_n} \to +\infty$. Hence $T < +\infty$. Notice now that $x \in K$ implies $xT \in S[M,\delta]$, and so $Kt \subset S(M,\varepsilon)$ for $t > T$, i.e., M is uniformly attracting, and the theorem is proved.

1.5.28 *THEOREM*

A compact positively invariant set $M \subset E$ *is asymptotically stable if and only if it is uniformly attracting.*

Proof. Let M be positively invariant and uniformly attracting. We shall prove that M is stable. Assume if possible that M is not stable. Then there is a sequence $\{x_n\}$, $x_n \to x \in M$, and a sequence $\{t_n\}$, $t_n \geqq 0$, such that $x_n t_n \in H(M,\varepsilon)$ for some $\varepsilon > 0$. Indeed $\varepsilon > 0$ may be chosen small to ensure that $S[M,\varepsilon]$ is compact, and $S[M,\varepsilon] \subset A(M)$. By uniform attraction there is a $T > 0$ such that $S[M,\varepsilon]t \subset S(M,\varepsilon)$ for $t > T$. Thus $t_n \leqq T$. There is then a subsequence $\{x_{n_k}\}$ of $\{x_n\}$, such that the corresponding subsequences $\{t_{n_k}\}$ and $\{x_{n_k} t_{n_k}\}$ converge. Let $t_{n_k} \to t$, and $x_{n_k} t_{n_k} \to y$. Then $y = xt \in M$, as M is positively invariant, and also $y \in H(M,\varepsilon)$ as $x_{n_k} t_{n_k} \in H(M,\varepsilon)$. This is impossible as $M \cap H(M,\varepsilon) = \emptyset$. Hence M is stable, and since M is an attractor, it is asymptotically stable. The converse of the theorem has already been proved (the previous theorem). The theorem is therefore proved.

1.5.29 *Remark.* The assumption that M is positively invariant is necessary. In fact, consider the following example (Figure 1.5.30).

1.5.30 *Figure.*

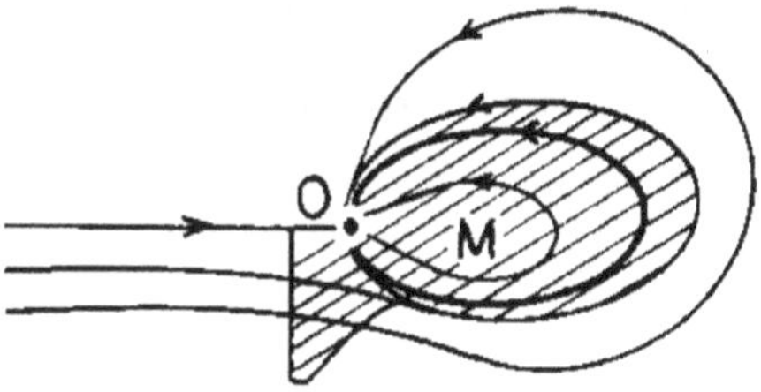

The shaded region represents the set M . The point 0 is an unstable attractor, (Example 1.4.9 iii) and M is uniformly attracting with a suitable time-parametrisation, but it is not stable.

The dependence of various concepts is illustrated below in a chart

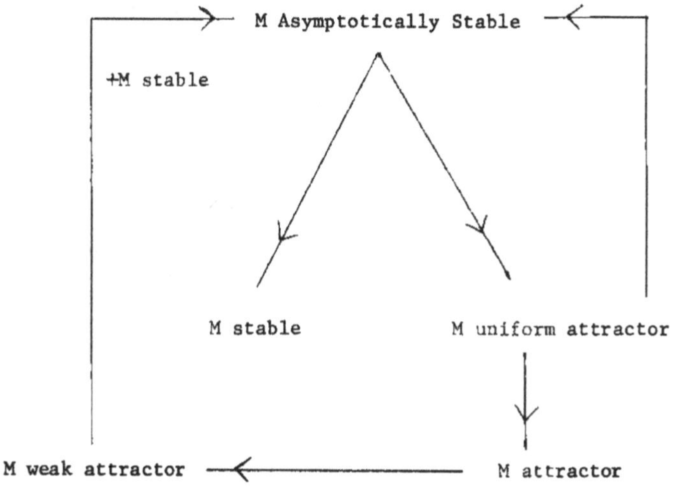

1.5.31 *Remark.* The definitions given and the theorems proved so far for compact sets, are meaningful and true under the slightly weaker hypothesis that $M \subset E$ is not a compact set, but a closed set with a compact vicinity.

1.5.32 *Examples.*

i) in Example 1.3.9 (Figure 1.3.10) choose any point p on the periodic trajectory γ . The set consisting of the point p is a *weak attractor*. This set has no other property listed in 1.5.6.

ii) Consider a planar dynamical system defined by the following diffential equations in polar coordinates.

$$\dot{r} = r(1-r)$$

$$\dot{\theta} = \sin^2(\theta/2) .$$

The phase portrait consists of two rest points $p_1 = (0,0)$, and $p_2 = (1,0)$ (Figure 1.5.33), a trajectory γ on the unit circle which together with the rest point p_2 forms a *path monogon* (the union of a trajectory γ and a rest point p such that $\Lambda^+(\gamma) = \Lambda^-(\gamma) = \{p\}$ will be generally called a path monogon). All orbits outside the unit circle have p_2 as their only positive limit point and their negative limit sets are empty. All trajectories in the interior of the unit circle (except the rest point p_1) have p_2 as their sole positive limit point and p_1 as their sole negative limit point. The point p_2 is an attractor with $A(p_2) = E^2 \setminus \{p_1\}$. It is not a uniform attractor, and is not stable.

1.5.33 *Figure.*

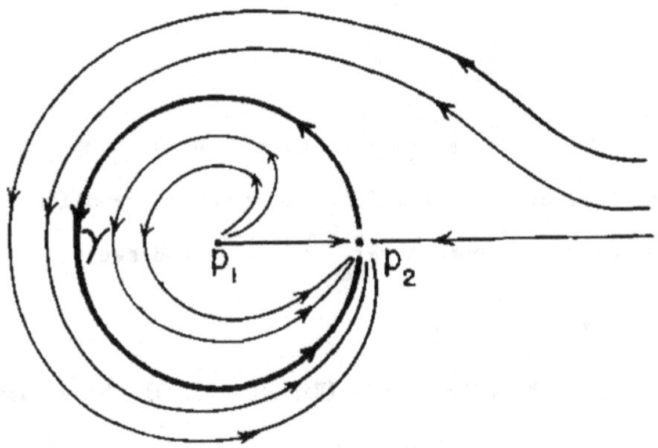

iii) In the above example (ii), the set M consisting of points on the unit disc is asymptotically stable. This set is also a uniform attractor. However, if we consider a set M^* consisting of points on the unit disc and another point not on the unit disc, then M^* is a uniform attractor, but it is not stable. A similar example can be built out of example 1.3.9 (i).

iv) Consider again a planar dynamical system defined by the following differential equations in cartesian coordinates.

$$\dot{x}_1 = x_2, \; \dot{x}_2 = -x_1 \quad .$$

The phase portrait consists of a rest point P -- the origin of coordinates and periodic trajectories which coincide with concentric circles with P as center. Any compact invariant set in this example is stable, but has none of the attractor properties. Thus, for example, the point P is stable.

1.5.34 *Figure*

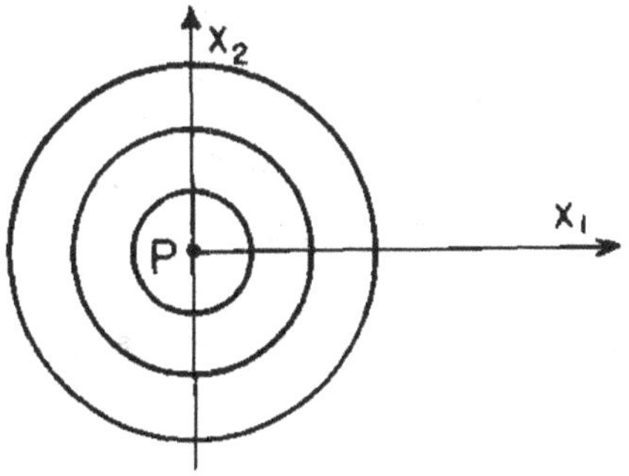

v) Consider finally a planar dynamical system given by the differential system in cartesian coordinates

$$\dot{x}_1 = x_2, \quad \dot{x}_2 = \sin^2\left(\frac{\pi}{x_1^2 + x_2^2}\right) x_2 - x_1 \; .$$

The phase portrait (Figure 1.5.35) consists of the rest point P -- the origin of coordinates, a sequence $\{\gamma_n\}$ of periodic trajectories which are circles with center P - $\gamma_n = \{(x_1, x_2): x_1^2 + x_2^2 = \frac{1}{n}\}$. All other trajectories are spirals. The point P is stable, but has no attractor property. No compact set except the point P is either stable or a weak attractor.

1.5.35 *Figure.*

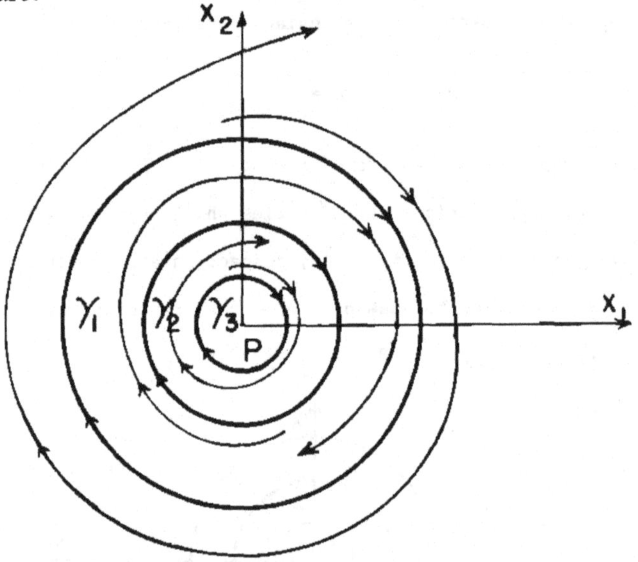

We shall now present further properties of stable and asymptotically stable compact sets.

1.5.36 *THEOREM*

A compact set $M \subset E$ *is stable if and only if each component of* M *is stable.*

The proof is given in 2.6.8.

1.5.37 *Remark.* Theorem 1.5.36 implies that our theory would have not been in any way restricted if instead of considering compact sets we would have limited ourselves to the case of continua.

We are now in the position of discussing the relative properties of and $A(M)$ and $C(A(M))$. This will be done in the next two theorems. Similar theorems in a much stronger form will be proved in Chapter 2.

1.5.38 *THEOREM*

Let M *be an asymptotically stable closed invariant set with a compact vicinity, then, if the set* $\overline{A(M)\setminus M}$ *is compact, the set* $C(A(M))$ *is completely unstable and* $C(M)$ *is its region of repulsion.*

Proof. The set $C(A(M))$ is closed and invariant as the set $A(M)\setminus M$ is open (by Theorem 1.5.11) and invariant (by Theorem 1.5.14) .

Notice now that for all $x \in A(M)$, $\Lambda^-(x) \subset C(A(M))$ and $\Lambda^-(x) \neq \emptyset$. Hence $C(A(M))$ is completely unstable and $C(M)$ its region of repulsion since $x \in M$ implies $\Lambda^-(x) \subset M$ as M is invariant.

1.5.39 *COROLLARY of Theorem* (1.5.38)

Let M *be a closed set with a compact vicinity. If* M *is completely unstable and invariant, then, if the set* $C(A^-(M))$ *has a compact vicinity, it is asymptotically stable and* $C(M)$ *is its region of attraction.*

1.5.40 *THEOREM*

Let M *be a positively invariant compact set, and let* $M^* \subset M$ *be the largest invariant set contained in* M *. Then,* M^* *is a stable attractor, relative to* M .

Proof. For any $x \in M$, $\Lambda^+(x) \neq \emptyset$ and compact, because $xR^+ \subset M$ and so

$x\overline{R^+}$ is compact. Again $\Lambda^+(x) \subset M^*$, because otherwise $\Lambda^+(x) \cup M^*$ will be a larger compact invariant subset of M. Hence M^* is an attractor. To see that it is stable, we must show that for $x \in M \setminus M^*$, $\Lambda^-(x) \cap M^* = \emptyset$. Assume the contrary, i.e. $\Lambda^-(x) \cap M^* \neq \emptyset$. Then there is a sequence $\{t_n\}$, $t_n \to -\infty$, with $xt_n \in M$, so that $(xt_n)R^+ \subset M$, as M is positively invariant. Since $xR = \bigcup (xt_n \in R^+;\ n = 1, 2, \ldots)$, we have $xR \subset M$. Notice now that $M^* \cup \overline{xR} = M^* \cup xR$ is a compact invariant set which is larger than M^*, which is a contradiction. Hence $\Lambda^-(x) \cap M^* = \emptyset$ for each $x \in M \setminus M^*$ and so M^* is stable.

The next theorem will further clarify the structure of asymptotically stable sets and of their regions of attraction.

1.5.41 *THEOREM*

If $M \in E$ *is a compact minimal set which is asymptotically stable, then for all* $x \in A(M)$ *the compact set* $\overline{xR^+} \in E$ *is asymptotically stable.*

Proof. As M is compact and minimal, we have for each $x \in A(M)$, $\Lambda^+(x) = M$. Otherwise, since $\Lambda^+(x) \subset M$, and $\Lambda^+(x)$ is closed and invariant, M will not be minimal if $\Lambda^+(x)$ is a proper subset. Now $\overline{xR^+} = xR^+ \cup \Lambda^+(x)$ $= xR^+ \cup M$ is compact, as $\Lambda^+(x)$ is compact. Therefore, definition 1.5.6 is applicable. Notice that $A(M)$ is open, and $\overline{xR^+}$ is a *compact* subset of it. Therefore $A(M)$ is a neighborhood of $\overline{xR^+}$. Now $y \in A(M) \setminus \overline{xR^+}$ implies that $y \in A(M)$ and therefore $\Lambda^+(y) \subset M \subset \overline{xR^+}$, and hence $\overline{xR^+}$ is an attractor. Again if $y \in A(M) \setminus \overline{xR^+}$, then $\Lambda^-(y) \cap M = \emptyset$, as M is asymptotically stable. But then $\Lambda^-(y) \cap \overline{xR^+} = \emptyset$, as $\Lambda^-(y) \cap A(M) \setminus M = \emptyset$. Thus $\overline{xR^+}$ is asymptotically stable. The theorem is proved.

The property of stability of a (compact) set M, defined in Theorem 1.5.19, is a rather weak property which cannot be characterized by the positive and negative limit sets of the points in a neighborhood of M. Such property can be characterized as a property of the first positive prolongation of M, as shown by Theorems 2.6.5 and 2.6.6.

We shall close this section by stating some important theorems on the stability properties of the first positive prolongation of compact attractors.

1.5.42 *THEOREM*

Let M *be a compact weak attractor. Then* $D^+(M)$ *is a compact asymptotically stable set. The region of attraction* $A(D^+(M))$ *of* $D^+(M)$ *coincides with the region of weak attraction* $A_\omega(M)$ *of* M. *Moreover,* $D^+(M)$ *is the smallest asymptotically stable set containing* M.

The proof of this theorem is given in 2.6.17.

1.5.43 *Notes and References*

Stability theory for dynamical systems was essentially developed by T. Ura [2] in the context of theory of prolongations. Early results and definitions can also be found in the book by Zubov [6]. The original defintions of stability and asymptotic stability for the case of differential equations are due to Liapunov. In his work, however, only local properties of equilibrium points are investigated. The concept of orbital stability (usually defined for limit cycles) found in many earlier works is a particular case of stability of sets (see, for instance, the book by L. Cesari [1]).

The concept of attraction seems to have been used by many authors, but a systematic study seems to have originated with the example of Mendelson [1].

The definition of weak attractor (1.5.6) is due to N. P. Bhatia [3].

Definition 1.5.16 is independent from stability. Our whole presentation of stability theory is motivated by this idea. This forces us to prove Theorem 1.5.19 proving that asymptotic stability implies stability; while usually asymptotic stability is defined as stability plus attraction. We have chosen this way of presenting asymptotic stability to clearly point out how this is a property of the positive and negative limit sets $\Lambda^+(S(M,\delta))$ and $\Lambda^-(S(M,\delta))$ only.

On the other hand, stability without attraction is not characterizable in terms of the properties of the limit sets above.

Theorem 1.5.27 is due to S. Lefschetz [2].

Theorem 1.5.28 is due to N. P. Bhatia, A. C. Lazer and G. P. Szegö [1].

1.6 *Liapunov Stability for Sets.*

In what follows the concepts and theorems developed so far will be extended to the general case of a set $M \subset E$. These extensions are by no means trivial. One of the major difficulties is the fact that the properties of the neighboring trajectories of M with respect to M are no longer characterisable in terms of their limit sets which may now be empty, even if the neighboring trajectories tend to M. In addition to this difficulty for non compact sets we are confronted with a very large number of possible stability properties which degenerate into a few basic properties for the case of closed sets with a compact vicinity. We shall present some of these different types of stability and instability without claiming that we shall exhaust all possible stability behaviors. The main reason for the study of these properties of non compact sets is that the stability properties of time-varying systems will be treated as a particular case of the Liapunov stability of non compact sets. To clear this point we shall define a time-varying dynamical system: $\mathcal{F}_t$ through a mapping $\pi: E \times R \times R \text{-->} E \times R$. This case is contained in the previously defined dynamical system by letting $E \times R = E^{n+1}$. Thus $\mathcal{F}_t$ is defined by the mapping $\pi: E^{n+1} \times R \text{-->} E^{n+1}$. As an illustration of the above remark, assume that the $\mathcal{F}_t$ has the invariant set (equilibrium state) $M = \{0\} \times R$. Its stability properties are equivalent to these of the invariant set $\{x: x_1 = x_2 = \ldots = x_n = 0\}$ in the space $E \times R = E^{n+1}$ with components x_i. These concepts will be fully explained and used in Section 3.4.

We shall now proceed with the definitions of the Liapunov-stability properties of sets in the space E.

1.6.1 *DEFINITION*

A set $M \subset E$ *is said to be (positively Liapunov)* ***stable,*** *if, given any* $\varepsilon > 0$ *, for each* $x \in M$ *there exists a* $\eta(\varepsilon, x)$ *such that* $S(x,\eta)R^+ \subset S(M,\varepsilon)$ *. This is equivalent to saying that given any* $\varepsilon > 0$ *there exists an open set* $O(M) \supset M$ *such that* $O(M)R^+ \subset S(M,\varepsilon)$ *.*

A set $M \subset E$ *is said to be (positively Liapunov)* ***uniformly stable***(*) *if, given any* $\varepsilon > 0$ *, there exists a* $\eta(\varepsilon)$ *such that* $S(M,\eta)R^+ \subset S(M,\varepsilon)$ *.*

From these definitions it obviously follows that

1.6.2 *THEOREM*

If a set $M \subset E$ *is uniformly stable, it is stable.*

On the other hand, it is easy to construct examples of sets which are stable but not uniformly stable.

1.6.3 *Example.*

Consider, for instance, the flow shown in Figure 1.6.4. This flow has the property that for all $x \in E^2$, $\lim\limits_{x_1 \to -\infty} (\gamma(x), \{x : x_2 = 0\}) \to 0$.

On the other hand the positive semitrajectory $\gamma^+(x)$ though all points $x = (x_1, x_2)$ with $x_1 \geq 0$ and x_2 arbitrary is a straight line parallel to the axis x_1 . Clearly then the set $\{x : x_2 = 0\}$ is stable, but not uniformly stable.

---

* Notice that Zubov [6] calls this property stability. We prefer to call it uniform stability to be consistent with the established terminology in the case of time-varying differential equations.

1.6.4 *Figure*

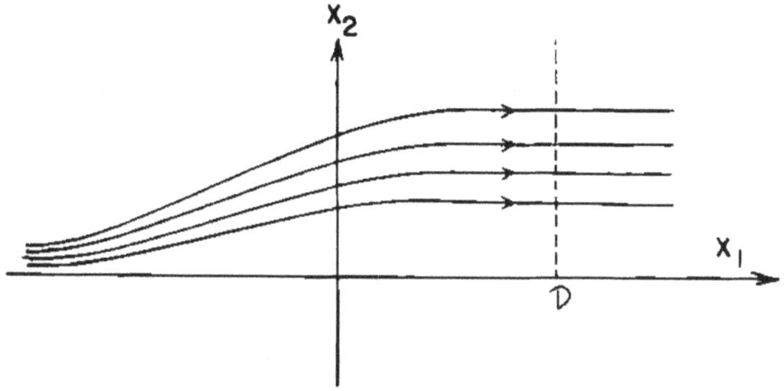

For the case of a compact set the property of stability and that of uniform stability coincide:

1.6.5 *THEOREM*

If a closed set $M \subset E$ *has a compact vicinity, then stability is equivalent to uniform stability.*

Proof. Given $\varepsilon > 0$, for each $x \in M$, there exists an $\eta(x) > 0$ such that $y \in S(x, \eta(x))$ implies $yR^+ \subset S(M,\varepsilon)$. Now ∂M is compact and the family of open sets $\{S(x, \eta(x)\}$, $x \in \partial M$ covers the compact set ∂M. Hence there is a finite subcovering $S(x_1, \eta(x_1))$, ..., $S(x_n, \eta(x_n))$ which covers ∂M. Notice now that $y \in M \cup S(x_1, \eta(x_1)) \cup \ldots \cup S(x_n, \eta(x_n)) = S(M)$ implies $yR^+ \subset S(M,\varepsilon)$. Since there is a $\eta > 0$ such that $S(M,\eta) \subset S(M)$ the theorem follows.

The next theorem is an extension of Theorem 1.5.24.

1.6.6 *THEOREM*

If the closed set $M \subseteq E$ *is stable, then it is positively invariant.*

Proof. Stability of M implies $MR^+ \subset S(M,\varepsilon)$ for all $\varepsilon > 0$. Hence

$$MR^+ \subset \cap \{S(M,\varepsilon),\ \varepsilon > 0\} = M$$

since M is a closed set. But $M \subset MR^+$ always holds, so that we have $MR^+ = M$ and M is positively invariant.

1.6.7 *Remark.* It is to be noted that the property of stability may be trivially satisfied if the set M is not closed. This is shown by:

1.6.8 *Example.* Let the boundary of the circle be a limit cycle (Figure 1.6.9) and let the orbits in the interior of the disc D approach it spirally. Let $x \in D$ be not a rest point. Then the set $\bar{D} \setminus \{x\}$ is still stable according to our definition 1.6.1. Note however that it has a compact vicinity, but it is not positively invariant.

1.6.9. *Figure*

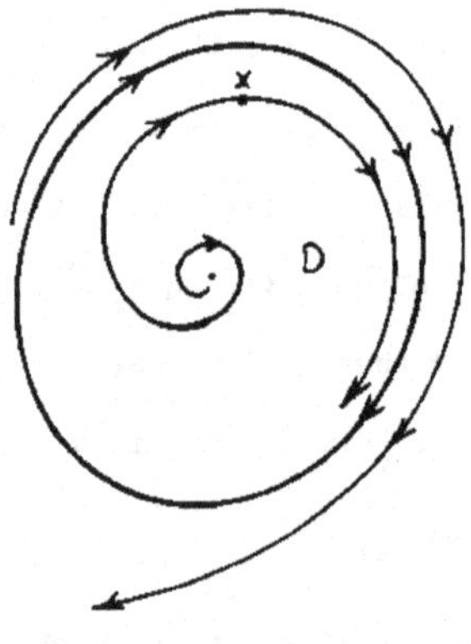

It is also noteworthy that the property of stability is not preserved for the closure of a set M , as shown by Example 1.6.8 although the property of uniform stability is preserved. In fact

1.6.10 *THEOREM*

If a set M *is uniformly stable, then* $\bar{M}$ *is also uniformly stable.*

Proof. The theorem is clear when one notices, that for any set M, $S(M,\eta) \equiv S(\bar{M},\eta)$.

The above theorems and examples indicate the role played by closedness in connection with stability properties.

The various nice properties which compact attractors always have, are not necessarily all present in the case of non compact sets. When the "attracting" property is constant on all points of M we can define weak attraction, attraction and uniform attraction, while in the case in which the attracting property of M varies from point to point, we shall call the same properties semi weak attraction and semi attraction. Essentially these properties are special forms of attraction relative to a set A(M) such that for any $\delta > 0$ $S(M,\delta) \not\subset A(M)$. In the case of a set with compact vicinity all these properties are equivalent and coincide with those given in the Definition 1.5.6.

1.6.11 *DEFINITION*

If for a set $M \subset E$ *there exists an open set* $O(M) \supset M$ *such that for each* $y \in O(M)$ *there is a sequence* $\{t_n\} : t_n \to +\infty$ *such that* $\rho(yt_n, M) \to 0$, M *is called* *semi weak attractor.*

If $O(M) \supset M$ *is such that for each* $y \in O(M)$ *it is* $\lim_{t \to +\infty} \rho(yt, M) = 0$, M *is called* *semi attractor.*

If for a set $M \subset E$ *there exists an* $\varepsilon > 0$ *such that for all* $y \in S(M,\varepsilon)$ *there is a sequence* $\{t_n\} : t_n \to +\infty$ *such that*

$\rho(yt_n, M) \to 0$, M *is called a* *weak attractor.*

If for a set $M \subset E$ *there exists an* $\varepsilon > 0$ *such that for all* $y \in S(M,\varepsilon)$ *it is* $\lim_{t \to +\infty} \rho(yt, M) = 0$, M *is called an* *attractor.*

If set $M \subset E$ *is such that for all* $\varepsilon > 0$ *there exists a* $\lambda(\varepsilon) > 0$ *and a* $\tau(\lambda, \varepsilon)$ *such that for all* $t \geq \tau$, $\rho(xt, M) < \varepsilon$ *for* $x \in S(M, \lambda)$, M *is called a* *uniform attractor.*

A set $M \subset E$ *is finally called* *equiattracting*(*) *if it is attracting and there exists a* $\lambda > 0$ *such that for each* ε, $0 < \varepsilon < \lambda$ *and* $T > 0$ *there exists a* $\tau > 0$ *with the property that for each* x, *such that* $\varepsilon \leq \rho(x,M) \leq \lambda$, $x[0,T] \cap S(M,\delta) = \emptyset$.

1.6.12 *DEFINITION*

The set

1.6.13 $$A_\omega(M) = \{x \in E : \{t_n\},\ t_n \to +\infty,\ xt_n \to M\}$$

is called the *region of weak attraction of the set* M. *The set*

1.6.14 $$A(M) = \{x \in E : \rho(xt, M) \to 0 \quad \text{as} \quad t \to +\infty\}$$

is called the *region of attraction of the set* M.

Notice that if M is an attractor, then

1.6.15 $$A_\omega(M) = A(\omega) \supset S(M,\tau) \quad \text{for some} \quad \tau > 0 .$$

The next theorem on the properties of $A(M)$ is a generalization of Theorem 1.5.14.

1.6.16 *THEOREM*

For any set M, $A(M)$ *is always invariant. If* M *is an attractor, then* $A(M)$ *is also open.*

---

*Notice that this property is equivalent to what Zubov [6] calls uniform attraction.

Proof. If $x \in A(M)$ and $\tau \in R$, then $\rho((x\tau)t, M) = \rho(x(\tau + t), M)$ $= \rho(x\tau', M) \to 0$ as $\tau' \to \infty$, where $\tau' = \tau + t$. Thus $x\tau \in A(M)$ and $A(M)$ is invariant.

As M is an attractor, there exists a $\delta > 0$ such that $S(M,\delta) \subset A(M)$. Now let $x \in A(M)\setminus S(M,\delta)$. We need to show, that there exists a $\mu > 0$ such that $S(x,\mu) \subset A(M)$. To see this, observe that, there is a $T > 0$ such that $\rho(xT,M) \leqq \frac{\delta}{2}$. Choose now $\varepsilon > 0$ such that $S(xT, \varepsilon) \subset S(M,\delta)$. Then $y \in S(xT, \varepsilon)$ implies $\rho(yt, M) \to 0$ as $t \to \infty$. Consider now the set $N = S(xT, \varepsilon)(-T) = \{y(-T) : y \in S(xT,\varepsilon)\}$. This set is a neighborhood of x. Note that $y \in N$ if and only if $yT \in S(xT, \varepsilon)$. Thus if $y \in N$, then $\rho(yt, M) \to 0$ as $t \to \infty$, and $A(M)$ is open which completes the proof.

It is easy to see that: a uniform attractor is an attractor, an attractor is a semi-attractor, and a weak attractor is a semi-weak attractor. Any other implication need not and does not hold in general. This is shown by the following examples.

1.6.17 *Example.*

i) <u>Semi Weak Attractor</u> Consider the flow shown in Figure 1.6.18.

1.6.18 *Figure.*

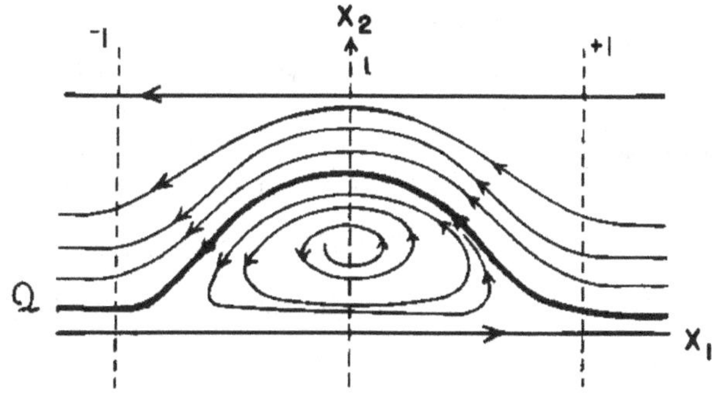

In the strip $-1 \leq x_2 \leq +1$ this flow has the property that the positive semi-trajectory $\Lambda^+(x)$ through all points $x = (x_1, x_2)$, with $x_1 \leq \frac{x_2-1}{x_2^2}$ and x_2 arbitrary and the negative semitrajectory $\gamma^-(x)$ through all points $x = (x_1,x_2)$ with $x_1 \geq \frac{1-x_2}{x_2^2}$ and x_2 arbitrary are straight lines parallel to the axis x_1. The flow for $x_1 \in (\frac{x_2-1}{x_2^2}, \frac{1-x_2}{x_2^2})$ is completed as shown in *Figure* 1.6.18. Where the separatrix Q is the trajectory $\gamma_g(x)$ with the property that for all $x \in E^2$, $x \notin Q$ implies $\rho(x,\{x:x_2=0\}) \neq 0$, while $\lim_{x_1 \to \pm\infty} \rho(\gamma_g(x),\{x:x_2=0\}) \to 0$.

Notice that then the positive limit set of all trajectories in the region G bounded by Q and the axis x_1 is the set $\Lambda^+(G) = Q \cup \{x:x_2=0\}$. Thus the set $\{x:x_2 = 0\}$ is a semi-weak attractor, but not a semi-attractor.

ii) Semi-attractor. Consider the flow shown in *Figure* 1.6.19.

1.6.19 *Figure*

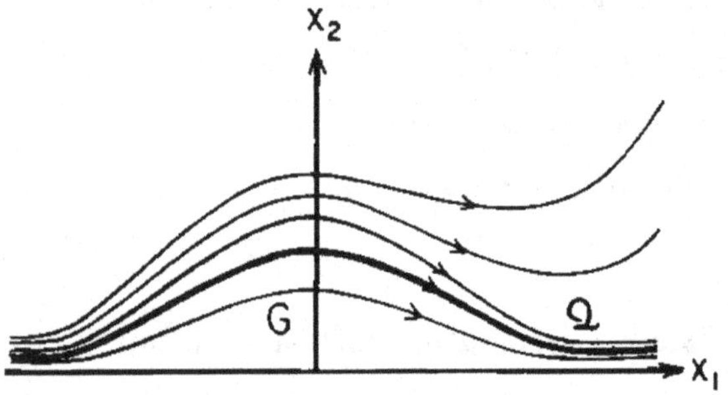

The trajectory Q has the same properties as the one in example i). The trajectories inside the region G bounded by Q and the axis x_1 are such that $y \in G$ implies that $\lim_{t \to +\infty} \rho(yt,\{x:x_2=0\}) = 0$. The set $\{x:x_2 = 0\}$ is a semi-attractor, but not a weak attractor.

iii) Weak attractor. Consider the flow shown in *Figure* 1.6.20.

1.6.20 *Figure*

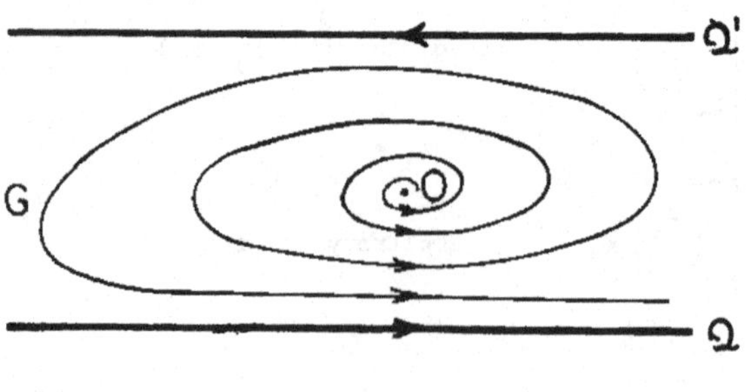

Let G be the infinite strip bounded by the parallel trajectories Q and Q^1. The flow may be for instance like the one defined in example 1.3.9 (ii).

The point $\{0\}$ is an equilibrium point. This flow has the property $\Lambda^+(G \setminus \{0\}) = \mathcal{Q} \cup \mathcal{Q}^1$. Then both $\mathcal{Q}$ and $\mathcal{Q}^1$ are weak attractors, but not attractors.

iv) Attractor. Consider the flow shown in *Figure* 1.6.21.

1.6.21 *Figure*

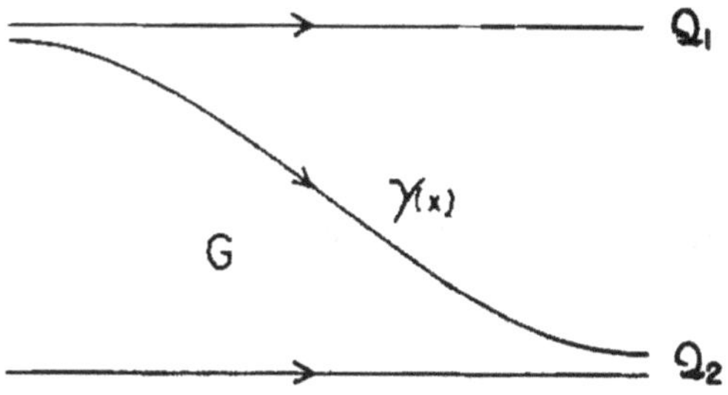

Let G be the infinite strip bounded by the parallel trajectories $\mathcal{Q}_1$ and $\mathcal{Q}_2$. Assume that for all $x \in G$ the flow has the same properties of the trajectory $\gamma(x)$ shown in the *Figure* 1.6.21, i.e.,

$\lim_{t \to -\infty} \rho(\mathcal{Q}_1, \gamma(x)) \to 0$ and $\lim_{t \to +\infty} \rho(\mathcal{Q}_2, \gamma(x)) \to 0$ uniformly. Then the set $M = \mathcal{Q}_1 \cup \mathcal{Q}_2$ is an attractor, but not a uniform attractor.*

In the case of compact sets, or closed sets with a compact vicinity one can prove that a semi-weak attractor is a weak attractor, and that a semi-attractor is an attractor. The proof rests on the fact that if M is compact, or is a closed set with a compact vicinity, then, if $O(M)$ is any open set containing M, one has for a sufficiently small $\varepsilon > 0$, $O(M) \supset S(M,\varepsilon)$, as is shown in Theorem 1.6.5. The proofs of these assertions are, therefore, omitted. Then:

1.6.22 *THEOREM*

If $M \subset E$ *is a closed set with a compact vicinity then semi weak attraction implies weak attraction and semi attraction implies attraction.*

It remains to be proven that in the case of compact sets the definitions 1.5.6 and 1.6.11 of an attractor are equivalent.

---

*Analytical examples can be found in a paper by Bhatia [1].

1.6.23 *THEOREM*

If $M \subset E$ *is compact the definition* 1.5.6 *of attractor is equivalent to definition* 1.6.11.

Proof: If (1.6.11) holds, then any sequence $\{xt_n\}, t_n \to +\infty$, and $x \in S(M,\delta)$ is in a compact set. Thus we may assume that it converges. Hence, for each $x \in S(M,\delta)$, $\Lambda^+(x) \neq \emptyset$. Notice further that $\rho(xt_n,M) \to 0$ as $t_n \to +\infty$. Thus if $xt_n \to y$, we have $\rho(y,M) = 0$, also $y \in M$ as M is closed, i.e., also $\Lambda^+(x) \subset M$.

Now assume that (1.5.6) holds. Assume, if possible, that $\rho(xt,M) \not\to 0$, as $t \to +\infty$. Then there is a sequence $\{\tau_n\}$, $\tau_n \to +\infty$ and $\rho(x\tau_n,M) \geq \varepsilon > 0$. We may assume that $\rho(x\tau_n,M) = \varepsilon$ for all n. As the set $\{y:\rho(y,M) = \varepsilon\}$ is compact, we can assume that $x\tau_n \to y$. Then $\rho(y,M) = \varepsilon$, so that $y \notin M$. But $y \in \Lambda^+(x) \subset M$, which is a contradiction, and proves the theorem.

If M is not compact, we can prove the following weaker version of Theorem 1.5.24.

1.6.24 *THEOREM*

If M *is a positively invariant closed set which is uniformly attracting, then it is stable.*

The proof follows from that of Theorem 1.5.24, when we notice that for any $\varepsilon > 0$ and $x \in M$, there is a $\delta_x > 0$ such that $yR^+ \subset S(M,\varepsilon)$ for $y \in S(x,\delta_x)$. Thus for $y \in O(M) = \bigcup_{x \in M} S(x,\delta_x)$, we have $yR^+ \subset S(M,\varepsilon)$, and since $O(M)$ is open, this implies stability of M.

By combining the five possible attracting properties with the two possible forms of stability we shall now define six different forms of asymptotic stability of sets. It is, in fact, easy to prove that

1.6.25 *THEOREM*

If a set $M \subset E$ *is (uniformly) stable and semi-weakly attracting, then it is semi-attracting. If a set* $M \subset E$ *is (uniformly) stable and weakly attracting, then it is attracting.*

Proof: We shall give the detailed proof only of the first statement; the proof of the second is similar.

If the assertion is not true there exists at least one sequence $\{t_n\}: t_n \to +\infty$, such that $\rho(yt_n, M) \not\to 0$ whereas there is a sequence $\{\tau_n\}$, $\tau_n \to +\infty$ such that $\rho(y\tau_n, x) \to 0$, for some $x \in M$. We may assume that $t_n \geq \tau_n$. Then the fact that $yt_n = y\tau_n(t_n - \tau_n)$ shows that definition 1.6.1 is contradicted and proves the theorem.

1.6.26 *DEFINITION*

If a set $M \subset E$ *is (uniformly) stable and semi-attracting it is called a (uniformly) stable semi-attractor.*

If a set $M \subset E$ *is (uniformly) stable and attracting it is called a (uniformly) stable attractor or an (asymptotically stable set).*

If a set $M \subset E$ *is (uniformly) stable and uniformly attracting it is called a (uniformly) stable uniform attractor or a (uniformly asymptotically stable set).*

We shall now give some examples of the various properties presented in definition 1.6.26.

1.6.27 *Examples*

i) Stable semi-attractor. Consider the flow shown in *Figure* 1.6.28. This flow

1.6.28 *Figure*

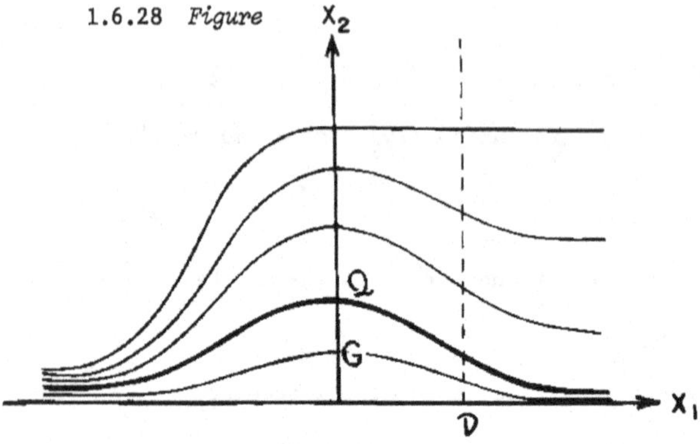

is essentially a variation of the flow shown in *Figure* 1.6.19. The only difference is that while the ax in 1.6.19 was not stable in 1.6.28 it is (positively Liapunov) stable. Stability is achieved by the property that now for $x \notin G$, $x_1 \geq D$ and x_2 arbitrary, the corresponding positive semi-trajectory $\gamma^+(x)$ is a straight line parallel to the axis x_1.

ii) Uniformly stable semi-attractor. Consider the flow shown in *Figure* 1.6.29.

1.6.29 *Figure*

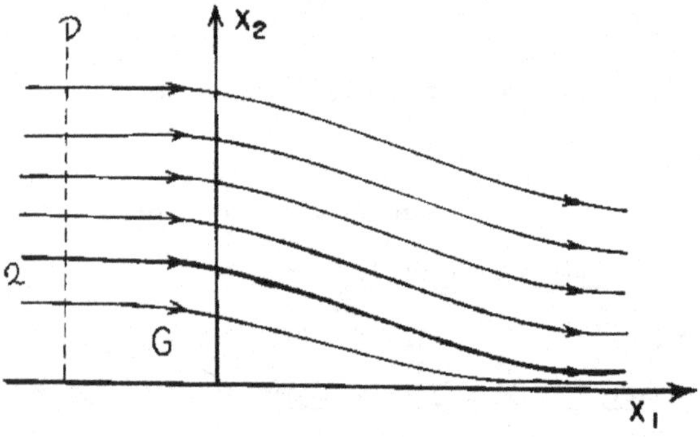

This flow has the property that for $x_1 \leq D$ and x_2 arbitrary the corresponding negative semi-trajectory $\gamma^-(x)$ is a straight line parallel to the axis x_1. Thus the region G bounded by the axis x_1 and the separatrix Q is an infinite strip in the direction $x_1 \to -\infty$. Clearly the set $\{x_2 = 0\}$ is uniformly (positively Liapunov) stable. The flow shown in *Figure* 1.6.29 has also the property that in the region $C(G)$, for $x_1 \geq (1 - x_2) / x_2^2$ the trajectories $\gamma(x)$ are straight lines parallel to the axis x_1. Hence the set $\{x_2 = 0\}$ is a uniformly stable semi-attractor.

iii) Stable attractor. Consider the flow shown in *Figure* 1.6.30. This flow is

1.6.30 *Figure*

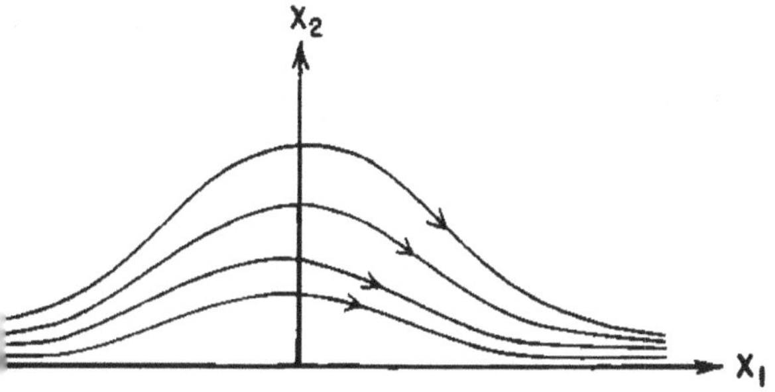

essentially a variation of the flows shown in *Figure* 1.6.4. Now the flow has the additional property that for all $x \in E^2$

$$\lim_{t \to +\infty} \rho(\gamma(x), \{x : x_2 = 0\}) \to 0.$$

Hence the set $\{x : x_2 = 0\}$ is a stable attractor.

iv) Uniformly stable attractor.

In the euclidean plane, consider the system

$$\dot{x}_1 = 1, \quad \dot{x}_2 = 0 \qquad \text{for} \quad x_1 \leq 0,$$

$$\dot{x}_1 = 1, \quad \dot{x}_2 = - \frac{2x_1x_2}{(1+x_1^2)} \qquad \text{for} \quad x_1 \geq 0.$$

The solution through any point (x_1^0, x_2^0) has the form

$$x_1 = t + x_1^0, \quad x_2 = \frac{1 + (x_1^0)^2}{1 + (t + x_1^0)^2} \quad x_2^0$$

for $t \geq -x_1^0$, and

$$x_1 = t + x_1^0, \quad x_2 = x_2^0 \qquad \text{for} \quad t \leq -x_1^0 \quad .$$

The x_1-axis is a uniformly stable attractor, but is not a uniform attractor.

In the proof of Theorems (1.5.11) and (1.5.14) no use has been made of the compactness of M. We may assume that these two theorems are true in the general case. The proof is left as exercise to the reader.

1.6.31 *THEOREM*

Let $M \subset E$ *be a closed attractor. Then* $A(M)$ *is open.*

1.6.32 *THEOREM*

Let $M \subset E$ *be a closed attractor. Then the set* $A(M)$ *is invariant.*

1.6.33 *THEOREM*

If a closed attractor $M \subset E$ *has a compact neighborhood the definition (1.6.26) is equivalent to the definitions (1.5.15) and (1.5.16).*

The proof follows immediately from Theorems (1.6.5), (1.6.22) and (1.6.23).

We shall now define and investigate a certain number of other properties of set; the instability properties. We shall first define two types of instability as the opposite of the two forms of stability defined in 1.6.1 then define various forms of negative attraction and complete instability. The classification that we give for these properties may not exhaust all possible behaviors.

1.6.34 *DEFINITION*

A set $M \subset E$ *is called*

i) _unstable_ *if it is not stable, i.e., if there exists an* $\varepsilon > 0$, *a point* $x \in M$, *a sequence* $\{x_n\}: x_n \in C(M); x_n \to x$ *and a sequence* $\{t_n\}; t_n \in R^+$ *such that* $\rho(x_n t_n, M) \geq \varepsilon$.

ii) _weakly unstable_ *if it is not uniformly stable, i.e., if there exists an* $\varepsilon > 0$, *a sequence* $\{x_n\}: x_n \in C(M)$; *and a sequence* $\{t_n\}: t_n \in R^+$ *such that* $\rho(x_n, M) \to 0$ *implies* $\rho(x_n t_n, M) \geq \varepsilon$.

1.6.35 *Remark*

It is important to point out that a set $M \subset E$ may be both stable and weakly unstable if it is stable, but not uniformly stable. A set with these properties is, for instance, the set M in the flow of *Figure* 1.6.4.
Again

1.6.36 *THEOREM*

If a compact set $M \subset E$ *is weakly unstable it is unstable.*

The difference between an unstable and a weakly unstable set lies in the different way with which a trajectory or a sequence of points leave the set M. If M is unstable, such a trajectory may possibly approach the set M again. We can then define a stronger form of instability when this does not happen, that is, if either the trajectory or the sequence of points will ultimately be bounded away from M. Thus

1.6.37 *DEFINITION*

A set $M \subset E$ *is called*

i) *ultimately unstable* *if there exists an* $\varepsilon > 0$, *a sequence* $\{x_n\}: x_n \in C(M), x_n \to x \in M$ *and a sequence* $\{t_n\}: t_n \in R^+$ *such that* $\rho(x_n(t_n + \tau), M) \geq \varepsilon$ *for all* $\tau \in R^+$.

ii) *ultimately weakly unstable* *if there exists an* $\varepsilon > 0$, *a sequence* $\{x_n\}: x_n \in C(M), x_n \to M$ *and a sequence* $\{t_n\}: t_n \in R^+$ *such that* $\rho(x_n(t_n + \tau), M) \geq \varepsilon$ *for all* $\tau \in R^+$.

Again it is easy to prove that

1.6.38 *THEOREM*

If $M \subset E$ *is a set with a compact vicinity then ultimate weak instability is equivalent to ultimate instability.*

We shall now introduce still stronger forms of instability and define properties of sets for which all trajectories and sequences in a certain neighborhood of it tend to leave. These definitions are made by requiring that all points of M have the property 1.6.34 i) or ii). It is, however, very important to point out that in this case the stronger form of the property 1.6.34 i) defines a weaker property than the stronger form of the property 1.6.34 ii).

1.6.39 *DEFINITION*

A set $M \subset E$ *is called*

i) *weakly completely unstable if there exists an* $\varepsilon > 0$ *such that for any sequence* $\{x_n\}: x_n \in C(M); x_n \to x \in M$, *there exist a sequence* $\{t_n\}; t_n \in R^+$ *such that* $\rho(x_n t_n, M) \geq \varepsilon$ *and ultimately weakly completely unstable if, in addition* $\rho(x_n(t_n + \tau), M) \geq \varepsilon$ *for all* $\tau \in R^+$.

ii) *completely unstable if there exists an* $\varepsilon > 0$ *such that for any sequence* $\{x_n\}: x_n \in C(M)$ $x_n \to M$, *there is a sequence* $\{t_n\}: t_n \in R^+$ *such that* $\rho(x_n t_n, M) \geq \varepsilon$ *and ultimately completely unstable if in addition* $\rho(x_n(t_n + \tau), M) \geq \varepsilon$ *for all* $\tau \in R^+$.

All the instability properties lised until now are the analogue of the "semi" properties for stability since they are essentially defined on open sets of M and not on spherical neighborhoods. It is immediate to prove that

1.6.40 *THEOREM*

A set $M \subset E$ *is ultimately completely unstable if and only if there exists an* $\varepsilon > 0$ *such that for all* $x \in S[M,\varepsilon] \setminus M$, *there exists a* $\tau(x) \in R^+$ *such that* $\rho(xt, M) > \varepsilon$ *for* $t \geq \tau(x)$.

Again it is easy to prove that

1.6.41 *THEOREM*

If $M \subset E$ *is a set with a compact vicinity then weak complete instability implies complete instability and ultimate weak complete instability implies ultimate complete instability.*

1.6.42 *Remark*

Obviously, by reversing the direction of motion on the trajectories all forms of stability and asymptotic stability will lead to some form of instability. It may happen that those negative asymptotic stability properties have even stronger instability properties than the one listed above since they characterize and classify the behavior of the flow also outside M. For practical reasons, however, these classifications are not very interesting in the case of instability.

1.6.43 *Examples*

i) Weak instability. Consider the flow shown in *Figure* 1.6.44. This flow has the

1.6.44 *Figure*

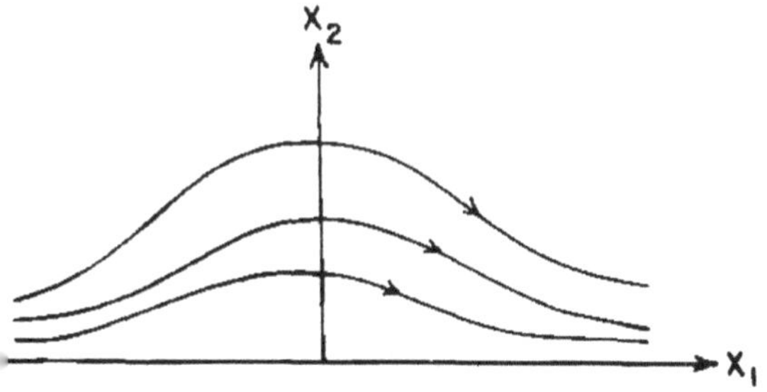

property that for all $x \in E^2$

$$\lim_{x_1 \to -\infty} \rho(\gamma(x), \{x: x_2 = 0\}) \to 0.$$

Thus the set $\{x: x_2 = 0\}$ is weakly unstable. On the other hand, it is neither unstable, nor ultimately weakly unstable.

ii) Instability. Consider the flow shown in *Figure* 1.6.45. This flow has an equilibrium

1.6.45 *Figure*

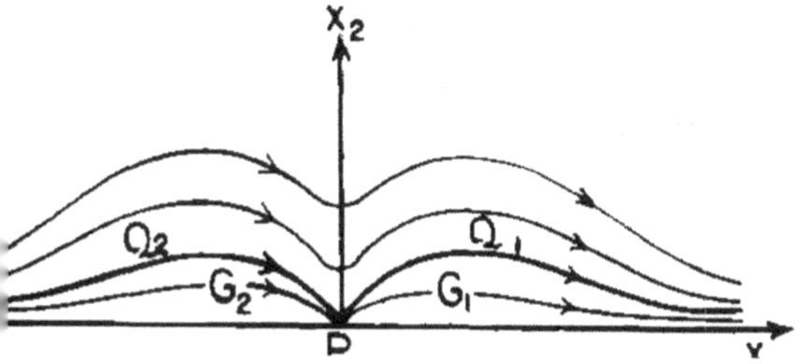

point P on the axis x_1. consider the separatrixes $\mathcal{Q}_1$ and $\mathcal{Q}_2$. Let's denote by G_1 and G_2 the region bounded by $\mathcal{Q}_1$ and $\mathcal{Q}_2$ and the axis x_1 respectively. This flow has the property that for all $x \in E^2$ it is $\lim\limits_{x_1 \to +\infty} \rho(\gamma(x),\{x: x_2 = 0\}) \to 0$ and for all $x \in C(G_1 \cup G_2)$ it is $\lim\limits_{x_1 \to -\infty} \rho(\gamma(x),\{x: x_2 = 0\}) \to 0$, while $x \in G_2$ implies $\Lambda^+(x) = \{p\}$ and $x \in G_1$ implies $\Lambda^-(x) = \{p\}$. The set $\{x: x_2 = 0\}$ is then unstable, but neither ultimately unstable nor completely unstable.

iii) Ultimate Weak Instability. Consider the flow shown in *Figure* 1.6.46. This flo

1.6.46 *Figure*

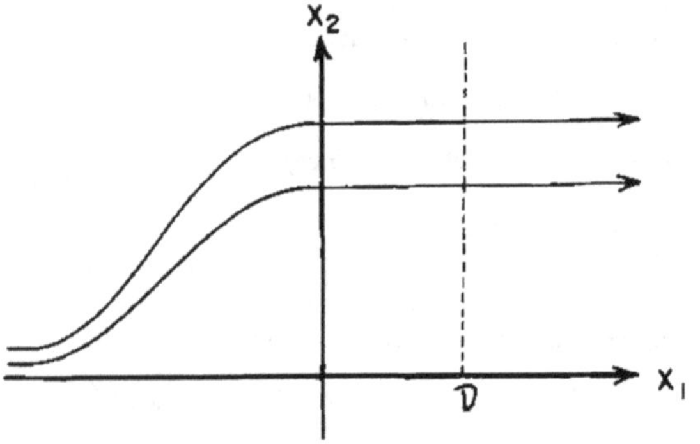

is essentially a variation of the flow shown in *Figure* 1.6.4 For $x_1 < 0$ the flow is the same as the one in *Figure* 1.6. Now for all $x \in E^2$, with $x_1 \geq \mathcal{D}$ and x_2 arbitrary the corresponding positive semi-trajectory $\gamma^+(x)$ is a straight line parallel to the axis x_1. Hence the set $\{x: x_2 = 0\}$ is ultimately weakly unstable.

iv) Ultimate Instability. Consider the flow shown in *Figure* 1.6.47. This flow is

1.6.47 *Figure*

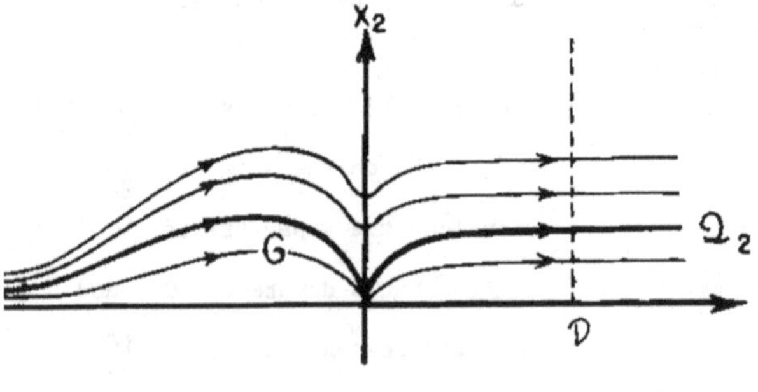

variation of the flow shown in *Figure* 1.6.45. It has the sam properties as $x_1 \to +\infty$ as th flow shown in *Figure* 1.6.45. The set $\{x: x_2 = 0\}$ is ultimatel unstable, but not completely ultimately unstable.

v) <u>Weak Complete Instability</u>. Consider the flow shown in *Figure* 1.6.48. This

1.6.48 *Figure*

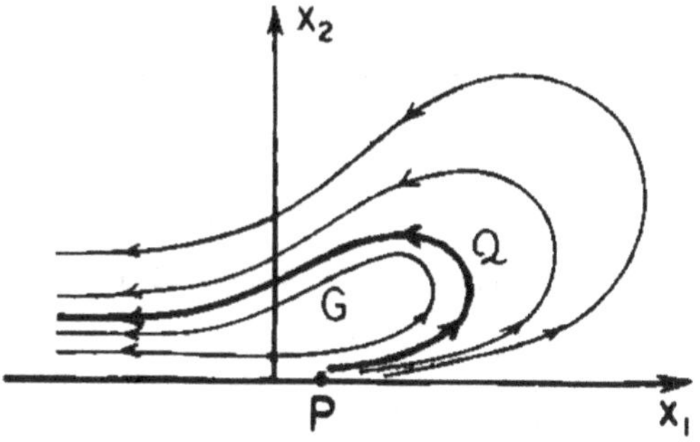

flow has an equilibrium point P on the axis x_1. Consider the separatrix Q. Denote with G the region bounded by Q and the axis x_1. This flow has the property that for all $x \in E^2$,

$$\lim_{t \to +\infty} \rho(\pi_x(t), \{x : x_2 = 0\}) \to 0,$$

for all $x \in G$, it is

$\lim_{t \to -\infty} \rho(\pi_x(t), \{x : x_2 = 0\}) \to 0$ and for $x \in \overline{C(G)}$, $\Lambda^-(x) = P$. The set $\{x : x_2 = 0\}$ is weakly completely unstable.

vi) <u>Ultimate Weak Complete Instability</u>. Consider the flow shown in *Figure* 1.6.49.

1.6.49 *Figure*

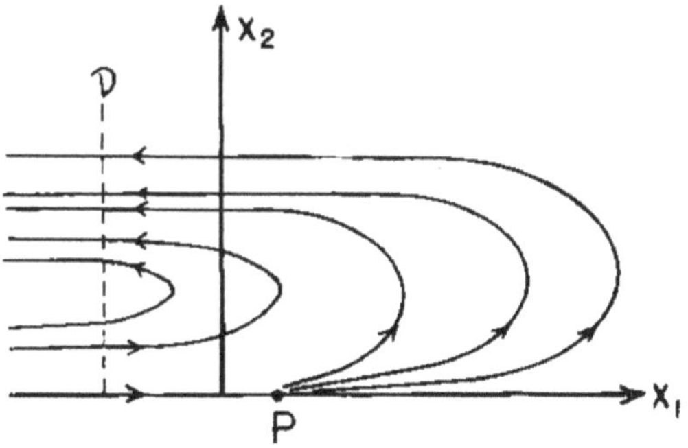

This flow is a variation of the flow shown in *Figure* 1.6.48. It has the same property for $x_1 > 0$ as the flow shown in *Figure* 1.6.48. For $x_1 \leq D$ the flow is modified in the usual way.

vii) Complete Instability. Consider the flow shown in *Figure* 1.6.50. This flow has the equilibrium point P and in addition all points of the semi-axis $x_1 > 0$ are equilibrium points We shall denote the semi-axis $x_1 > 0$ with M. Consider the separatrix $\mathcal{Q}$. Denote with G the (shaded) region bounded by $\mathcal{Q}$ and the set M. This flow has the following properties. For $x \in G$, the trajectories $\gamma(x)$ are closed bounded curves clustering around P and filling the set $I(G)$. For $x \in \overline{C(G)} \setminus M$, $\Lambda^-(x) = \{0\}$ and $\lim_{t \to +\infty} \rho(\pi_x(t), M) \to 0$. The set M is completely unstable.

1.6.50 *Figure*

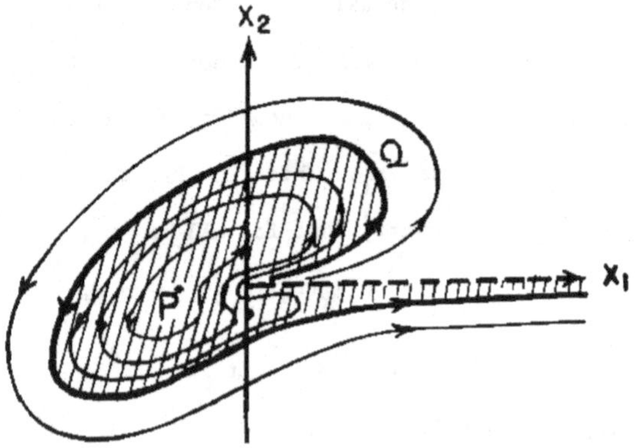

viii) Ultimate Complete Instability. Consider the flow shown in *Figure* 1.6.51. The set $B_1 \cup B_2$ is ultimately complet pletely unstable.

1.6.51 *Figure*

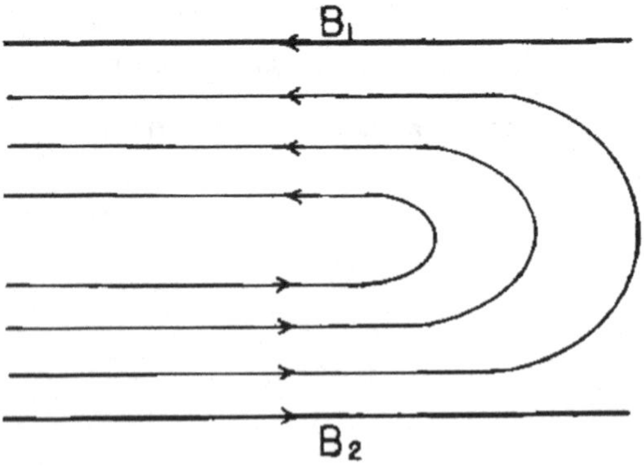

1.6.52 *Notes and References*

Some of the above given definitions have been presented under slightly different forms by Zubov [6] and by Bhatia [1].

1.7 *Stability and Liapunov functions.*

In this section we shall formulate some necessary and sufficient conditions for a closed set $M \subset E$ to be stable, uniformly stable, stable semi-attracting and asymptotically stable in terms of the existence of certain scalar function $v = \phi(x)$ defined on a suitable neighborhood of M. In this section we are interested in deriving necessary and sufficient conditions for the above mentioned properties which require weak properties of the scalar function $v = \phi(x)$. In Section 2. we shall, on the other hand, be interested in giving necessary and sufficient conditions for the above mentioned properties of a very sharp type, i.e., by using scalar functions $v = \phi(x)$ of a very special type.

1.7.1 *THEOREM*

Necessary and sufficient for stability of a closed set $M \subset E$ *is the existence of a function* $\phi(x)$ *defined on a positively invariant open set* W *such that* $W \supset S(x, \delta(x))$ *for all* $x \in M$ *and some* $\delta = \delta(x) > 0$ *and having the following properties:*

i) For every $\varepsilon > 0$*, there exists a* $\delta > 0$ *such that* $x \in W \setminus S(M,\varepsilon)$ *implies* $\phi(x) \geq \delta$*,*

ii) For every $\varepsilon > 0$ *and a compact set* $K \subset E$*, there exists a* $\delta > 0$ *such that* $\phi(x) \leq \varepsilon$ *for* $x \in W \cap S(M,\delta) \cap K$*,*

iii) $\phi(xt) \leq \phi(x)$*, for* $x \in W$ *and* $t > 0$.

Proof. The conditions are sufficient, because for any given $\varepsilon > 0$ such that $W \setminus S(M,\varepsilon) \neq \emptyset$, and any compact set $K \subset E$, we can choose $\delta > 0$ such that

$$\sup\{\phi(x): x \in W \cap S(M,\delta) \cap K\} < \inf\{\phi(x): x \in W \setminus S(M,\varepsilon)\},$$

and $S(M,\delta) \cap K \subset W$.

We can then assert that $x \in S(M,\delta) \cap K$ implies $\gamma^+(x) \subset S(M,\varepsilon)$. For by hypothesis (iii)

$$\phi(xt) \leq \phi(x) < \inf\{\phi(x): x \in W \setminus S(M,\varepsilon)\}.$$

This implies $xt \in S(M,\varepsilon)$ for $t > 0$, i.e., that M is stable.

To prove the necessity, set

$$W = \{x: \gamma^+(x) \subset S(M,\nu)\}.$$

where $\nu > 0$ is arbitrary. The set W is positively invariant, open and such that $M \subset W$. The first assertion is obvious. We prove that W is open and such that $M \subset W$. Let $x \in M$ then there exists a $\mu > 0$ such that $\overline{S(x,\mu)}$ is compact. By hypothesis, for $\nu > 0$ and the compact set $\overline{S(x,\mu)}$, there exists $\delta > 0$ such that $y \in \overline{S(x,\mu)} \cap S(M,\delta)$ implies $\gamma^+(y) \subset S(M,\nu)$. Now let $\eta = \min(\mu,\delta)$. Then as $x \in M$, $S(x,\eta) \subset \overline{S(x,\mu)} \cap S(M,\delta)$ and consequently $S(x,\eta) \subset W$, i.e., for all $x \in M$ there exists $\eta(x) > 0$ such that $S(x, \eta(x)) \subset W$. For $x \in W$ define now

1.7.2 $$\phi(x) = \sup\{\rho(x\tau,M): \tau \geq 0\}.$$

Then

$$\phi(xt) = \sup\{\rho(x(t+\tau),M): \tau \geq 0\}$$

$$= \sup\{\rho(x\tau,M): \tau \geq t\} \leq \phi(x), \quad \text{if } t > 0,$$

i.e., $\phi(x)$ has the property (iii).

Since $\phi(x) \geq \rho(x,M)$, $x \in W \setminus S(M,\varepsilon)$ implies $\phi(x) \geq \varepsilon$, so that $\phi(x)$ has property (i).

Lastly, for any $\varepsilon > 0$, choose $\delta > 0$ such that $x \in S(M,\delta)$ implies $\gamma^+(x) \subset S(M,\varepsilon)$. Then $\phi(x) \leq \varepsilon$ for $x \in S(M,\delta)$ and in particular for $x \in S(M,\delta) \cap W$. This is property (ii). Q.E.D.

1.7.3 *COROLLARY*

A sufficient condition for stability of a closed set $M \subset E$ *is the existence of a continuous function* $\phi(x)$ *defined in some* $S(M,\delta)$, $\delta > 0$, *and a continuous monotonic increasing function* $\alpha(r), \alpha(0) = 0$, *defined for* $0 \leq r < \delta$, *such that*

(i) $\alpha(\rho(x,M)) \leq \phi(x), \phi(x) = 0$, *for* $x \in M$.

(ii) $\phi(xt) \leq \phi(x)$ *for* $t > 0$ *and* $x \in S(M,\delta)$.

We now give a similar theorem for uniform stability

1.7.4 *THEOREM*

Necessary and sufficient for the uniform stability of a closed set $M \subset E$ *, is the existence of a function* $\phi(x)$, *defined on a positively invariant set* $W \supset S(M,\delta)$, $\delta > 0$, *and having the following properties.*

i) For every $\varepsilon > 0$, *there exists a* $\delta > 0$ *such that* $\phi(x) \geq \delta$ *for* $x \in W \setminus S(M,\varepsilon)$;

ii) For every $\varepsilon > 0$, *there exists a* $\delta > 0$ *such that* $\phi(x) \leq \varepsilon$ *for* $x \in W \cap S(M,\delta)$;

iii) $\phi(xt) \leq \phi(x)$ *for* $x \in W$ *and* $t > 0$.

Proof. The conditions are clearly sufficient. We prove the necessity. For a fixed $\varepsilon_0 > 0$ define $W = \{x : \gamma^+(x) \subset S(M, \varepsilon_0)\}$. This set W is positively invariant and such that $W \supset S(M,\delta), \delta > 0$. In fact, by the definition of uniform stability there exists a $\delta_0 > 0$ such that $x \in S(M,\delta_0)$ implies $\gamma^+(x) \subset S(M,\varepsilon_0)$. Thus $S(M,\delta_0) \subset W$ and $W \supset S(M,\delta)$, $\delta > 0$. Again if $x \in W$ and $t > 0$, then $xt \in W$, for $\gamma^+(xt) \subset \gamma^+(x) \subset W$.

Thus W is positively invariant.

Now, for $x \in W$ define $\phi(x)$ as in 1.7.2. This $\phi(x)$ has all the properties (i),(ii),(iii), which can be verified as in the Theorem 1.7.1.

1.7.5 *COROLLARY*

A sufficient condition for the uniform stability of a closed set $M \subset X$ *is the existence of a continuous function* $\phi(x)$ *defined in some* $S(M,\delta)$ *for* $\delta > 0$, *and two continuous monotonic increasing functions* $\alpha(\mu)$ *and* $\beta(\mu)$, $\alpha(0) = \beta(0) = 0$, *defined for* $0 \leq \mu < \delta$, *such that*

(i) $\alpha(\rho(x,M)) \leq \phi(x) \leq \beta(\rho(x,M))$ *for* $x \in S(M,\delta)$ *and*

(ii) $\phi(xt) \leq \phi(x)$ *for* $t > 0$.

1.7.6 *Remark*

It is to be noted that the theorem 1.7.4 does not predict the existence of a continuous function $\phi(x)$ possessing the properties mentioned in case of uniform stability. Notice, for instance, that no continuous function $\phi(x)$ satisfying the conditions of Theorem 1.7.4 can exist for the stable rest point p in Example 1.5.32 (v).

The situation that even for a compact set M which is stable (and hence uniformly stable), a continuous function satisfying conditions of Theorem 1.7.4 need not exist, has led to the introduction of a host of stronger concepts of stability, each lying somewhere between stability and asymptotic stability. This has been made possible by the general theory of prolongations. These we shall discuss in later sections, but let it be mentioned that the concept of stability, for which the existence of a continuous $\phi(x)$ satisfying conditions of Theorem 1.7.4 is guaranteed, is called absolute stability. We may, however, note that even in the case of ordinary stability, any function $\phi(x)$ satisfying conditions of Theorem 1.7.4, is continuous at all points of M, and that, in general, a function $\phi(x)$, continuous along the trajectories of the dynamical system in a neighborhood N of M, does always exist.

We now prove a theorem on stable semi-attractors.

1.7.7 *THEOREM*

A necessary and sufficient condition, that a closed set $M \subset E$ *be a stable semi-attractor is the existence of a continuous function* $\phi(x)$ *defined in an open invariant set* W *such that* $W \supset S(x, \delta(x))$ *for all* $x \in M$ *and some* $\delta = \delta(x) > 0$, *which satisfies the following conditions:*

i) *there is a continuous monotonic increasing function* $\alpha(\mu)$, $\alpha(0) = 0$, *defined for* $\mu \geq 0$, *such that*

$$\phi(x) \geq \alpha(\rho(x,M)) \quad \textit{for} \quad x \in W,$$

ii) *for every* $\varepsilon > 0$ *and a compact set* K, *there is a* $\delta > 0$ *such that* $\phi(x) \leq \varepsilon$ *for* $x \in W \cap K \cap S(M,\delta)$,

iii) $\phi(xt) < \phi(x)$ *for* $x \in W \setminus M, t > 0$,

iv) $\phi(xt) \to 0$ *as* $t \to \infty$.

Proof. That the conditions are sufficient is clear, we prove the necessity.

We set W = interior of $A(M)$. This is an open invariant set with $M \subset A(M)$. Let now for $x \in W$

$$\phi(x) = \sup \{\rho(x\tau,M) : \tau \geq 0\}.$$

Then $\phi(x)$ has properties (i), (ii) and (iii). We prove that it is also continuous. For this purpose, we define $W_\varepsilon = \{x \in W : \gamma^+(x) \subset S(M,\varepsilon)\}$. This W_ε is open, positively invariant, has the property $M \subset W_\varepsilon$ and further has the important property that for each $x \in W$, there exists a $T > 0$ such that $xT \in W_\varepsilon$. Let now $x \in W$ and $\phi(x,M) = \lambda$. There exists a $T > 0$ such that $xT \in W_\mu$, where $\mu = \lambda/4$. We can choose $\sigma > 0$ such that $S(xT,\sigma) \subset W_\mu$ and $\overline{S(xT,\sigma)}$ is compact. Then $S(xT,\sigma)(-T) = N$ is a neighborhood of x. We can choose, then a $\eta > 0$ such that $\eta < (\lambda/4)$ and $S(x,\eta) \subset N$. Then if $y \in S(x,\eta)$,

$$|\phi(x) - \phi(y)| = |\sup_{\tau\geq 0} \rho(x\tau,M) - \sup_{\tau\geq 0} \rho(y\tau,M)|$$

$$\leq \sup_{0\leq\tau\leq T} |\rho(x\tau,M) - \rho(y\tau,M)|$$

$$\leq \sup_{0\leq\tau\leq T} \rho(x\tau,y\tau) \quad .$$

This shows, however, that $\phi(y) \to \phi(x)$ as $y \to x$. Thus $\phi(x)$ is continuous. To get a $\phi(x)$ with property (iii) we can set

$$\Phi(x) = \int_0^\infty e^{-\tau}\phi(x\tau)d\tau ,$$

which has all the properties (i-iv). Q.E.D.

We now prove the following theorem on asymptotic stability.

1.7.8 *THEOREM*

A closed set $M \subset E$ *is asymptotically stable if and only if there exists a continuous scalar function* $\phi(x)$ *defined on an open invariant set* $W \supset S(M,\delta)$, $\delta > 0$ *having the properties:*

i) There exist two continuous monotonic increasing functions $\alpha(\mu)$ *and* $\beta(\mu)$, *defined for* $\mu \geq 0$, $\alpha(0) = \beta(0) = 0$, *such that*

$$\alpha(\rho(x,M)) \leq \phi(x) \leq \beta(\rho(x,M)) \quad \textit{for} \quad x \in W,$$

ii) $\phi(xt) < \phi(x)$ *for* $x \in W \setminus M, t > 0$,

iii) $\phi(xt) \to 0$ *as* $t \to \infty$ *for each* $x \in W$.

Proof. The conditions are obviously sufficient. We prove the necessity. By hypothesis, the region of attraction $A(M)$ is such that $A(M) \supset S(M,\delta)$, $\delta > 0$ and it is open and invariant by Theorem 1.6.16. For $x \in W = A(M)$ define

$$\phi(x) = \sup\{\rho(x\tau,M): \tau \geq 0\}.$$

Clearly $\phi(x) \geq \rho(x,M)$, so that we can set $\alpha(\mu) = \mu$. To see the existence of $\beta(\mu)$, we note first that $\phi(x) \equiv 0$ for $x \in M$. Now for any $\varepsilon > 0$, define $\delta(\varepsilon) = \sup \{\delta > 0 : x \in S(M,\delta)$ implies $\gamma^+(x) \subset S(M,\varepsilon)\}$. Then $\delta(\varepsilon)$ is positive for $\varepsilon > 0$, is nondecreasing and $\delta(0) = 0$. We can choose a continuous monotonic increasing function $\delta^*(\varepsilon)$, such that $\delta^*(0) = 0$ and $\delta(\varepsilon) \geq \delta^*(\varepsilon)$. Let now $\beta(\mu)$ be the inverse function of $\delta^*(\varepsilon)$, then we have $\phi(x) \leq \beta(\rho(x,M))$. Lastly note that $\phi(x)$ has the property (iii) for all $x \in W$.

We now prove that $\phi(x)$ is also continuous. Let $\rho(x,M) = \lambda > 0$ for a given $x \in W \setminus M$. Let $\delta > 0$ be a number which corresponds to $(\lambda/4)$ by the hypothesis of stability, i.e., $x \in S(M,\delta)$ implies $\gamma^+(x) \subset S(M,\lambda/4)$. Let $T > 0$ be chosen such that $\rho(xT,M) < \delta$. Then $\rho(xt,M) < (\lambda/4)$ for $t \geq T$. Choose now $\mu > 0$, such that $S(xT,\mu) \subset S(M,\delta)$, and $\overline{S(xT,\mu)}$ is compact (this is possible as $S(M,\delta)$ is open). Then the set $S(xT,\mu)(-T) = \{y(-T) : y \in S(xT,\mu)\} = N$ is a neighborhood of x. Further $z \in N$ implies $zT \in S(xT,\mu)$ and consequently $zt \in S(M,\lambda/4)$ for $t \geq T$. Thus

$$\begin{aligned} |\phi(x) - \phi(y)| &= |\sup_{\tau \geq 0} \rho(x\tau,M) - \sup_{\tau \geq 0} \rho(y\tau,M)| \\ &\leq \sup_{0 \leq \tau \leq T} |\rho(x\tau,M) - \rho(y\tau,M)| \\ &\leq \sup_{0 \leq \tau \leq T} \rho(x\tau,y\tau) \quad \text{if } y \in N \cap S(x,(\lambda/4)). \end{aligned}$$

But $\sup_{0 \leq \tau \leq T} \rho(x\tau,y\tau) \to 0$ as $y \to x$, so that $\phi(y) \to \phi(x)$ as $y \to x$, implying that $\phi(x)$ is continuous. This $\phi(x)$ may not have the property (ii), although $\phi(xt) \leq \phi(x)$ for $t > 0$ is satisfied. To have a $\phi(x)$ which also has the property (ii), we can set $\Phi(x) = \int_0^\infty e^{-\tau}\phi(x\tau)d\tau$, for $x \in W$. This $\Phi(x)$ has the properties (i), (ii) and (iii).

To see, that it has the property (ii), note that, $\phi(x\tau) \to 0$ as $\tau \to \infty$, and $0 < \phi(x\tau) \leq \phi(x)$ for $x \in W \setminus M$, $\tau > 0$.

Now examine the difference

$$\Phi(x) - \Phi(xT) = \int_0^\infty e^{-\tau}[\phi(x\tau) - \phi(x(T + \tau))]d\tau, \quad T > 0.$$

By the properties of $\phi(x)$, there is an interval (t_1,t_2) such that $\phi(x\tau) - \phi(x(T + \tau)) > 0$ for $\tau \in (t_1,t_2)$. Then

$$\Phi(x) - \Phi(xT) \geq \int_{t_1}^{t_2} e^{-\tau}[\phi(x\tau) - \phi(x(T + \tau))]d\tau > 0.$$

This proves the theorem completely.

1.7.9 *Remarks*

For compact sets M, we do not need condition (iii) as will be proved next. That condition (iii) is essential for noncompact sets can be seen from the following example.

1.7.10 *Example*

Consider the dynamical system defined by the differential equations

$$\dot{x} = 1, \dot{y} = -2xy/(1 + x^2)(2 + x^2) \quad \text{for} \quad x \geq 0$$

$$= 2xy/(1 + x^2) \quad \text{for} \quad x \leq 0,$$

in the euclidean plane. The x-axis is stable, but not asymptotically stable. If the x-axis is denoted by M, then we may define $\phi(x,y) = |y| = \rho((x,y),M)$. This function has the properties (i) and (ii) required in the above theorem, but not the property (iii).

1.7.11 *THEOREM*

A necessary and sufficient condition for the compact set $M \subset E$ *to be asymptotically stable is that there exists a continuous scalar function* $\phi(x)$ *defined in a positively invariant neighborhood of* M *and such that there exists two continuous strictly increasing functions* $\alpha(\nu)$ *and* $\beta(\nu)$, *defined for* $\nu \geq 0$, $\alpha(0) = \beta(0) = 0$, *such that*

i) $\alpha(\rho(x,M)) \leq \phi(x) \leq \beta(\rho(x,M))$ *for* $x \in N$

ii) $\phi(xt) < \phi(x)$ *for* $x \in N \setminus M$, $t > 0$

Proof. Sufficiency. Let $\varepsilon > 0$. The set $S[M,\varepsilon]$ is compact. The stability part follows from 1.7.4. Choose $\delta > 0$ such that

1.7.12 $$\beta(\delta) \geq \alpha(\varepsilon)$$

If $x \in S(M,\delta)$, then $xR^+ \subset S(M,\varepsilon)$. For, if not, then there is a $\tau > 0$ such that $x\tau \in \partial S(M,\varepsilon)$. Hence

$$\alpha(\varepsilon) = \alpha(\rho(x\ ,M)) < \phi(x\tau) < \phi(x) \leq \beta(\delta).$$

which contradicts 1.7.12. This proves stability of M. It must now be proved that M is an attractor. If M is not an attractor then there exists an $x \subset S(M,\delta)$ such that $\lim_{t \to +\infty} \rho(xt,M) \neq 0$. Then there exists a $\eta > 0$ and a sequence $\{t_n\}$: $t_n \to +\infty$ such that

$$\eta \leq \rho(xt_n,M) < \varepsilon$$

Since $S[M,\varepsilon] \setminus S(M,\eta)$ is compact, the sequence $\{xt_n\}$ has a convergent sub sequence. Thus there is a point $y \in \Lambda^+(x)$ with $y \notin M$. There exists therefore a $\tau > 0$ such that by condition (ii)

1.7.13 $$\phi(y\tau) < \phi(y)$$

However, since $y\tau \in \Lambda^+(x)$ and $y \in \Lambda^+(x)$ there are sequences $\{t_n\}$ and $\{t'_n\}$

such that $xt_n \to y$ and $xt_n' \to y\tau$. We might assume, without loss of generality that $t_n > t_n'$ for each n. Then

$$\phi(xt_n) < \phi(xt_n')$$

and proceeding to the limit, since ϕ is continuous we obtain that

$$\phi(y) \leq \phi(y\tau)$$

which contradicts 1.7.13. Thus it must be $\lim_{t\to+\infty} \rho(xt,M) = 0$ for each $x \in S(M,\delta)$. This completes the proof of sufficiency.

Necessity: The set $A(M)$ is an open and invariant neighborhood of M. For $x \in A(M)$ set.

1.7.14 $$\phi(x) = \sup\{\rho(x\tau,M): \tau \geq 0\}$$

Clearly $\phi(x) \geq \rho(x,M)$, so that we may take $\alpha(\mu) = \mu$. Since M is compact we need only prove that $\phi(x)$ is continuous, which can be done exactly as in the previous theorem, then the existence of $\beta(\mu)$ will follow from Theorem 0.3.2 .

The scalar function 1.7.14 may not have the property (ii), although $\phi(xt) \leq \phi(x)$ for $t > 0$ is satisfied, and $\phi(x\tau) \to 0$ as $\tau \to \infty$.

To have a $\phi(x)$ satisfying all the properties set

$$\Phi(x) = \int_0^\infty e^{-\tau}\phi(x\tau)\,d\tau, \qquad \text{for } x \in A(M) \quad .$$

This scalar function $\Phi(x)$ has the properties (i) and (ii). The proof is the same as in the previous theorem.

This proves the theorem completely.

1.7.15 *Remark*

Notice that any $\phi(x)$ satisfying the above theorem has the property that $\phi(xt) \to 0$ as $t \to \infty$, although this is not explicitly assumed in the hypothesis.

1.7.16 *Notes and References*

Almost all results presented in this section are due to N. P. Bhatia [1]. Few similar results can be found in the book by Zubov [6] and in a paper by Roxin [3]. The use of the function $\phi(x)$ for characterizing stability properties was introduced by Liapunov [1]. See 1.12.13 and 3.6.32.

1.8 *Topological methods.*

1.8.1 *DEFINITION*

Let N, M *be open sets with* $M \subset N \subset E$. *Let* N *be positively invariant for a flow* $\mathcal{F}$ *on* E. *A point* $x \in \partial M \cap N$ *is called an* *egress point* *(or an ingress point) of* M *if there exists an* $\varepsilon > 0$ *such that* $x(-\varepsilon, 0) \subset M$ *(or* $x(0,\varepsilon) \subset M$). *If in addition,* $x(0,\eta) \cap \bar{M} = \emptyset$ *(or* $x(-\eta,0) \cap \bar{M} = \emptyset$), *then* x *is called a* *strict egress point* *(or strict ingress point). Sometimes a point* $x \in \partial M \cap N$ *may be called a non-egress point, if it is not an egress point. The sets of egress points and strict egress points of* M *will be denoted respectively by* M_e *and* M_{se}.

1.8.2 *LEMMA*

If $\partial M \cap N$ *is either empty or consists only of non-egress points, then* M *is positively invariant.*

Proof. If M is not positively invariant, then there is an $x \in M$, and a $t \in R^+$, with $xt \notin M$. Let $\tau = \inf\{t \in R^+; xt \notin M\}$. Then, by continuity of π, $x(0,\tau) \subset M$, and $x\tau \in \partial M \cap N$, because $xR^+ \subset N$ (note N is positively invariant). Setting $y = x\tau$, we note that $x(0,\tau) = x\tau(-\tau,0) = y(-\tau,0)$, showing that y is an egress point of M, which is a contradiction. Q.E.D.

1.8.3 *DEFINITION*

Let U *be a topological space and* $V \subset U$. *A continuous mapping* $f: U \to V$ *is called a* *retraction* *of* U *onto* V *if* $f(u) \in V$ *for all* $u \in U$ *and* $f(v) = v$ *for all* $v \in V$. *When there exists a retraction of* U *onto* V, *the set* V *is called a* *retract* *of* U.

1.8.4 *THEOREM*

Let $M, N, M \subseteq N \subseteq E$, *be open sets such that* N *is positively invariant for a flow* $\mathcal{F}$ *on* E. *Let* $M_e = M_{se}$, *i.e., all egress points of* M *are strict egress points. Let* S *be a non-empty subset of* $M \cup M_e$ *such that*

$S \cap M_e$ *is a retract of* M_e, *but is not a retract of* S. *Then there exists at least one point* $x \in S \cap M$ *such that* $xR^+ \subset M$.

Proof. Suppose that the theorem is false. Then for each $x \in S \setminus M_e$ there

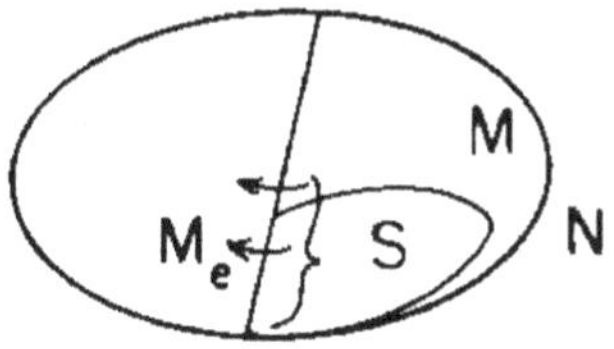

is a t_x such that $t_x > 0$ and $x[0,t_x) \subset M$ and $xt_x \in M_e$. Consider the map $f:S \to M_e$ defined by: $f(x) = xt_x$ if $x \in S \setminus M_e$ and $f(x) = x \in S \cap M_e$. This map is continuous since the map π defining the flow $\mathcal{F}$ is continuous and $M_e = M_{se}$. To see this let $x \in S \cap M_e$. Let $y \in S$. Let $\varepsilon > 0$ be sufficiently small, but arbitrary. Then $xt \in S \cap M_e$ for $0 \leq t \leq t_x - \varepsilon$, and $xt \notin \bar{M}$ for $t_x < t \leq t_x + \varepsilon$. Set $\eta = \min\{\rho(x(t_x + \varepsilon),\bar{M}), \rho(x(t_x - \varepsilon),\partial\bar{M})\}$. Then there is a $\delta > 0$ such that for $\rho(x,y) < \delta$, we have $\rho(xt,yt) < \eta/2$ for $0 \leq t \leq t_x + \varepsilon$, i.e., also $y(t_x + \varepsilon) \notin \bar{M}$ and $y(t_x - \varepsilon) \in M$. Hence $t_x - \varepsilon \leq t_y \leq t_x + \varepsilon$. Therefore t_x is a continuous function of $x \in S$. Hence also $f:S \to M_e$ is continuous for $x \in S \setminus M_e$. A similar argument holds for $x \in S \cap M_e$.

If now $g:M_e \to S \cap M_e$ is a retract of M_e onto $S \cap M_e$, then the composite map gf is a retraction of S onto $S \cap M_e$.

The existence of such a retraction contradicts the hypothesis, so that the theorem is proved.

1.8.5 *Remark*

If M is compact the only non-trivial condition of the theorem is that $M_e = M_{se}$. In this case the result of the theorem is rather weak as will be shown in the next section.

1.8.6 *Notes and References*

Theorem 1.8.4 is due to Ważewski [3]. This theorem is the cornerstone of the so-called topological methods for the study of properties of flows. Additional results along these lines are due to F. Albrecht [1] and to A. Pliss [1].

1.9 *Topological properties of attractors.*

1.9.1 *DEFINITION*

We shall say that a compact set $M \subset E$ *has* <u>*strong stability properties*</u> *if it is either asymptotically stable or completely unstable. We shall denote with* $A^s(M)$ *the open invariant set in which these properties hold, namely either* $A(M)$ *or* $A^-(M)$.

In this section and in the next we shall discuss some relationships which exist between the topological properties of closed sets having certain stability properties and the topological properties of the regions of the space E in which such stability properties hold. In particular we shall devote our attention to the case of closed sets with strong stability properties and discuss the relationship between the topological properties of a closed set $M \subset E$ and those of $A^s(M)$. The case of strong stability properties is not only the most interesting, but also the easiest to solve since in this case the existing continuous Liapunov functions have many properties which are very useful in the proof of the various results. In this section we shall limit ourselves to the discussion of global properties with respect to equilibrium points. In the next we shall present the very few results available for the case of local properties, while we postpone the more general discussion of the properties of sets to Section 2.8. Since for the proof of these results a more involved mathematical machinery is required, we urge the reader of this section to read at least the statement of the theorems presented in the advanced section.

Most of the theorems that we shall present are given for the case of weak attractor. Obviously they hold for asymptotically stable sets and, with the due changes (by inverting the direction of motion on the trajectories), for completely unstable sets.

All the results that we shall present are at a very early development stage; they are incomplete and admit further improvement. It is only because we think that those problems right now are among the most important problems in stability theory that we expose the reader to these preliminary results and incomplete theories.

1.9.2 *THEOREM*

Let $M \subset E$ *be a compact minimal set, and let* M *be a global weak attractor. Then* M *is a rest point.*

Proof. By Theorem 1.5.42, $D^+(M)$ is globally asymptotically stable. Let $x_0 \in E$ be arbitrary but fixed. Choose $\alpha > 0$ sufficiently large so that $D^+(M) \subset S(x_0,\alpha)$. Choose further $\varepsilon > 0$ sufficiently small such that $S(D^+(M),\varepsilon) \subset S[x_0,\alpha]$. By Theorem 1.5.27, $D^+(M)$ is uniformly attracting, and hence there is a $T > 0$ such that $\pi(x,t) \in S(D^+(M),\varepsilon) \subset S[x_0,\alpha]$, whenever $x \in S[x_0,\alpha]$ and $t \geq T$. For each $\tau \geq T$ define the map $\pi_\tau : X \to X$ by $\pi_\tau(x) = \pi(x,\tau)$. Then π_τ is continuous and $\pi_\tau(S[x_0,\alpha]) \subset S[x_0,\alpha]$. Thus by the Brouwer fixed point theorem $S[x_0,\alpha]$ contains a fixed point of the map π_τ, i.e., there is an $x \in S[x_0,\alpha]$ such that $\pi_\tau(x) = x = \pi(x,\tau)$. Hence $\pi(x,t) = \pi(\pi(x,\tau),t) = \pi(x,t+\tau)$ for all $t \in R$, and so $\gamma(x)$ is a periodic trajectory. Notice that $x \in M$, for otherwise, if $x \notin M$, then $\gamma(x) \cap M = \emptyset$, for M is invariant. On the other hand, since $\gamma(x)$ is a periodic trajectory, we have $\gamma(x) \equiv \Lambda^+(x)$, and as $x \in A_\omega(M)$, we must have $\Lambda^+(x) \cap M \neq \emptyset$, i.e., $\gamma(x) \cap M \neq \emptyset$. This contradiction proves that $\gamma(x) \subset M$. Since $\overline{\gamma(x)} = \gamma(x)$, we must have $\gamma(x) \equiv M$, as M is minimal. Thus M is a periodic trajectory with period τ. If now M is not a rest point, then it will have a least period say τ_0, and all other periods must be the numbers $m\,\tau_0$, where m is an integer. However, we have in fact shown that all numbers $\tau \geq T$ are periods of M. This is a contradiction and so M is a rest point, and the theorem is proved.

1.9.3 *Remark*

An important implication of the above theorem is that if M is a compact minimal set, and is not a rest point, M cannot be globally weakly attracting, or in particular, globally asymptotically stable. Thus the trajectory of a periodic motion, or the closure of the trajectory of an almost periodic or recurrent motion cannot be globally weakly attracting.

1.9.4 *COROLLARY*

Let $M \subset E$ *be a compact minimal set with global strong stability properties, then* M *is a rest point.*

The following theorem is a generalization of one of the principal results of the Poincare-Bendixson Theory of planar dynamical systems described by differential equations viz., every periodic trajectory contains in its interior a rest point. This is clear when we notice that a periodic trajectory and its interior form an invariant set homeomorphic to the unit disc.

The proof of this theorem is an elementary application of the Brouwer fixed point theorem and of the following lemma.

1.9.5 *LEMMA*

Let $M \subset X$ *be a compact positively invariant set. Let* $\{\gamma_n\}$ *be a sequence of periodic trajectories with periods* T_n, *such that* $\gamma_n \subset M$, *and* $T_n \to 0$. *Then* M *contains a rest point.*

Proof. Consider any sequence of points $\{x_n\}$, with $x_n \in \gamma_n$, $n = 1,2,\ldots$. We may assume without loss of generality that $x_n \to x \in M$, as M is compact. We will demonstrate that x is a rest point. For suppose that this is not the case.

Then there is a $\tau > 0$, such that $x \neq \pi(x,\tau)$. Let $d(x,\pi(x,\tau)) = \alpha(> 0)$. The spheres $S(x,\frac{\alpha}{4})$, and $S(\pi(x,\tau),\frac{\alpha}{4})$ are disjoint. Now choose $T, 0 < T < \tau$, such that $\rho(x,\pi(x,t)) \leq \frac{\alpha}{8}$ for $0 \leq t \leq T$. By continuity of π there is a $\delta > 0$ such that $\rho(x,y) < \delta$ implies $\rho(\pi(x,t),\pi(y,t)) < \frac{\alpha}{8}$ for $0 \leq t \leq \tau$. Notice in particular that if $\rho(x,y) < \delta$, then

$$\rho(x,\pi(y,t)) \leq \rho(x,\pi(x,t)) + \rho(\pi(x,t),\pi(y,t))$$

$$< \frac{\alpha}{8} + \frac{\alpha}{8} = \frac{\alpha}{4} \quad \text{if} \quad 0 \leq t \leq T,$$

and $\rho(\pi(x,\tau),\pi(y,\tau)) < \frac{\alpha}{4}$.

Now for sufficiently large n we have $T_n < T$ and $d(x,x_n) < \delta$. Hence $\rho(x,\pi(x_n,t)) < \frac{\alpha}{4}$ for $0 \leq t \leq T_n < T$. And as γ_n is periodic of period T_n, we have $\rho(x,\pi(x_n,t)) < \frac{\alpha}{4}$ for all $t \in R$. This is impossible, because we must have $\rho(x,\pi(x_n,\tau)) \geq \rho(x,\pi(x,\tau)) - \rho(\pi(x,\tau),\pi(x_n,\tau)) = \alpha - \frac{\alpha}{4} = \frac{3}{4}\alpha$. This contradiction proves that the point $x \in M$ is a rest point.

1.9.6 *THEOREM*

Let $M \subset E$ *be a compact positively invariant set, which is homeomorphic to the unit closed ball in* E. *Then* M *contains a rest point.*

Proof. Consider any sequence $\{\tau_n\}$, $\tau_n > 0$, $\tau_n \to 0$. Consider the sequence of maps $\{\pi_n\}$, $\pi_n : E \to E$, $\pi_n(x) = \pi(x,\tau_n)$. As π is continuous, so is each one of the maps π_n. Further, as M is positively invariant each π_n is a continuous map of M into itself. Thus by the Brouwer Fixed Point Theorem, M contains a fixed point of each one of the maps π_n. Let $x_n \in M$ be a fixed point of the map π_n. Then since $x_n = \pi_n(x_n) = \pi(x_n, \tau_n)$, the trajectory $\gamma(x_n) = \gamma_n$ is a rest point or a periodic trajectory with a period τ_n, and as

M is positively invariant $\gamma_n \subset M$. By the above lemma, M contains a rest point, and the theorem is proved.

1.9.7 *Remark*

Theorem 1.9.6 is not in general true in any compact space. Consider for example a dynamical system defined on a torus. There is a periodic trajectory γ which is not contractible to a point.

1.9.8 *Figure*

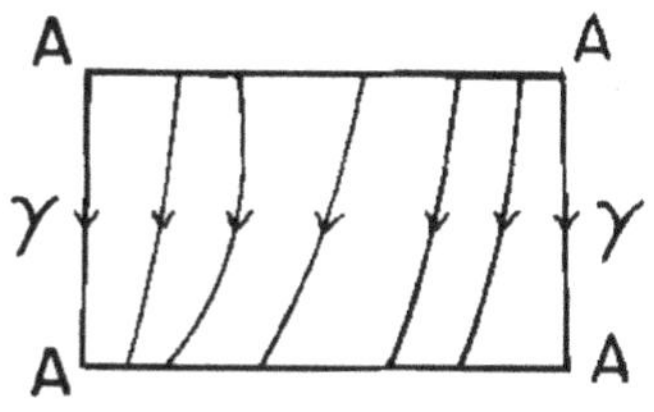

All other trajectories have γ as their positive as well as negative limit sets (see Figure 1.9.8). Notice that in this case

$D^+(\gamma) = A(\gamma) = X$ (the torus)

The following theorem holds in general.

1.9.9 *THEOREM*

Let X *be a compact invariant set. Let* $M \subset X$ *be compact and weakly attracting with* $A_\omega(M) = X$. *Then* $D^+(M) = X$.

Proof. Let if possible $X \neq D^+(M)$. Let $x \in X \setminus D^+(M)$. Now $\Lambda^-(x) \neq \emptyset$, as X is compact, and $\Lambda^-(x) \cap M = \emptyset$. For if $\Lambda^-(x) \cap M \neq \emptyset$, then $x \in D^+(M)$. To see this, note that if $\Lambda^-(x) \cap M \neq \emptyset$, then there is a sequence $\{\tau_n\}$, $\tau_n \to -\infty$, such that $\pi(x,\tau_n) \to y \in M$. Note that $x = \pi(\pi(x,\tau_n),-\tau_n)$, and since $-\tau_n \to +\infty$, $x \in D^+(y) \subset D^+(M)$. Thus $\Lambda^-(x) \cap M = \emptyset$. Now recall that $\Lambda^-(x)$ is non-empty closed and invariant, so that for any $z \in \Lambda^-(x)$, we have $\Lambda^+(z) \subset \Lambda^-(x)$. Hence, $\Lambda^+(z) \cap M = \emptyset$. This contradicts the assumption that M is globally weakly attracting and proves the theorem.

1.9.10 *COROLLARY*

Let X *be a compact invariant set. Let* $M \subset X, M \neq X$ *be a weak attractor with* $A_\omega(M) = X$. *Then* M *is not stable.*

1.9.11 *Notes and References*

Most of the results presented in this section are derived in the work by Bhatia, Lazer and Szegö [1].Theorem 1.9.6 is also mentioned by Petrovskii [1].

1.10 *From periodic motions to Poisson stability*

In this section we shall be mostly concerned with those properties of a motion which are generalizations of the concept of periodic motions. For a detailed investigation and a complete study of some of the most important properties of the concepts that we are going to introduce now, the reader is referred to Chapter 2.

In order of decreasing strength the concepts that we shall present are: periodicity, almost periodicity, recurrence and Poisson stability. It will be seen that each one of the above concepts imply the following one. All these are properties of motions. With exception of the case of periodic motion, no geometrical characterization of the trajectories defined by the motions with the weaker properties is possible. For the sake of completeness we shall start from the definition of periodic motions.

1.10.1 *DEFINITION*

A motion π_x *which for all* $t \in R$ *and some* $\tau \neq 0 \in R$ *has the property* $\pi_x(t + \tau) = \pi_x(t)$ *is called* *periodic.*

1.10.2 *DEFINITION*

A motion π_x *is said to be* *almost periodic* *if for all* $\varepsilon > 0$ *there exists a relatively dense set of numbers* $\{\tau_n\}$, *called displacements, such that* $\rho(xt, x(t + \tau_n)) < \varepsilon$ *for all* $t \in R$ *and* τ_n.

Notice that the set $\{\tau_n\}$ of the definition 1.10.2 does not depend on x.

Obviously, periodicity implies almost periodicity, while the converse is not true.

1.10.3 *DEFINITION*

A motion π_x *is said to be* *recurrent* *if for every* $\varepsilon > 0$ *there exists a* $\tau = \tau(t) > 0$ *such that for all* $t \in R$ $\quad xR \subset S(x[t-\tau, t+\tau], \varepsilon)$.

The property of recurrence can be expressed as almost periodicity if the set $\{\tau_n\}$ is made to depend on x. It can be proved that almost periodicity implies recurrence and that there exist recurrent motions which are not almost periodic. Many theorems of the relative properties of compact minimal sets and recurrent motions are given in Section 2.9.

1.10.4 *DEFINITION*

A motion π_x *is called positively Poisson-stable* (P^+*-stable) if* $x \in \Lambda^+(x)$, *negatively Poisson-stable* (P^-*-stable) if* $x \in \Lambda^-(x)$ *and Poisson-stable* (P*-stable) if both* $x \in \Lambda^+(x)$ *and* $x \in \Lambda^-(x)$ *holds.*

Again, the property of Poisson stability may be defined as a weak form of the property 1.10.2 where the set $\{\tau_n\}$ may depend upon x and does not need to be relatively dense.

Obviously, if π_x is $P^\cdot$-stable, then for all $t \in R$ also π_{xt} is $P^\cdot$-stable. Poisson stability can, therefore, also be defined as a property of trajectories and their limit sets. In fact, it is easy to show that if x is P^+-stable then

1.10.5 $$xR \cap \Lambda^+(x) \neq \emptyset$$

and that if x is P^--stable, then

1.10.6 $$xR \cap \Lambda^-(x) \neq \emptyset$$

We shall then prove that

1.10.7 *THEOREM*

If x *is* P^+*-stable*

1.10.8 $$\Lambda^-(x) \subset \Lambda^+(x) = \overline{xR}$$

If x *is* P^-*-stable*

1.10.9 $$\Lambda^+(x) \subset \Lambda^-(x) = \overline{xR}$$

and if x *is* P-*stable, then*

1.10.10 $$\Lambda^+(x) = \Lambda^-(x) = \overline{xR}.$$

Proof. We shall prove 1.10.8. The proof of 1.10.9 and 1.10.10 is analogous. Because of 1.10.5, from the closedness of $\Lambda^+(x)$, if x is P^+-stable, then $\overline{xR} \subset \Lambda^+(x)$. From the definition of limit sets, $\Lambda^-(x) \subset \overline{xR^-} \subset \overline{xR}$, hence, $\Lambda^-(x) \subset \Lambda^+(x)$, and $\Lambda^+(x) \subset \overline{xR^+} \subset \overline{xR}$. Thus $\Lambda^+(x) = \overline{xR}$.

The following theorem on Poisson-stable motion is very simple and its proof is left as an exercise.

1.10.11 *THEOREM*

A motion π_x *is* P^+-*stable if and only if the trajectory* xR *is not positively asymptotic.*

It can be proved that recurrence implies Poisson stability and there exist Poisson-stable motions which are not recurrent.

1.10.12 *Notes and References*

The definition of a recurrent motion given in this section is due to G. D. Birkhoff [1, Vol. 1, pg. 660). See also 2.10.17.

1.11 *Stability of motions*

Liapunov stability and asymptotic stability of sets are properties of a given set with respect to the neighboring trajectories. Thus Liapunov stability and asymptotic stability are purely geometrical properties of the oriented trajectory: the set $xR^+ \cup xR^-$. We shall now extend these concepts to the case of motions. It is important to point out that the concept of stability of a motion π_x defined by $\mathcal{F}$ and $x \in E$ is completely different from the concept of stability of the set xR. The stability of a motion π_x can be defined as follows.

1.11.1 *DEFINITION*

A motion π_x *is said to be* *(positively Liapunov) stable* *if for every* $\varepsilon > 0$ *there exists a* $\eta(\varepsilon) > 0$ *such that*

for all $y \in E$ *with* $\rho(x,y) < \eta$

1.11.2

$$\rho(xt,yt) < \varepsilon \quad \textit{for all} \quad t \in R^+$$

If the property 1.11.2 *is true for all* $t \in R^-$, *or for all* $t \in R$ *the motion* π_x *is said to be* *negatively (Liapunov) stable*, *or* *(Liapunov) stable*, *respectively.*

Similarly to the case of stability of non-compact sets one can define a stronger form of stability of a motion, namely uniform stability, in the following way:

1.11.3 *DEFINITION*

A motion π_x *is said to be* *(positively Liapunov) uniformly stable*, *if, given any* $\varepsilon > 0$ *there exists* $\eta(\varepsilon) > 0$ *such that for all* $y \in E$ *with* $\rho(x\tau,y) < \eta$ $\quad \rho(x(\tau + t),yt) < \varepsilon$ *for all* $t \in R^+$ *and* $\tau \in R$.

Similarly one can define negatively Liapunov uniformly stable motions and Liapunov uniformly stable motions. From this definition follows a rather interesting result which is presented in the next theorem. Similar results can be given for the case of negatively stable and stable motions.

1.11.4 *THEOREM*

A motion π_x *is positively (uniformly) stable if and only if every motion* π_y *with* $y \in xR$ *is positively (uniformly) stable.*

From the definitions 1.11.1 and 1.11.3 it clearly follows that

1.11.5 *THEOREM*

If a motion π_x *is (uniformly) stable then the corresponding trajectory* xR^+ *is also (uniformly) stable.*

It is on the other hand easy to show that if π_x is stable the closure of the positive semi-trajectory $\overline{xR^+}$ need not be stable. Consider for example the continuous flow $\mathcal{F}$ shown in Figure 1.11.6. Clearly the closed segment M, limited by the equilibrium points z and y is unstable. On the other hand, one can define motions $\pi_x, x \in M$ which are stable (but not uniformly stable).

1.11.6 *Figure*

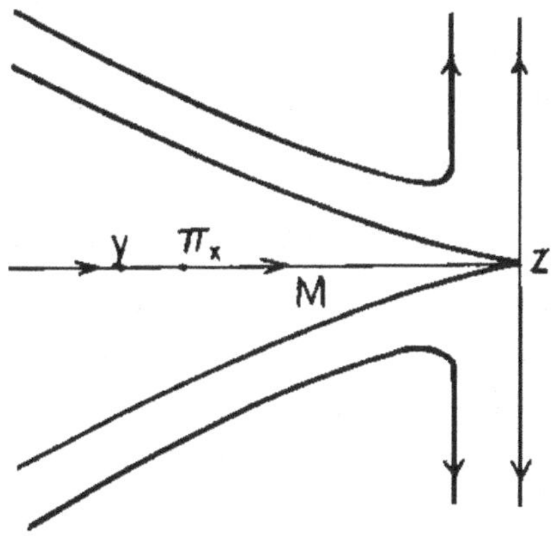

The converse of Theorem 1.11.5 does not hold, in fact (uniform) stability of a trajectory does not imply stability of the motions defined on it. This fact can be shown by many examples. For instance, consider the flow $\mathcal{F}$ on the plane represented by Figure 1.11.7. The trajectory through each point is a circle with its center in the origin of the plane.

1.11.7 *Figure*

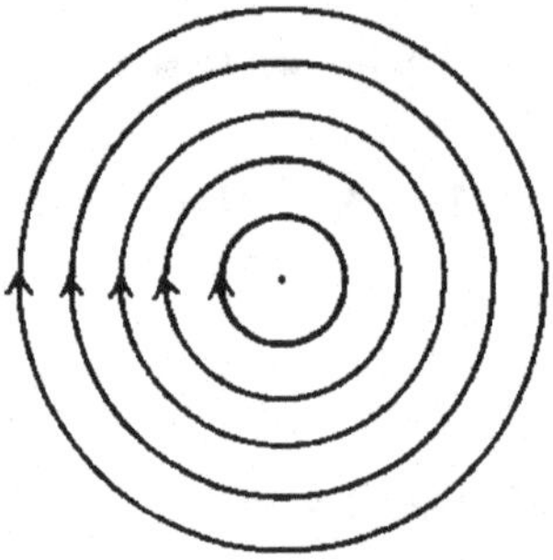

Obviously the origin as well as all circular trajectory are uniformly stable.

Assume that the tangential velocity of rotation defined by the motion on each trajectory is a constant, it follows that the angular velocity is decreasing as the radius of the circles is increasing. Thus the periodic motions are not stable.

In the case that for $x \in E$, the corresponding set $\overline{xR}$ is compact, or even in the case of almost periodic motion some stronger connections between the stability properties of motions and those of the corresponding trajectories will be shown (Section 2.9).

In the particular case of an equilibrium point the two concept: stability of motion and stability of trajectory, coincide. This is the reason for the not clear distinction between stability of sets and stability of motion in the classical literature.

1.11.8 *THEOREM*

The *equilibrium motion* π_x $[\pi_x(t) = x$ *for all* $t \in R]$, *is stable if and only if the set* $\{x\} = xR$ *is stable.*

Proof. Let the equilibrium motion π_x be stable. Then, given any $\varepsilon > 0$, there is a $\delta > 0$ such that $\rho(x,y) < \delta$ implies $\rho(xt,yt) < \varepsilon$ for all $t \in R^+$. Since $xt = x$ for all $t \in R$, we get $\rho(x,yt) < \varepsilon$. Clearly then $y \in S(\{x\},\delta)$ implies $yR^+ \subset S(\{x\},\varepsilon)$, i.e., the set $\{x\}$ is stable.

Now let the invariant set $\{x\}$ be stable. Then, given any $\varepsilon > 0$, there is a $\delta > 0$ such that $\rho(x,y) < \delta$ implies $yR^+ \subset S(x,\varepsilon)$. Also in particular $yt \in S(x,\varepsilon), t \in R^+$. Since, however, $xt = x$ for all $t \in R$, we have $yt \in S(xt,\varepsilon)$, i.e., $\rho(xt,yt) < \varepsilon$ for $t \in R^+$, the motion π_x is stable. This completes the proof.

In the literature, it is also given, for the case of motions, a stability property stronger than the one given in definition 1.11.1, namely stability with respect to a set.

1.11.9 *DEFINITION*

A motion π_x *is called* *(positively Liapunov) stable with respect to the set* $B \subset E$ *if for every* $\varepsilon > 0$ *there exists* $\eta(\varepsilon) > 0$ *such that for every* $y \in B$ *with* $\rho(x,y) < \eta$ $\rho(xt,yt) < \varepsilon$ *for all* $t \in R^+$, *and* *(positively Liapunov) uniformly stable with respect to the set* $B \subset E$ *if for every* $\varepsilon > 0$ *there exists* $\eta(\varepsilon) > 0$ *such that for every* $y \in B$ *with* $\rho(x\tau,y) < \delta$ $\rho(x(\tau + t),yt) < \varepsilon$ *for all* $t \in R^+$ *and* $\tau \in R$.

Similarly one can define negatively Liapunov stable and Liapunov stable motion with respect to B. If B is a neighborhood of x, then the concept of stability of π_x and stability of π_x with respect to B coincide. Notice, however, that if $B = xR^+$ this need not be true since there may exist motions π_x which are not stable with respect to xR^+. For instance, in the case of Figure 1.11.10 where xR^+ is a straight half line

1.11.10 *Figure*

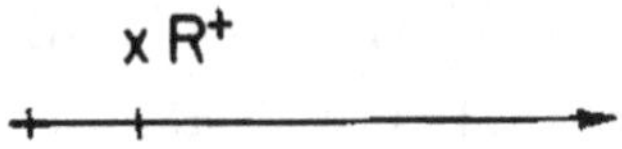

and the motions is accelerating on it.

It will be proved in Chapter 2 that for almost periodic motion this situation cannot arise.

Similarly to that done for the case of stability one can define attracting motions as:

1.11.11 *DEFINITION*

A motion π_x *is said to be* <u>*attracting*</u> *if there exists a* $\delta > 0$ *such that* $\rho(x,y) < \delta$ *implies* $\rho(xt,yt) \to 0$ *for* $t \to +\infty$.

Clearly the property of attraction of a motion π_x implies the property of attraction of the corresponding trajectory $x R^+$.

As in the case of trajectories one can also define asymptotic stability.

1.11.12 *DEFINITION*

A motion π_x *is said to be* <u>*asymptotically stable*</u> *if it is both stable and attracting.*

Notice that for the case of motions having noncompact trajectory closures one can define at least as many properties as the ones defined in Section 1.6. Since we shall not use these properties it is pointless to define them in detail. Their definition is very simple.

1.11.13. *Notes and References*

The original definition of stability of motions is due to A. M. Liapunov [1]. The presentation given here is adopted from Nemytskii and Stepanov. It must be emphasized that the stability of a given motion was the only form of stability considered by Liapunov [1] as well as from many other authors like Chetaev [5], Malkin [8], Hahn,[2] etc. See also 3.6.32.

DYNAMICAL SYSTEMS IN METRIC SPACES

2.1 *Definition of a dynamical system and related notation.*

A dynamical system or continuous flow on X is the triplet (X,R,π), where $\pi: X \times R \to X$ is a map from the product space $X \times R$ into X satisfying the following axioms:

2.1.1 $\pi(x,0) = x$ for every $x \in X$,

2.1.2 $\pi(\pi(x,t_1),t_2) = \pi(x,t_1 + t_2)$ for every $x \in X$, and $t_1, t_2 \in R$,

2.1.3 π is continuous.

The above three axioms are usually referred to as the Identity, Homomorphism and Continuity Axioms, respectively.

In the sequel we shall generally delete the symbol π. Thus the image $\pi(x,t)$ of a point $(x,t) \in X \times R$ will be written simply as xt. The identity and the homomorphism axioms then read

2.1.1' $x0 = x$ for every $x \in X$, and

2.1.2' $xt_1(t_2) = x(t_1 + t_2)$ for all $x \in X$ and $t_1, t_2 \in R$.

Notice also that the continuity axiom is equivalent to:

2.1.3' If $\{x_n\}$, $\{t_n\}$ are sequences in X and R respectively such that $x_n \to x$, $t_n \to t$, then $\{x_n t_n\}$ is a sequence in X such that $x_n t_n \to xt$.

In line with the above notation, if $M \subset X$ and $A \subset R$, we set

$MA = \{xt : x \in M$ and $t \in A\}$. If either M, or A, is a singleton, (a set containing exactly one element) i.e., $M = \{x\}$, or $A = \{t\}$, we simply write xA, or Mt for $\{x\}A$, or $M\{t\}$, respectively.

For a given dynamical system on X, the space X is generally called the phase space, and the map π as the phase map (of the dynamical system).

The phase map π determines two other maps when one of the variables x or t is fixed. Thus for a fixed $t \in R$ the map $\pi^t : X \to X$ determined by $\pi^t(x) = xt$ is called a transition. For each $t \in R$, π^t is a homeomorphism of X onto itself. Again for a fixed $x \in X$ the map $\pi_x : R \to X$ determined by $\pi_x(t) = xt$ is called a motion (through x).

2.2 *Elementary Concepts: Trajectories and their Limit Sets.*

For any $x \in X$, the trajectory (or orbit), the positive semi-trajectory, and the negative semi-trajectory are the sets given respectively by

2.2.1 $$\gamma(x) = \{xt: t \in R\} ,$$

2.2.2 $$\gamma^+(x) = \{xt: t \in R^+\} , \quad \text{and}$$

2.2.3 $$\gamma^-(x) = \{xt: t \in R^-\} .$$

We shall reserve in the sequel the symbols γ, γ^+, γ^- for the maps from X to 2^X defined respectively by 2.2.1, 2.2.2, and 2.2.3.

A subset $M \subset X$ will be called *invariant*, positively invariant, or negatively invariant if the condition $\gamma(M) = M$, $\gamma^+(M) = M$ or $\gamma^-(M) = M$ is satisfied, respectively.

2.2.4 *DEFINITION*

A subset $M \subset X$ *is called* ***minimal***, *if it is non-empty, closed, and invariant, and no proper subset of* M *has these properties.*

2.2.5 *DEFINITION*

For any $x \in X$, *the* ***positive*** *or* ***omega limit set***, *and the* ***negative*** *or* ***alpha limit set*** *are the sets given respectively by* $\Lambda^+(x) = \{y \in X$: *there is a sequence* $\{t_n\}$, $t_n \to +\infty$, *such that* $xt_n \to y\}$; *and* $\Lambda^-(x) = \{y \in X$: *there is a sequence* $\{t_n\}$, $t_n \to -\infty$, *such that* $xt_n \to y\}$.

Examples of limit sets are given in Section 1.3.

2.2.6 *Exercises*

i) Show that $\gamma(x) = \gamma(xt)$ for every $t \in R$.

ii) $\gamma(\gamma(x)) = \gamma(x)$, $\gamma^+(\gamma^+(x)) = \gamma^+(x)$, and $\gamma^-(\gamma^-(x)) = \gamma^-(x)$. Thus $\gamma(x)$, $\gamma^+(x)$, and $\gamma^-(x)$ are respectively, invariant, positively invariant, and negatively invariant.

iii) Show that

$$\Lambda^+(x) = \bigcap\left\{ \overline{\gamma^+(xt)} : t \in R\right\} , \qquad \text{and}$$

$$\Lambda^-(x) = \bigcap\left\{ \overline{\gamma^-(xt)} : t \in R\right\} .$$

iv) Show that

$$\Lambda^+(x) = \Lambda^+(xt) , \quad \text{and} \quad \Lambda^-(x) = \Lambda^-(xt) , \quad t \in R .$$

2.2.7 *Note.* The relations (iii) in 2.2.6 are frequently used to define the positive and negative limit sets.

2.2.8 *Exercises*

i) Show that $\overline{\gamma^+(x)} = \gamma^+(x) \cup \Lambda^+(x)$.

ii) $\Lambda^+(x)$ is closed and invariant.

2.2.9 *THEOREM*

If the space X *is locally compact, then a positive limit set* $\Lambda^+(x)$ *is connected whenever it is compact. Further, whenever a positive limit set is not compact, then none of its components is compact.*

Proof. Let $\Lambda^+(x)$ be compact, and let it be not connected. Then $\Lambda^+(x) = P \cup Q$, where P, Q are non-empty, closed, disjoint sets. Since $\Lambda^+(x)$ is compact, so are P and Q . Further, since X is locally compact, there is an $\varepsilon > 0$ such that $S[P,\varepsilon]$, $S[Q,\varepsilon]$ are compact and disjoint. Now let $y \in P$ and $z \in Q$. Then there are sequences $\{t_n\}$,

$\{\tau_n\}$, $t_n \to +\infty$, $\tau_n \to +\infty$, such that $xt_n \to y$, and $x\tau_n \to z$. And we may assume without loss of generality, that $xt_n \in S(P,\varepsilon)$, $x\tau_n \in S(Q,\varepsilon)$, and $\tau_n - t_n > 0$ for all n. Since the trajectory segments $x[t_n, \tau_n]$, $n = 1, 2, \ldots$, are compact connected sets, they clearly intersect $H(P,\varepsilon)$ and $H(Q,\varepsilon)$. Thus, in particular, there is a sequence $\{T_n\}$, $t_n < T_n < \tau_n$, such that $xT_n \in H(P,\varepsilon)$ which is compact. We may therefore assume that $xT_n \to \tilde{y}$, and as $T_n \to +\infty$, we have $\tilde{y} \in \Lambda^+(x)$. However, $\tilde{y} \notin P \cup Q$, which is a contradiction. This establishes the first part of the theorem.

To prove the second part of the theorem we need the following topological theorem, which we give without proof.

2.2.10 *TOPOLOGICAL THEOREM.*

Let S *be a Hausdorff continuum (a compact connected Hausdorff space),* *let* U *be an open subset of* S, *and let* C *be a component of* U. *Then* $\bar{U}\setminus U$ *contains a limit point of* C.

2.2.11 *Proof of the 2nd Part of Theorem* 2.2.9

Notice that the space X is a locally compact Hausdorff space, and everything that has been said above goes through in such a space. Now X possesses a one-point compactification. So let $\tilde{X} = X \cup \{\omega\}$ be the one-point compactification of X by the ideal point ω. Extend the dynamical system (X,R,π) on X to a dynamical system $(\tilde{X},R,\tilde{\pi})$ on $\tilde{X}$, where $\tilde{\pi}$ is given by $\tilde{\pi}(x,t) \equiv \pi(x,t)$ for $x \in X$, $t \in R$, and $\tilde{\pi}(\omega,t) = \omega$ for all $t \in R$. If now for $x \in \tilde{X}$, $\tilde{\Lambda}^+(x)$ denotes the positive limit set of x, then clearly $\tilde{\Lambda}^+(x) = \Lambda^+(x) \cup \{\omega\}$, whenever $x \in X$ and $\Lambda^+(x)$ is not compact. However, $\tilde{\Lambda}^+(x)$ is compact, as $\tilde{X}$ is compact, and by the first part of the theorem it is connected. $\tilde{\Lambda}^+(x)$ is therefore a Hausdorff continuum. Further $\Lambda^+(x)$ is an open set in $\tilde{\Lambda}^+(x)$.

Now $\tilde{\Lambda}^+(x) - \Lambda^+(x) = \{\omega\}$, and so by Theorem 2.2.10 every component of $\Lambda^+(x)$ has ω as a limit point, and so is not compact. This proves the theorem completely. Similarly to what was done in Chapter 1, one can easily prove that:

2.2.13 *THEOREM*

If $\overline{\gamma^+(x)}$ *is compact, then* $\Lambda^+(x)$ *is compact. Further if* X *is locally compact and* $\Lambda^+(x)$ *is compact, then* $\overline{\gamma^+(x)}$ *is compact.*

2.2.14 *Exercises*

Let X be locally compact. Then

i) If $\Lambda^+(x)$ is compact, then $\rho(xt,\Lambda^+(x)) \to 0$ as $t \to +\infty$,

ii) Give an example to show that (i) is false if $\Lambda^+(x)$ is not compact.

iii) Let X be not locally compact, then give an example showing that Theorem 2.2.9 is false.

iv) Show that if $M \subset X$ is invariant, then the sets $\overline{M}$, ∂M, $C(M)$, $I(M)$ are also invariant.

v) Show that if $M \subset X$ is connected and $\Lambda^+(M)$ is compact, then $\Lambda^+(M)$ is connected provided that X is locally compact.

2.2.15 *Notes and References*

Alternative definitions of limit sets have been proposed by S. Lefschetz [2] and T. Ura [2]. For instance, Lefschetz uses the definition $\Lambda^+(\gamma) = \Lambda^+(x) = \bigcap\{\overline{\gamma^+(y)};\ y \in \gamma^+(x)\}$ and $\Lambda^-(\gamma) = \Lambda^-(x) = \bigcap\{\overline{\gamma^-(x)};\ y \in \gamma^-(x)\}$. Ura gives a slightly more general definition which is essentially the same as the one by Lefschetz in the case (X,R,π), but can be used to define limit sets of general topological transformation groups (T,G,π).

Theorem 2.2.10 can be found, for instance, in the book by Hocking and Young at pg. 37.

2.3 *The (first) (positive) prolongation and the prolongational limit set*

2.3.1 *DEFINITION*

For any $x \in X$, *the (first) positive prolongation and the (first) negative prolongation are the sets given respectively by*

$$D^+(x) = \{y \in X : \exists \{x_n\} \subset X \text{ and } \{t_n\} \subset R^+ \text{ such that } x_n \to x \text{ and } x_n t_n \to y\}$$

$$D^-(x) = \{y \in X : \exists \{x_n\} \subset X \text{ and } \{t_n\} \subset R^- \text{ such that } x_n \to x \text{ and } x_n t_n \to y\}.$$

2.3.2 *Note*

The reason that the prolongations defined above are called first prolongations is that there are others with which we shall deal in a later section. This fact is, however, insignificant for most applications. Since we shall mostly deal with the properties and application of the notion of the positive prolongation, we shall delete the adjective positive.

Various examples of $D^+(x)$ are given in 1.4.7

2.3.3 *Exercises*

Show that for any $x \in X$,

i) $D^+(x) = \bigcap \{\overline{\gamma^+(S(x,\alpha))} : \alpha > 0\}$,

ii) $D^-(x) = \bigcap \{\overline{\gamma^-(S(x,\alpha))} : \alpha > 0\}$,

iii) $D^+(x) \supset \overline{\gamma^+(x)}$, and

iv) $D^-(x) \supset \overline{\gamma^-(x)}$.

2.3.4 *THEOREM*

For any $x \in X$, $D^+(x)$ *is closed and positively invariant.*

The proof is left as an exercise.

2.3.5 *THEOREM*

Let X *be locally compact. Then for any* $x \in X$, $D^+(x)$ *is connected whenever it is compact. Further, if* $D^+(x)$ *is not compact, then none of its components is compact.*

The proof follows exactly the same lines as that of Theorem 2.2.9 and is, therefore, omitted.

2.3.6 *DEFINITION*

The (first) positive prolongational limit set, and the (first) negative prolongational limit set of any $x \in X$ *are the sets given respectively by*

$$J^+(x) = \{y \in X: \exists \{x_n\} \subset X, \{t_n\} \subset R, \text{ such that } x_n \to x, t_n \to +\infty \text{ and } x_n t_n \to y\}$$

$$J^-(x) = \{y \in X: \exists \{x_n\} \subset X, \{t_n\} \subset R, \text{ such that } x_n \to x, t_n \to -\infty \text{ and } x_n t_n \to y\}$$

2.3.7 *Example*

In Example 1.4.7 (i) for any p in the x_1-axis, $J^+(p)$ is the x_2-axis. In Example 1.4.7. (ii) for any $p \in \gamma_{-1}$, $J^+(p) = \gamma_0 \cup \gamma_{-2}$.

2.3.8 *Exercises*

Show that for any $x \in X$

i) $J^+(x) = \bigcap \{D^+(xt): t \in R\}$, and

$J^-(x) = \bigcap \{D^-(xt): t \in R\}$.

ii) $J^+(x) = J^+(xt)$, $J^-(x) = J^-(xt)$ $t \in R$.

iii) $\gamma(J^+(x)) = J^+(x)$, i.e., $J^+(x)$ is invariant.

2.3.9 *Note*

The relations (i), (ii) in Exercises 2.3.3 are frequently used to define $D^+(x)$ and $D^-(x)$. The relations (i) in Exercise 2.3.8 are frequently used to define $J^+(x)$, $J^-(x)$ once $D^+(x)$, $D^-(x)$ have been defined.

2.3.10 *Exercises*

Show that

i) $D^+(x) = \gamma^+(x) \cup J^+(x)$ and $D^-(x) = \gamma^-(x) \cup J^-(x)$.

ii) $J^+(x)$, $J^-(x)$ are closed and invariant.

2.3.11 *THEOREM*

Let X *be locally compact. Then* $J^+(x)$ *is connected whenever it is compact. Further, if* $J^+(x)$ *is not compact, then none of its components is compact.*

The proof of the first part although similar to that of Theorem 2.2.9 will be made to depend on the following lemma. Proof of the second part will be omitted.

2.3.12 *LEMMA*

For any $x \in X$, $J^+(x) \subset D^+(\omega)$, *whenever* $\Lambda^+(x) \neq \emptyset$ *and* $\omega \in \Lambda^+(x)$. *In particular, if* $\omega \in \Lambda^+(x)$, *and* $y \in J^+(x)$, *then there exist sequences* $\{x_n\}$ *in* X, $\{t_n\}$ *and* $\{\tau_n\}$ *in* R^+, *with* $x_n \to x$, $t_n - \tau_n > 0$ *for each* n, $t_n \to +\infty$, $\tau_n \to +\infty$, $x_n\tau_n \to \omega$, *and* $x_n t_n \to y$.

Proof. Given $\omega \in \Lambda^+(x)$, and any $y \in J^+(x)$, there exist sequences $\{\tau_n'\}$, $\tau_n' \to +\infty$, $x\tau_n' \to \omega$, and $\{t_n'\}$ and $\{x_n\}$, $x_n \to x$, $t_n' \to +\infty$, $x_n t_n' \to y$. We can assume, if necessary by choosing subsequences, that $t_n' - \tau_n' > 0$ for each n. Consider for each fixed k, k = 1,2,..., the sequence $\{x_n\tau_k'\}$. By the continuity axiom $x_n\tau_k' \to x\tau_k'$, k = 1,2,... . We may, therefore, assume without loss of generality that for each fixed k, $\rho(x_n\tau_k', x\tau_k') \leq \frac{1}{k}$ for $n \geq k$. This shows that $x_n\tau_n' \to \omega$, because $\rho(\omega, x_n\tau_n') \leq \rho(\omega, x\tau_n') + \rho(x\tau_n', x_n\tau_n') \leq \rho(\omega, x\tau_n') + \frac{1}{n}$. The sequences $\{\tau_n'\}$, $\{t_n'\}$, $\{x_n\}$ have then the required properties. Now notice that $x_n t_n' = x_n\tau_n'(t_n' - \tau_n')$, and $x_n t_n' \to y$, $x_n\tau_n' \to \omega$, and $t_n' - \tau_n' > 0$. Hence $y \in D^+(\omega)$. As $y \in J^+(x)$ was arbitrary, we have $J^+(x) \subset D^+(\omega)$, and the lemma is proved.

2.3.13 *Proof of the first part of Theorem 2.3.11*

Notice that $\Lambda^+(x) \subset J^+(x)$ holds always. And if X is locally compact, then $\Lambda^+(x) \neq \emptyset$, whenever $J^+(x)$ is compact. To see this assume that $\Lambda^+(x) = \emptyset$.

Since X is locally compact, and $J^+(x)$ is compact, we can find an $\varepsilon > 0$ such that $S[J^+(x),\varepsilon]$ is compact. If $\Lambda^+(x) = \emptyset$, there exists a $T > 0$ such that $\gamma^+(xT) \cap S[J^+(x),\varepsilon] = \emptyset$. This is so for, otherwise, there will be sequence $\{xt_n\}$, $t_n \to +\infty$, such that $xt_n \in S[J^+(x),\varepsilon]$, and as $S[J^+(x),\varepsilon]$ is compact, the sequence $\{xt_n\}$ will have a limit point $\omega \in \Lambda^+(x)$ contradicting $\Lambda^+(x) = \emptyset$. Notice also that $x \notin J^+(x)$, for then $\gamma^+(x) \subset J^+(x)$, and as $J^+(x)$ is compact, $\Lambda^+(x)$ will not be empty. It is thus clear that there is an $\alpha > 0$ such that $S[J^+(x),\alpha]$ is compact, and $\gamma^+(x) \cap S[J^+(x),\alpha] = \emptyset$. Now let $y \in J^+(x)$. Then there are sequences $\{x_n\}$, $\{t_n\}$, $x_n \to x$, $t_n \to +\infty$, and $x_nt_n \to y$. We may assume that $x_n \notin S[J^+(x),\alpha]$, and $x_nt_n \in S(J^+(x),\alpha)$ for all n. But then there is a sequence $\{\tau_n\}$, $0 < \tau_n < t_n$, such that $x_n\tau_n \in H(J^+(x),\alpha)$. Since $H(J^+(x),\alpha)$ is compact, we may assume without loss of generality that $x_n\tau_n \to z \in H(J^+(x),\alpha)$. We claim that $\{\tau_n\}$ is bounded, because otherwise it will have an unbounded subsequence so that we will have $z \in J^+(x)$, whereas $J^+(x) \cap H[J^+(x),\alpha] = \emptyset$ However, if $\{\tau_n\}$ is bounded, it has a convergent subsequence, and we may assume without loss of generality that $\tau_n \to t$. But then by the continuity axiom $x_n\tau_n \to xt \in \gamma^+(x)$, i.e., $z = xt \in \gamma^+(x)$. This is impossible as $\gamma^+(x) \cap S[J^+(x),\alpha] = \emptyset$. Thus $\Lambda^+(x)$ cannot be empty. To complete the proof of the theorem, assume that $J^+(x)$ is not connected. Then we have $J^+(x) = P \cup Q$, where P,Q are compact, non-empty and disjoint. As $\Lambda^+(x)$ is non-empty and compact, it is connected (Theorem 2.2.9) and so $\Lambda^+(x)$ is a subset of only one of the sets P and Q, say $\Lambda^+(x) \subset P$. Choose $y \in Q$, and $\omega \in \Lambda^+(x) \subset P$. Further, choose $\varepsilon > 0$ such that $S[P,\varepsilon] \cap S[Q,\varepsilon] = \emptyset$, and $S[P,\varepsilon]$, $S[Q,\varepsilon]$ are compact (this is possible as X is locally compact, and P,Q are compact and disjoint). Now, by Lemma 2.3.12, there are sequences $\{x_n\}$, $\{t_n\}$, $\{\tau_n\}$, $x_n \to x$, $t_n - \tau_n > 0$, $\tau_n \to +\infty$, $x_n\tau_n \to \omega$, and $x_nt_n \to y$. We may assume that $x_n\tau_n \in S(P,\varepsilon)$, and $x_nt_n \in S(Q,\varepsilon)$ for all n. But then there is a sequence $\{t_n'\}$, $\tau_n < t_n' < t_n$, such that $x_nt_n' \in H(P,\varepsilon)$, and indeed $t_n' \to +\infty$. Since $H(P,\varepsilon)$ is compact, we may assume that $x_n t'_n \to z \in H(P,\varepsilon)$. But then $z \in J^+(x)$, while $z \notin P \cup Q$. This contradiction proves the theorem.

The first part of the above proof contains the following lemma which we give below for future reference.

2.3.14 *LEMMA*

If X *is locally compact, and if* $J^+(x) \neq \emptyset$ *is compact, then* $\Lambda^+(x) \neq \emptyset$ *and is indeed compact.*

2.3.15 *Exercise*

Give an example to show that Lemma 2.3.14 does not hold in general metric spaces X.

2.3.16 *Exercises*

i) If $M \subset X$ is non-empty and compact, then $D^+(M)$ is closed.

ii) If $M \subset X$ is connected, and $D^+(M)$ is compact, then $D^+(M)$ is connected if X is locally compact.

2.3.17 *Example*

Consider the Example 1.4.7 i) modified by deleting the origin of the coordinates from the plane. Then for each p in X (i.e., the plane without the origin of coordinates) $\Lambda^+(p) = \emptyset$. If p is a point in the x_1-axis, then $J^+(p)$ has two components, viz. the positive and negative parts of the x_2-axis. Both are indeed non-compact.

In Example 1.4.7 ii) note that $\Lambda^+(p) = \emptyset$ for all points p in the plane, whereas $J^+(p) \neq \emptyset$ for all $p \in \gamma_{-1}$.

2.3.18 *Notes and References* (see also 1.4.14)

The definition of prolongation is due to T. Ura [2]. He adopts the relations 2.3.3 (i) and 2.3.3 (ii) as definitions.

Theorem 2.3.5 is essentially due to N. P. Bhatia [3]. The concept of prolongational limit sets is due to J. Auslander, N. P. Bhatia and P. Seibert.

Theorem 2.3.11 is due to N. P. Bhatia [3].

2.4 *Self-intersecting trajectories*

2.4.1 *DEFINITION*

A point $x \in X$ *such that* $xt = x$ *for all* $t \in R$, *is called a* <u>*rest point*</u> *(or a critical point, or an equilibrium point).*

2.4.2 *DEFINITION*

For any $x \in X$, *the trajectory* $\gamma(x)$ *(and also the motion* π_x *through* x*) is called periodic with a period* τ *whenever* $x(t + \tau) = xt$ *for all* $t \in R$.

Notice that a rest point $x \in X$ is a periodic orbit having every number $\tau \in R$ as a period.

However, the following lemma holds.

2.4.3 *LEMMA*

If $\{x\} \neq \gamma(x)$, *i.e.,* x *is not a rest point, and if* $\gamma(x)$ *is periodic, then there exists a least positive number* T, *such that* T *is a period of* $\gamma(x)$ *and if* τ *is any period, then* $\tau \in \{kT : k = \pm 1, \pm 2, \ldots\}$.

Proof. Notice that if τ is a period, then so is $-\tau$, because if

$$xt = x(t + \tau) \qquad \text{for all } t \in R,$$

then by the homomorphism axiom

$$x(t - \tau) = xt(-\tau) = x(t + \tau)(-\tau) = x(t + \tau - \tau) = xt,$$

showing that $-\tau$ is a period. Thus the periodic trajectory has positive numbers as periods. Notice further that if $\tau > 0$ is a period, then for any integer k

$$x(t + k\tau) = xt \qquad \text{for all } t \in R,$$

so that $k\tau$ is a period. Now let P be the set of positive periods of the periodic trajectory $\gamma(x)$. If there is no least positive period, then there is a sequence

$\{\tau_n\}$, $\tau_n \in P$, and $\tau_n \to 0$ (because, if τ_1, τ_2 are periods, then $\tau_1 \pm \tau_2$ are periods). By the continuity axiom, given $\varepsilon > 0$ there is a $\delta > 0$ such that $\rho(x,xt) < \varepsilon$ if $|t| < \delta$. Further, if $\tau_n > 0$ is a period, then obviously $\gamma(x) = x[0,\tau_n]$. As $\tau_n \to 0$, we notice that $\gamma(x) = x[0,\tau_n] \subset S(x,\varepsilon)$ for large n. Hence $\gamma(x) = \bigcap \{S(x,\varepsilon); \varepsilon > 0\} = \{x\}$, i.e., x is a rest point contrary to the hypothesis. This proves the lemma.

2.4.4 *DEFINITION*

For any $x \in X$, *the trajectory* $\gamma(x)$ *is said to be self-intersecting, if there exist* $t_1, t_2 \in R$, $t_1 \neq t_2$, *such that* $xt_1 = xt_2$.

2.4.5 *THEOREM*

If for any $x \in X$, $\gamma(x)$ *is self-intersecting, then either* x *is a rest point or* $\gamma(x)$ *is periodic.*

Notice that rest points and periodic trajectories are self-intersecting. The above theorem shows that these are the only self-intersecting trajectories. The proof of the above theorem is trivial and is, therefore, omitted.

2.4.6 *Exercise*

A self-intersecting trajectory is a compact minimal set.

2.5 *Lagrange and Poisson stability*

2.5.1 *DEFINITION*

For any $x \in X$, *the motion* π_x *is said to be* *positively Lagrange stable* *if* $\overline{\gamma^+(x)}$ *is compact. Further, if* $\overline{\gamma^-(x)}$ *is compact, then the motion* π_x *is called* *negatively Lagrange stable*. *It is said to be* *Lagrange stable* *if* $\overline{\gamma(x)}$ *is compact.*

2.5.2 *Remark*

If $X = E$, then the above statements are equivalent to the sets $\gamma^+(x)$, $\gamma^-(x)$, $\gamma(x)$ being bounded, respectively.

2.5.3 *Exercises*

(i) If X is locally compact, then a motion π_x is positively Lagrange stable if and only if $\Lambda^+(x)$ is a non empty compact set.

(ii) If a motion π_x is positively Lagrange stable, then $\Lambda^+(x)$ is compact and connected.

(iii) If a motion π_x is positively Lagrange stable, then $\rho(xt,\Lambda^+(x)) \to 0$ as $t \to +\infty$.

It will be useful to compare the statements in the above exercise with Theorem 2.2.9 and the Exercise 2.2.16 (i) and (ii).

2.5.4 *DEFINITION*

A motion π_x *is said to be* *positively (negatively) Poisson stable* *if* $x \in \Lambda^+(x)$ $(x \in \Lambda^-(x))$. *It is said to be* *Poisson stable* *if it is both positively and negatively Poisson stable, i.e., if* $x \in \Lambda^+(x) \cap \Lambda^-(x)$.

2.5.5 *Exercise*

(i) A motion π_x is positively Poisson stable if and only if $\gamma(x) \subset \Lambda^+(x)$.

(ii) A motion π_x is positively Poisson stable if and only if $\Lambda^-(x) \subset \Lambda^+(x) = \overline{\gamma(x)}$

(iii) If π_x is positively Poisson stable then for any $t \in R$, the motion π_{xt} is positively Poisson stable.

2.5.6 *Exercise*

A self-intersecting trajectory is Lagrange stable and Poisson stable. Indeed the following theorem holds.

2.5.7 *THEOREM*

A motion π_x *is positively Poisson stable if and only if* $\overline{\gamma^+(x)} = \Lambda^+(x)$.

The proof is trivial (see Exercise 2.5.5) and is left as an exercise.

In view of the above theorem it is interesting to inquire about the consequences of the condition $\gamma^+(x) = \Lambda^+(x)$. The answer is contained in the following theorem.

2.5.8 *THEOREM*

$\gamma^+(x) = \Lambda^+(x)$ *if and only if either* x *is a rest point or* $\gamma(x)$ *is a periodic trajectory.*

Proof. Let $\gamma^+(x) = \Lambda^+(x)$. If x is a rest point, the relation holds and there is nothing to prove. Suppose x is not a rest point. Indeed $x \in \Lambda^+(x)$ and as $\Lambda^+(x)$ is invariant we see that $\gamma^+(x) = \Lambda^+(x) = \gamma(x)$. Thus for each $\tau < 0$, $x\tau \in \gamma^+(x)$, and, therefore, there is a $\tau' \geq 0$ such that $x\tau = x\tau'$. Hence by the homomorphism axiom $xt = x(t + \tau' - \tau)$ for all $t \in R$, showing that the trajectory $\gamma(x)$ is periodic with a period $\tau' - \tau (> 0)$. The converse holds triviall and the theorem is proved.

2.5.9 *Remark*

It is to be noted that if $\gamma^+(x) = \Lambda^+(x)$ then the motion π_x is indeed

Poisson stable. It is, therefore, appropriate to inquire whether there exist motions which are Poisson stable but are not periodic (i.e., also not a rest point). We give below an example of a motion which is Poisson stable but is neither a rest point nor a periodic motion.

2.5.10 *Example*

Consider a dynamical system defined on a torus by means of the planar differential system

$$\frac{d\phi}{dt} = f(\phi,\theta), \qquad \frac{d\theta}{dt} = \alpha f(\phi,\theta),$$

where $f(\phi,\theta) \equiv f(\phi+1, \theta+1) \equiv f(\phi+1, \theta) \equiv f(\phi,\theta+1)$, and $f(\phi,\theta) > 0$ if ϕ and θ are not both zero (mod 1), $f(0,0) = 0$. Let $\alpha > 0$ be irrational. It is easily seen that the trajectories of this system on the torus consist of a rest point p corresponding to the point $(0,0)$. There is exactly one trajectory γ_1 such that $\Lambda^-(\gamma_1) = \{p\}$, and exactly one trajectory γ_2 such that $\Lambda^+(\gamma_2) = \{p\}$. For any other trajectory γ, $\Lambda^+(\gamma) = \Lambda^-(\gamma) =$ the torus. Further $\Lambda^+(\gamma_1) = \Lambda^-(\gamma_2) =$ the torus. In this example, therefore, the trajectory γ_1 is positively Poisson stable, but not negatively Poisson stable. The trajectory γ_2 is negatively Poisson stable, but not positively Poisson stable. All other trajectories are Poisson stable. Note that no trajectory except the rest point p is periodic.

The following theorem sheds some light on a positively Poisson stable motion π_x when $\gamma^+(x) \neq \Lambda^+(x)$.

2.5.11 *THEOREM*

Let X *be a complete metric space. Let a motion* π_x *be positively Poisson stable, and let it not be a rest point or a periodic motion. Then the set* $\Lambda^+(x) \setminus \gamma(x)$ *is dense in* $\Lambda^+(x)$, *i.e.,* $\overline{\Lambda^+(x) \setminus \gamma(x)} = \Lambda^+(x) = \overline{\gamma(x)}$.

Proof. Since π_x is positively Poisson stable, we have $\Lambda^+(x) = \overline{\gamma(x)}$. To see that

$\overline{\Lambda^+(x) \setminus \gamma(x)} = \Lambda^+(x)$, it is sufficient to show that if $y \in \gamma(x)$ and $\varepsilon > 0$ is arbitrary, then there is a point $z \in \Lambda^+(x) \setminus \gamma(x)$ such that $z \in S(y,\varepsilon)$.
To see this notice that since $y \in \Lambda^+(x) \equiv \Lambda^+(y)$, there is a monotone increasing sequence $\{t_n\}$, $t_n \to +\infty$, such that $yt_n \to y$. Choose $\tau_1 > t_1$ such that $y\tau_1 \in S(y,\varepsilon)$. Then $y\tau_1 \notin y[-t_1,t_1]$ (otherwise π_x will be periodic). Also $\delta_1 = \rho(y\tau_1, y[-t_1,t_1]) > 0$. Set $\varepsilon_1 = \min\{\frac{\varepsilon}{2}, \varepsilon - \rho(y,y\tau_1), \frac{\delta_1}{2}\}$. Then $S(y\tau_1,\varepsilon_1) \subset S(y,\varepsilon)$ and $\overline{S(y\tau_1,\varepsilon_1)} \cap y[-t_1,t_1] = \emptyset$. Having defined $y\tau_{n-1}$ and ε_{n-1}, choose $\tau_n > t_n$ such that $y\tau_n \in S(y\tau_{n-1},\varepsilon_{n-1})$ (possible because of positive Poisson stability of π_x). Then define $\varepsilon_n = \min\{\frac{\varepsilon_{n-1}}{2}, \varepsilon_{n-1} - \rho(y\tau_{n-1},y\tau_n), \frac{\delta_{n-1}}{2}\}$, where $\delta_n = \rho(y\tau_n, y[-t_n,t_n])$. Note that $\delta_n > 0$ as the motion is not periodic. Clearly $S(y\tau_n,\varepsilon_n) \subset S(y\tau_{n-1},\varepsilon_{n-1})$, and $\overline{S(y\tau_n,\varepsilon_n)} \cap y[-t_n,t_n] = \emptyset$. The sequence $\{y\tau_n\}$ has the property that $\rho(y\tau_n,y\tau_{n-1}) < \varepsilon_{n-1} \leq \frac{\varepsilon}{2^{n-1}}$ for $n = 1,2,\ldots$. $\{y\tau_n\}$ is, therefore, a Cauchy sequence which converges to a point z as the space X is complete. Since $y\tau_n \in \gamma(x)$, and $\tau_n \to +\infty$, we have $z \in \Lambda^+(x)$. Further $\rho(y,y\tau_n) < \varepsilon$, so that $\rho(y,z) \leq \varepsilon$. Notice further that $z \notin \gamma(x)$. For, otherwise, if $z \in \gamma(x) \equiv \gamma(y)$, we will have $z = y\tau$. But there is an n such that $t_n > |\tau|$, so that $z \in y[-t_n,t_n]$. However, $z \in \overline{S(y\tau_n,\varepsilon_n)}$, and by construction $\overline{S(y\tau_n,\varepsilon_n)} \cap y[-t_n,t_n] = \emptyset$, i.e., $z \notin y[-t_n,t_n]$. This contradiction proves that $z \notin \gamma(x)$ and the theorem is proved.

It is now clear that

2.5.12 *THEOREM*

If X *is complete, then a necessary and sufficient condition that* $\gamma(x)$ *be periodic is that* $\gamma(x) = \Lambda^+(x)$ $[= \Lambda^-(x)]$.

2.5.13 *Remark*

Theorem 2.5.12 is not true if X is not complete. This can be shown for instance by constructing an almost periodic motion on a torus and then delete from the space all points which do not belong to the trajectory defined by that motion.

Obviously, $\gamma(x) = \Lambda^+(x)$, but $\gamma(x)$ is not periodic.

2.5.14 *THEOREM*

A motion π_x *is positively Poisson-stable if and only if for every* $\varepsilon > 0$ *there exist a* $t \geq 1$ *such that* $xt \in S(x,\varepsilon)$.

The proof is left as an exercise to the reader.

2.5.15 *Notes and References*

This section has essentially been adopted from the book by Nemytskii and Stepanov.

2.6 *Attraction, stability, and asymptotic stability of compact sets*

2.6.1 *DEFINITIONS*

A compact set $M \subset X$ *is said to be a* weak attractor, *if there is an* $\varepsilon > 0$ *such that* $\Lambda^+(x) \cap M \neq \emptyset$ *whenever* $x \in S(M,\varepsilon)$;

an attractor, *if there is an* $\varepsilon > 0$ *such that* $\Lambda^+(x) \neq \emptyset$, *and* $\Lambda^+(x) \subset M$ *for all* $x \in S(M,\varepsilon)$;

a uniform attractor, *if it is an attractor and is such that given any* $\delta > 0$ *and a compact set* K *with the property that* $\Lambda^+(x) \neq \emptyset$, $\Lambda^+(x) \subset M$ *for all* $x \in K$, *there exists a* $T = T(K,\delta) \geq 0$ *with* $Kt \subset S(M,\delta)$ *for all* $t > T$;

stable, *if given any* $\varepsilon > 0$ *there is a* $\delta > 0$ *such that* $\gamma^+(S(M,\delta)) \subset S(M,\varepsilon)$;

asymptotically stable, *if it is both stable and an attractor; and finally*

unstable, *if it is not stable.*

2.6.2 *Remark*

The concepts of attraction and stability are in general independent of each other as we shall presently see. However, under certain circumstances attraction and uniform attraction do imply stability. Further, if a stable set is a weak attractor, then it is an attractor and hence asymptotically stable, and an asymptotically stable set is a uniform attractor. Thus the combination of stability with any one of the attractor properties yields asymptotic stability. For details see Section 1.5 .

2.6.3 *DEFINITION*

Given any set $M \subset X$, *set*

$$A_\omega(M) = \{x \in X : \Lambda^+(x) \cap M \neq \emptyset\}, \qquad \text{and}$$

$$A(M) = \{x \in X : \Lambda^+(x) \neq \emptyset, \text{ and } \Lambda^+(x) \subset M\}.$$

The sets $A_\omega(M)$, *and* $A(M)$ *are respectively called* *the region of weak attraction*, *and the* *region of attraction* *of the set* M.

Note that $A_\omega(M) \supset A(M)$ holds always.

2.6.4 *Exercise*

Show that the sets $A_\omega(M)$ and $A(M)$ are invariant.

The implications of the various stability properties defined in 2.6.1 and the elementary properties of compact sets having one of these stability properties have been discussed at length in Section 1.5. We shall now present some more results.

2.6.5 *THEOREM*

If a compact set $M \subset X$ *is stable, then* $D^+(M) = M$.

Proof. Let, if possible, $D^+(M) \neq M$. Then there is a point $y \in D^+(M) \setminus M$. Let $\rho(y,M) = \delta > 0$. Since $y \in D^+(M)$, there is an $x \in M$ with $y \in D^+(x)$, and hence sequences $\{x_n\}$, $\{t_n\}$, $x_n \to x$, $t_n \geq 0$, $x_n t_n \to y$. In view of Theorem 1.5.24 we may assume that $x_n \notin M$, $x_n \in S(M, \frac{\delta}{2})$, $x_n t_n \notin S[M, \frac{\delta}{2}]$. This shows that for every α, $0 < \alpha \leq \frac{\delta}{2}$, $\gamma^+(S(M,\alpha)) \not\subset S(M, \frac{\delta}{2})$, i.e., M is not stable. This proves the theorem.

The converse of the above theorem is not in general true. However, in locally compact metric spaces X we do have

2.6.6 *THEOREM*

If X *is locally compact, then a compact set* $M \subset X$ *is stable if and only if* $D^+(M) = M$.

Proof. Let $D^+(M) = M$, and suppose if possible that M is not stable. Then there is an $\varepsilon > 0$, a sequence $\{x_n\}$, and a sequence $\{t_n\}$, with $t_n \geq 0$,

$\rho(x_n,M) \to 0$, and $\rho(x_n t_n,M) \geq \varepsilon$. We may assume without loss of generality that $\varepsilon > 0$ has been chosen so small that $S[M,\varepsilon]$ and hence $H(M,\varepsilon)$ is compact (this is possible as X is locally compact). Further, we may assume that $x_n \to x \in M$. We can now choose a sequence $\{\tau_n\}$, $0 \leq \tau_n \leq t_n$, such that $x_n\tau_n \in H(M,\varepsilon)$, $n = 1,2,\ldots$. Since $H(M,\varepsilon)$ is compact, we may assume that $x_n\tau_n \to y \in H(M,\varepsilon)$. Then clearly $y \in D^+(x) \subset D^+(M)$, but $y \notin M$. This contradiction shows that M is stable. The converse has already been proved in the previous theorem and so the proof is completed.

The following example shows that Theorem 2.6.6 does not hold in general metric spaces.

2.6.7 *Example*

Consider Example 1.5.32 (ii) (see Figure 1.5.33). Let the space X be the set $X \setminus \gamma$. This space with the usual euclidean distance is not locally compact (note that the point p_2 does not have any compact neighborhood). Note also that now $D^+(p_2) = p_2$, but p_2 is not stable. The trajectories in the present example are the same as in Example 1.5.32 ii) except that the trajectory γ has been deleted.

The following exercise contains yet another characterization of a stable compact set M.

2.6.7b *Exercises*

i) If M is stable, then $J^-(X \setminus M) \cap M = \emptyset$.

ii) If X is locally compact, then a compact set M is stable if and only if $J^-(X \setminus M) \cap M = \emptyset$.

iii) Show that if $x, y \in X$, then $x \in J^+(y)$ if and only if $y \in J^-(x)$. Further $x \in D^+(y)$ if and only if $y \in D^-(x)$.

We shall now present an interesting property of the components of stable compact sets.

2.6.8 *THEOREM*

Let $M \subset X$ *be compact and let* X *be locally compact. Then* M *is stable if and only if every component of* M *is stable.*

Proof. Note that if M is compact, then every component of M is compact. Further if M is positively invariant, so is every one of its components. Now let $M = \bigcup\{M_i : i \in I\}$ where I is an index set, and M_i are components of M.

Let each M_i be stable, i.e., $D^+(M_i) = M_i$. Then $D^+(M) = \bigcup D^+(M_i) = \bigcup M_i = M$ and M is stable. To see the converse, let $D^+(M) = M$, i.e., M is stable. Let M_i be a component of M. Then $D^+(M_i)$ is a compact connected set, and $D^+(M_i) \subset M$. Since $D^+(M_i) \supset M_i$ and M_i is a component we have $D^+(M_i) = M_i$ and M_i is stable. The theorem is proved.

2.6.9 *Remark*

Theorem 2.6.8 is not true if X is not locally compact.

We shall now prove that in any dynamical system there do not exist compact stable sets which are weak attractors but not attractors. Before doing so, we shall prove a number of preliminary lemmas.

2.6.10 *LEMMA*

For any given set $M \subset X, x \in A_\omega(M)$ *implies* $\Lambda^+(x) \subset D^+(M)$.

Proof. Since $x \in A_\omega(M)$, we have $\Lambda^+(x) \cap M \neq \emptyset$. Choose any $y \in \Lambda^+(x) \cap M$, and let $z \in \Lambda^+(x)$ be arbitrary. Then there are sequences $\{t_n\}$, $\{\tau_n\}$, $t_n \to +\infty$, $\tau_n \to +\infty$ such that $xt_n \to y$, and $x\tau_n \to z$ (since $y, z \in \Lambda^+(x)$). We may assume without loss of generality that $\tau_n - t_n > 0$ for each n. Setting

$xt_n \equiv y_n$, $n = 1,2,\ldots$, we notice that $x\tau_n = xt_n(\tau_n - t_n) = y_n(\tau_n - t_n)$. Since $y_n \to y \in M$, $\tau_n - t_n > 0$, and $y_n(\tau_n - t_n) = x\tau_n \to z$, we have $z \in D^+(y) \subset D^+(M)$. Thus $\Lambda^+(x) \subset D^+(M)$ as $z \in \Lambda^+(x)$ was arbitrary. This proves the lemma.

2.6.11 *LEMMA*

Let $M \subset X$ *be compact. If* M *is a weak attractor, then* $A_\omega(M)$ *is an open invariant set containing* M. *If* M *is an attractor, then* $A(M)$ *is an open invariant set containing* M.

Proof. We shall only prove the first statement, as the proof of the second is the same. By definition of weak attractor, there is an $\varepsilon > 0$ such that $A_\omega(M) \supset S(M,\varepsilon)$. Now let $x \in A_\omega(M)$. Then since $\Lambda^+(x) \cap M \neq \emptyset$, there is a $T > 0$ such that $xT \in S(M,\varepsilon)$. Since $S(M,\varepsilon)$ is open, there is a $\delta > 0$ such that $S(xT,\delta) \subset S(M,\varepsilon) \subset A_\omega(M)$. Consider now the inverse image of the open set $S(xt,\delta)$ by means of the transition π^T. Since π^T is continuous, the inverse image $S(xT,\delta)(-T)$ is open and contains x. Note that for any $y \in S(xT,\delta)(-T)$, $yT \in S(xT,\delta)$, and, therefore, $\Lambda^+(y) \cap M \neq \emptyset$, as $\Lambda^+(y) = \Lambda^+(yT)$, and $\Lambda^+(yT) \cap M \neq \emptyset$ because $yT \in S(M,\varepsilon)$. Hence $S(xT,\delta)\ (-T) \subset A_\omega(M)$. This shows that $A_\omega(M)$ is open. $A_\omega(M)$ is indeed always invariant and the Lemma is proved.

We are now ready to prove our promised theorem.

2.6.12 *THEOREM*

Let M *be a compact stable set. If* M *is a weak attractor, then it is an attractor and hence is an asymptotically stable set.*

Proof. Since M is stable, we have $D^+(M) = M$. If $x \in A_\omega(M)$, then we have by

Lemma 2.6.10 $\Lambda^+(x) \subset D^+(M) = M$. This shows that $A_\omega(M) \subset A(M)$, and as $A(M) \subset A_\omega(M)$ holds always, we have $A_\omega(M) = A(M)$. As M is a weak attractor $A_\omega(M)$ is a neighborhood of M. Thus M is an attractor. The theorem is proved.

We shall next characterize the property of asymptotical stability of a compact set M in terms of $J^+(x)$.

2.6.13 *THEOREM*

Let M *be compact and positively invariant. Then* M *is asymptotically stable, if and only if there is a* $\delta > 0$ *such that* $x \in S(M,\delta)$ *implies* $J^+(x) \neq \emptyset$ *and* $J^+(x) \subset M$.

Proof. Let M be asymptotically stable. Since M is stable, we have $M = D^+(M)$. There is further an $\eta > 0$ such that $x \in S(M,\eta)$ implies $\Lambda^+(x) \neq \emptyset$ and $\Lambda^+(x) \subset M$. This shows that for $x \in M$, $J^+(x) \neq \emptyset$ and $J^+(x) \subset M$. Now let $x \in S(M,\eta) \setminus M$. Let, if possible, $J^+(x) \not\subset M$. There is then a $y \in J^+(x)$, $y \notin M$. Choose $z \in \Lambda^+(x)$. Since $\Lambda^+(x) \neq \emptyset$, $\Lambda^+(x) \subset M$, we have $D^+(z) \subset M$. By Lemma 2.3.12 $J^+(x) \subset D^+(z)$. But as $M = D^+(M)$, we have $D^+(z) \subset D^+(M) = M$, also $J^+(x) \subset M$.

Now let for $x \in S(M,\delta)$, $J^+(x) \neq \emptyset$, $J^+(x) \subset M$. Clearly $J^+(x)$ is compact. Hence $\Lambda^+(x) \neq \emptyset$ and further $\Lambda^+(x) \subset J^+(x) \subset M$. This shows that M is an attractor. Now let $x \in M$, then $D^+(x) = \gamma^+(x) \cup J^+(x)$. Since $J^+(x) \subset M$ and M is positively invariant, we have $D^+(x) \subset M$. Hence $D^+(M) = M$, i.e., M is stable. The theorem is proved.

2.6.14 *Exercise*

If a compact set M is asymptotically stable, then $J^+(A(M)) \subset M$.

We shall close this section by proving a very interesting property of compact weak attractors in locally compact metric spaces.

2.6.15 *THEOREM*

Let X *be locally compact. Let* M *be a compact weak attractor. Then*

$D^+(M)$ *is a compact asymptotically stable set, with* $A(D^+(M)) \equiv A_\omega(M)$. *Moreover,* $D^+(M)$ *is the smallest asymptotically stable set containing* M.

For the proof we shall need the following lemma.

2.6.16 *LEMMA*

Let X *be locally compact. Let* M *be a compact weak attractor, and let* $\alpha > 0$. *Then there is* $T > 0$ *such that*

$$D^+(M) \subset S[M,\alpha] \cdot [0,T] \equiv \pi(S[M,\alpha],[0,T]),$$

where π *is the map defining the dynamical system.*

Proof. Choose ε, $0 < \varepsilon \leq \alpha$, such that $S[M,\varepsilon]$ is a compact subset of $A_\omega(M)$. For $x \in H(M,\varepsilon)$, define $\tau_x = \inf\{t > 0: xt \in S(M,\varepsilon)\}$; since $x \in A_\omega(M)$, τ_x is defined. Set $T = \sup\{\tau_x : x \in H(M,\varepsilon)\}$. We claim that $T < +\infty$. If this is not the case, there is a sequence $\{x_n\}$ in $H(M,\varepsilon)$ for which $\tau_{x_n} \to +\infty$. We may assume that $x_n \to x \in H(M,\varepsilon)$. Let $\tau > 0$ such that $x\tau \in S(M,\varepsilon)$. For sufficiently large n, we have then $\tau_{x_n} < \tau$, which contradicts $\tau_{x_n} \to +\infty$. Now let $y \in D^+(M) \setminus S[M,\varepsilon]$. Then there are sequences $\{x_n\}$, $\{t_n\}$ with $x_n \to x \in M$, and $t_n \geq 0$ such that $x_n t_n \to y$. Then for all sufficiently large n there is a τ_n, $0 < \tau_n < t_n$ such that $x_n \tau_n \in H(M,\varepsilon)$, and $xt \notin S[M,\varepsilon]$ for $\tau_n < t \leq t_n$. By the first part of this proof $0 < t_n - \tau_n < T$. Then $x_n t_n = x_n \tau_n (t_n - \tau_n) \in S[M,\varepsilon][0,T]$. Therefore, $y \in S[M,\varepsilon][0,T]$, since this set is closed. The lemma is proved.

2.6.17 *Proof of Theorem* 2.6.15

Notice that if $S[M,\varepsilon]$ is compact, then $S[M,\varepsilon][0,T]$ is compact for any $T > 0$. Thus $D^+(M)$ being a closed subset of the compact set $S[M,\varepsilon][0,T]$ (by the above lemma) is compact. Further, as $\varepsilon > 0$ is chosen such that $S[M,\varepsilon] \subset A_\omega(M)$ we have $D^+(M) \subset S[M,\varepsilon][0,T] \subset A_\omega(M)$. Thus $A_\omega(M)$ is an open invariant set containing

$D^+(M)$, and is, therefore, a neighborhood of $D^+(M)$. By Lemma 2.6.10 $x \in A_\omega(M)$ implies $\Lambda^+(x) \neq \emptyset$, and $\Lambda^+(x) \subset D^+(M)$. Therefore, $D^+(M)$ is an attractor. Notice that $A_\omega(M) = A(D^+(M))$, for if there is an $x \in A(D^+(M))$, then there is a $t > 0$ such that $xt \in A_\omega(M)$ (this being a neighborhood of $D^+(M)$), and since $\Lambda^+(x) \equiv \Lambda^+(xt)$, we have $x \in A_\omega(M)$. To show that $D^+(M)$ is stable, let $x \in D^+(M)$. Since $x \in A_\omega(M)$, we can choose an $\omega \in \Lambda^+(x) \cap M$. Then $J^+(x) \subset D^+(\omega) \subset D^+(M)$ by Lemma 2.3.12. Thus $D^+(x) = \gamma^+(x) \cup J^+(x) \subset D^+(M) \cup D^+(M) = D^+(M)$, for $D^+(M)$ is positively invariant. This shows that $D^+(D^+(M)) = D^+(M)$, i.e., $D^+(M)$ is stable (2.6.6). We have thus proved that $D^+(M)$ is asymptotically stable. Finally, let M^* by any compact set such that $M \subset M^* \subset D^+(M)$. Then $D^+(M) \subset D^+(M^*) \subset D^+(D^+(M)) = D^+(M)$, and so $D^+(M^*) = D^+(M)$. If M^* is stable, then $M^* = D^+(M^*) = D^+(M)$. Thus $D^+(M)$ is the smallest stable (also asymptotically stable) set containing M. The theorem is proved.

2.6.18 *Exercises*

i) Let M be a compact invariant set. Let M be a weak attractor. Let X be locally compact. Then $\Lambda^-(y) \cap M \neq \emptyset$ for every $y \in D^+(M)$.

ii) Let X be locally compact, and M a compact invariant weak attractor. Then M is a negative weak attractor if and only if $D^+(M) \equiv A_\omega(M)$.

2.6.19 *THEOREM*

Let X *be a locally compact and locally connected metric space. Let* $M \subset X$ *be a compact asymptotically stable set. Then* M *has a finite number of components, each of which is asymptotically stable.*

2.6.20 *Remark*

Theorem 2.6.19 is not true if the space does not have the properties listed above. Consider for the case of a dynamical system defined only on a compact sequence of points tending to one point. The compact set is asymptotically stable and so are its isolated components. But the limit point (a component) is not asymptotically stable and there are an infinite number of components.

2.6.21. *Notes and References*

The first systematic application of the notion of a prolongation to attractors seems to have been done by Auslander, Bhatia, and Seibert. Most results in this study were shown to be valid for weak attractors by Bhatia [3] who introduced this later notion. Theorem 2.6.19 is essentially due to Desbrow, who proves it for a connected, locally connected, locally compact metrizable space X. We observe that connectedness of the space is not required, but local connectedness is essential. A trivial counter example was given in the last remark.

2.7 *Liapunov functions and asymptotic stability of compact sets.*

The basic feature of the stability theory à la Liapunov is that one seeks to characterize the stability or instability properties of a given set of the phase space in terms of the existence of certain types of scalar functions (i.e., real valued functions) defined in suitable sets (usually neighborhoods of the given set) of the phase space. Such functions are generally required to be monotone along the trajectories of the given dynamical system. Any such function which guarantees a stability or instability property of a set is termed as a Liapunov function for that set. In what follows, we shall present some very strong results. By this we mean theorems on necessary and sufficient conditions for asymptotic stability of compact sets based upon the existence of continuous real-valued functions of very special types. These functions will indeed characterize the behavior of the dynamical system much better than the functions presented in Section 1.7.

The simplest and perhaps the best known result on asymptotic stability is

2.7.1 *THEOREM*

A compact set $M \subset X$ *is* *asymptotically stable* *if and only if there exists a continuous scalar function* $v = \Phi(x)$ *defined in a neighborhood* N *of* M *such that*

i) $\Phi(x) = 0$ *if* $x \in M$ *and* $\Phi(x) > 0$ *if* $x \notin M$;

ii) $\Phi(xt) < \Phi(x)$ *for* $x \notin M, t > 0$ *and* $x[0,t] \subset N$.

Remark. This theorem is similar to Theorem 10 in Auslander and Seibert [2]. The minor changes being necessitated as we have not assumed M or N to be positively invariant as is the case in [2]. The corresponding minor changes in the proof of sufficiency can be made and so we omit this part of the proof. Since we have a different proof of necessity we give it below. The difference lies in the fact, that in [2] the authors prove the existence of a suitable function in a relatively compact positively invariant neighborhood of M, whereas we show that the same method as in [2] yields a function with desired properties defined on the whole region of attraction $A(M)$ of M. Since $A(M)$ need not be relatively compact we need a different proof.

2.7.2 *Proof of necessity in Theorem* 2.7.1. Let M be asymptotically stable and let $A(M)$ be its region of attraction. For each $x \in A(M)$ define

$$\phi(x) = \sup\{\rho(xt,M): t \geq 0\}.$$

Indeed $\phi(x)$ is defined for each $x \in A(M)$, because if $\rho(x,M) = \alpha$, then there is a $T > 0$ with $x[T, +\infty) \subset S(M,\alpha)$. Thus

$$\phi(x) \equiv \sup\{\rho(xt,M): 0 \leq t \leq T\}.$$

As $\rho(xt,M)$ is a continuous function of t, $\phi(x)$ is defined. This $\phi(x)$ has the properties: $\phi(x) = 0$ for $x \in M$, and $\phi(x) > 0$ for $x \notin M$, and $\phi(xt) \leq \phi(x)$ for $t \geq 0$. This is clear when we remember that M is stable and hence positively invariant and that $A(M)$ is invariant. So that if $\phi(x)$ is defined for any $x \in A(M)$ it is defined for all xt with $t \in R$. We further claim that this $\phi(x)$ is continuous in $A(M)$. Indeed stability of M implies continuity of $\phi(x)$ on M. For $x \in M$ we can prove the continuity of $\phi(x)$ as follows. For $x \notin M$, set $\rho(x,M) = \alpha (> 0)$ and choose ε, $0 < \varepsilon < \frac{\alpha}{4}$, such that $S[x,\varepsilon]$ is a compact subset of $A(M)$; this is possible as X is locally compact and $A(M)$ is open. Since M is a uniform attractor (1.5.28), there is a $T > 0$ such that $S[x,\varepsilon]t \subset S(M, \frac{\alpha}{4})$ for all $t \geq T$. Thus for $y \in S[x,\varepsilon]$ we have

$$\begin{aligned}\phi(x) - \phi(y) &= \sup\{\rho(xt,M):t \geq 0\} - \sup\{\rho(yt,M):t \geq 0\} \\ &= \sup\{\rho(xt,M):0 \leq t \leq T\} - \sup\{\rho(yt,M): 0 \leq t \leq T\}.\end{aligned}$$

So that

$$|\phi(x) - \phi(y)| \leq \sup\{|\rho(xt,M) - \rho(yt,M)|: 0 \leq t \leq T\}$$

$$\leq \sup\{\rho(xt,yt): 0 \leq t \leq T\}.$$

The continuity axiom implies that the right hand side of the above inequality tends to zero as $y \to x$, for T is fixed for $y \in S[x,\varepsilon]$. $\phi(x)$ is therefore continuous in $A(M)$. The above function indeed may not be strictly decreasing along parts of trajectories in $A(M)$ which are not in M and so may not satisfy (ii). Such a function can be obtained by setting

$$\Phi(x) = \int_0^\infty \phi(xt)\exp(-\tau)d\tau$$

That $\Phi(x)$ is continuous and satisfies (i) in $A(M)$ is clear. To see that $\Phi(x)$ satisfies (ii), let $x \notin M$ and $t > 0$. Then indeed $\Phi(xt) \leq \Phi(x)$ holds, because $\phi(xt) \leq \phi(x)$ holds. To rule out $\Phi(xt) = \Phi(x)$, observe that in this case we must have $\phi(x(t + \tau)) \equiv \phi(x\tau)$ for all $\tau \geq 0$. Thus, in particular, letting $\tau = 0,t,2t,\ldots$ we get $\phi(x) = \phi(x(nt))$, $n = 1,2,3,\ldots$. But asymptotic stability of M implies that for $x \in A(M)$, $\rho(xt,M) \to 0$ as $t \to \infty$. Thus $\phi(x(nt)) \to 0$ as $n \to \infty$, as $\phi(x)$ is continuous. This shows that $\phi(x) = 0$. But as $x \notin M$, we must have $\phi(x) > 0$, a contradiction. We have thus proved that $\Phi(xt) < \Phi(x)$ for $x \notin M$ and $t > 0$. The theorem is proved.

2.7.3 *Remark*

Theorem 2.7.1 says nothing about the extent of the region of attraction of M. Thus if a function $\Phi(x)$ as in Theorem 2.7.1 is known to exist in a neighborhood N of M, we need not have either $N \subset A(M)$ or $A(M) \subset N$. We will give an example to elucidate this point. (The observation is indeed well known, but examples are woefully lacking in the literature). In particular this means that the above theorem <u>cannot</u> immediately be stated as a theorem on global asymptotic stability:

A compact set M is said to be globally asymptotically stable if it is asymptotically stable and $A(M) \equiv X$.

2.7.4 *Example*

Consider a dynamical system defined in the real euclidean plane by the differential equations

2.7.5 $\dot{x} = f(x,y), \qquad \dot{y} = g(x,y)$,

where

$g(x,y) = -y$ for all (x,y), and

2.7.6 $f(x,y) = x$ if $x^2y^2 \geq 1$; $\qquad f(x,y) = 2x^3y^2 - x$ if $x^2y^2 < 1$.

These equations are integrable by elementary means and the phase portrait is as in *Figure* 2.7.7

2.7.7 *Figure*

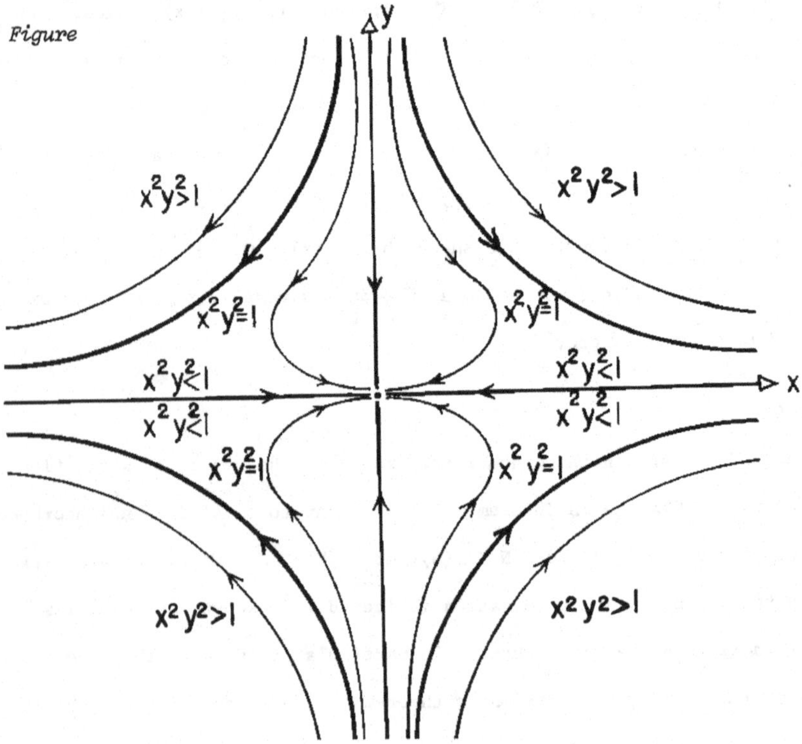

The origin (0,0) is asymptotically stable, with the set $\{(x,y): x^2y^2 < 1\}$ as its region of attraction. Consider now the function

2.7.8 $$\Phi(x,y) = y^2 + \frac{x^2}{1 + x^2} \quad .$$

This function satisfies conditions of Theorem 2.7.1 in the whole euclidean plane. To see this one may find the derivative $\dot{\Phi}(x,y) = \frac{\partial\Phi}{\partial x} f(x,y) + \frac{\partial\Phi}{\partial y} g(x,y)$ of $\Phi(x)$ for the given system as is standard practice for differential equations. It can easily be verified that $\dot{\Phi}(x,y) < 0$ for all $(x,y) \neq (0,0)$, which implies that this $\Phi(x)$ satisfies conditions of Theorem 2.7.1 in every neighborhood of the origin. But not every neighborhood of the origin is contained in the region of attraction, nor does it contain the region of attraction.

Since Theorem 2.7.1 guarantees asymptotic stability of M, it is clear that $A(M) \cap N$ is a neighborhood of M. How far can the extent of this neighborhood be determined by a $\Phi(x)$ function, as in Theorem 2.7.1, is the subject of the following proposition. The following proposition and remark will be obvious to those familiar with the work of LaSalle [3]. They are included because they are needed in our investigation and some of the points here do not seem to have been clarified by LaSalle.

2.7.9 *LEMMA*

Let $\Phi(x)$ *be a function defined in a neighborhood* N *of* M *and having properties as in Theorem* 2.7.1. *Then if* $K \subset N$ *is any compact positively invariant set, we have* $K \subset A(M)$. *Further for all sufficiently small* $\alpha > 0$ *the set* $K_\alpha = \{x \in N: \Phi(x) \leq \alpha\}$ *has a compact positively invariant subset* P_α *such that* $P_\alpha \cap \overline{K_\alpha \setminus P_\alpha} = \emptyset$ *and* P_α *is a neighborhood of* M, *further* $P_\alpha \subset A(M)$. *In particular if* K_α *is compact, then* $K_\alpha \subset A(M)$.

Proof. If $x \in K$, then $\Lambda^+(x) \neq \emptyset$ and $\Lambda^+(x) \subset K \subset N$. For any $y \in \Lambda^+(x)$ and $t > 0$, we have $\Phi(yt) \leq \Phi(y)$ by (i) and (ii) of Theorem 2.7.1. Further, there are sequences $\{t_n\}$ and $\{\tau_n\}$ $t_n \to +\infty$, $\tau_n \to +\infty$, and $xt_n \to y$, $x\tau_n \to yt$ (as $yt \in \Lambda^+(x)$ also). We may assume that $t_n \geq \tau_n$ for each n. Thus we have $\Phi(xt_n) \leq \Phi(x\tau_n)$ for each n. As $\Phi(x)$ is continuous we can proceed to the limit and get $\Phi(y) \leq \Phi(yt)$. Thus $\Phi(y) = \Phi(yt)$. By (ii) Theorem 2.7.1 then, we have $y \in M$, i.e., $\Lambda^+(x) \subset M$. Hence $K \subset A(M)$. To prove the second part, let $\varepsilon > 0$ be chosen such that $S[M,\varepsilon]$ is a compact subset of $I(N)$. Set $m(\varepsilon) = \min\{\Phi(x) : x \in H(M,\varepsilon)\}$. Clearly $m(\varepsilon) > 0$. Now let $0 < \alpha < m(\varepsilon)$, and set $P_\alpha = K_\alpha \cap S[M,\varepsilon]$. We claim that P_α is a set predicted in the lemma. Clearly $P_\alpha \neq \emptyset$ as $P_\alpha \supset M$. P_α is compact, for if $\{x_n\}$ is a sequence in P_α, then we may assume that it is convergent, because $\{x_n\}$ is in the compact set $S[M,\varepsilon]$. If now $x_n \to x$, then clearly $x \in S[M,\varepsilon]$ on one hand, and $\Phi(x) \leq \alpha$, because $\Phi(x_n) \leq \alpha$ for each n, showing that $x \in K_\alpha$. It follows that $x \in (K_\alpha \cap S[M,\varepsilon]) = P_\alpha$. That P_α is positively invariant may be seen by first observing that if $x \in P_\alpha$, and $xR^+ \not\subset S(M,\varepsilon)$, then there is a $t > 0$ such that $xt \in H(M,\varepsilon)$ and $x[0,t] \subset S[M,\varepsilon] \subset N$. By (ii) Theorem 2.7.1 we get $\Phi(xt) \leq \Phi(x) \leq \alpha$, and $xt \in H(M,\varepsilon)$ implies $\Phi(xt) \geq m(\varepsilon) > \alpha$, a contradiction. Hence $xR^+ \subset S(M,\varepsilon) \subset N$. It follows by (ii) Theorem 2.7.1 that $\Phi(xt) \leq \Phi(x) \leq \alpha$ for each $t \geq 0$, which shows that $xR^+ \subset K_\alpha$. Thus $xR^+ \subset K_\alpha \cap S(M,\varepsilon) \subset P_\alpha$. That is, P_α is positively invariant. By the first part of the lemma we have $P_\alpha \subset A(M)$. That P_α is a neighborhood of M follows from the continuity of Φ and the fact that $\Phi(x) = 0$ for $x \in M$. Finally to see that $P_\alpha \cap \overline{K_\alpha \setminus P_\alpha} = \emptyset$, observe that $\bar{K}_\alpha \cap H(M,\varepsilon) = \emptyset$. The last part follows by observing that if K_α is compact, then it is positively invariant.

2.7.10 *Remark*

For any given $\varepsilon, \alpha > 0$ chosen as above, the set P_α defined in the above proof is the largest, compact positively invariant set in K_α. For if $S \subset K_\alpha$ were a larger set, then $S \setminus P_\alpha \subset C(S[M,\varepsilon])$, so that $\overline{S \setminus P_\alpha} \cap P_\alpha = \emptyset$. It follows that $S \setminus P_\alpha$ is compact and positively invariant. This is impossible unless $S \setminus P_\alpha = \emptyset$. For, if $S \setminus P_\alpha$ were a non-empty subset of N, by the above lemma, $S \setminus P_\alpha \subset A(M)$. Therefore, $(S \setminus P_\alpha)R^+ \cap P_\alpha \neq \emptyset$ as P_α is a neighborhood of M. But $(S \setminus P_\alpha) R^+ = S \setminus P_\alpha$, and $(S \setminus P_\alpha) \cap P_\alpha = \emptyset$.

2.7.11 *THEOREM*

Let $M \subset X$ *be a compact asymptotically stable set. Then its region of attraction* $A(M)$ *contains a countable dense subset. In particular, if* $A(M) = X$, *then* X *is separable.*

Proof. Take any $\Phi(x)$ defined in a neighborhood N of M and satisfying conditions of Theorem 2.7.1. Choose ε, $\alpha > 0$ as in the proof of Theorem 2.7.9 and construct P_α as before. As P_α is a compact positively invariant neighborhood of M, and $P_\alpha \subset A(M)$,

$$A(M) = P_\alpha R = P_\alpha R^+ \cup P_\alpha R^- = P_\alpha \cup P_\alpha R^- = P_\alpha R^-.$$

Thus $A(M) = \bigcup\{P_\alpha[-n,0] : n = 1,2,\ldots\} \equiv P_\alpha R^-$.

Since each $P_\alpha[-n,0]$ is compact and thus contains a countable dense subset, the result follows.

2.7.12 *Remark*

When we observe that $P_\alpha[-n,0]$ is for each positive integer n a compact and hence relatively compact positively invariant neighborhood of M, we can use the method of proof of Auslander and Seibert [2] to get our result on the necessity part of Theorem 2.7.1.

The above proposition motivates the following definition.

2.7.13 *DEFINITION*

A continuous scalar function $\Phi(x)$ *defined on a set* $N \subset X$ *will be said to be* *uniformly unbounded* *on* N *if given any* $\alpha > 0$ *there is a compact set* $K \subset N$, $K \neq N$, *such that* $\Phi(x) \geq \alpha$ *for* $x \notin K$.

2.7.14 *THEOREM*

Let $M \subset X$ *be a compact asymptotically stable set and let* M *be invariant. Then there exists a continuous uniformly unbounded function* $\Phi(x)$ *defined on* $A(M)$ *such that*

i) $\Phi(x) = 0$ *for* $x \in M$, *and* $\Phi(x) > 0$ *for* $x \notin M$,

ii) $\Phi(xt) = e^{-t}\,\Phi(x)$ *for all* $x \in A(M)$ *and* $t \in R$.

Proof: Consider any function $\phi(x)$ defined in any neighborhood N of M and satisfying conditions of Theorem 2.7.1. Choose $\varepsilon, \alpha > 0$ as in Theorem 2.7.9, and consider the set P_α. Note that $\partial P_\alpha = \{x \in S[M,\varepsilon] : \phi(x) = \alpha\}$. We claim that for every point $x \in A(M)\setminus M$, there is a unique $\tau(x) \in R$ such that $x\tau(x) \in \partial P_\alpha$ The uniqueness of $\tau(x)$ follows from the fact that if $x\tau(x) \in \partial P_\alpha$, then $\phi(x(\tau(x) + t)) < \phi(x\tau(x)) = \alpha$ for $t > 0$. As P_α is positively invariant and $\phi(y) = \alpha$ for $y \in \partial P_\alpha$, we conclude that $x(\tau(x) + t)$ is in the interior of P_α for all $t > 0$. This establishes uniqueness of $\tau(x)$ for all x for which $\tau(x)$ is defined. That $\tau(x)$ is defined for all $x \in A(M)\setminus P_\alpha$ follows from the fact that P_α is a neighborhood of M, so that any trajectory $\gamma(x) \subset A(M)$ with $x \notin P_\alpha$ must intersect ∂P_α. For $x \in \partial P_\alpha$ indeed $\tau(x)$ is defined and $\tau(x) = 0$. For $x \in (I(P_\alpha)\setminus M)$, if there is no $\tau(x)$ such that $x\tau(x) \in \partial P_\alpha$, then $\gamma(x) \subset (I(P_\alpha)\setminus M) \subset P_\alpha$. P_α being compact

$\Lambda^-(x) \neq \emptyset$, $\Lambda^-(x) \subset P_\alpha$, but $\Lambda^-(x) \cap M = \emptyset$ (otherwise M will be unstable). Now $\Lambda^-(x)$ is compact and invariant, so that if $y \in \Lambda^-(x)$ we have $\Lambda^+(x) \neq \emptyset$, and $\Lambda^+(y) \subset \Lambda^-(x)$. Then on one hand we have $\Lambda^+(y) \cap M = \emptyset$ and on the other hand $\Lambda^+(y) \subset M$ as $y \in A(M)$. This contradicts $\Lambda^+(y) \neq \emptyset$. Thus $\gamma(x) \cap \partial P_\alpha \neq \emptyset$, and $\tau(x)$ is defined for each $x \in A(M) \setminus M$. Note that $A(M) \setminus M$ is invariant, as M and A(M) are both invariant. For $x \in A(M) \setminus M$ and $\tau \in R$ observe now that

$$\tau(xt) \equiv \tau(x) - t .$$

This follows from the fact that any trajectory $\gamma(x)$ in $A(M) \setminus M$ intersects ∂P_α at exactly one point. Thus $xt(\tau(xt)) = x\tau(x)$, i.e., by the homomorphism axiom

2.7.15 $$x(t + \tau(xt)) = x(\tau(x)) \quad .$$

As $\gamma(x)$ can neither be periodic nor a rest point, we have $t + \tau(xt) = \tau(x)$. This shows further that $\tau(xt)$ is a continuous function of t and $\tau(xt) \to \pm \infty$ as $t \to \mp \infty$. We now claim that $\tau(x)$ is continuous on $A(M) \setminus M$. For any $x \in A(M) \setminus M$, and $\varepsilon > 0$, the point $y \equiv x(\tau(x) + \varepsilon) \in I(P_\alpha)$. There is therefore a neighborhood N_y of y such that $N_y \subset P_\alpha$. Then $N^+ = N_y(-\tau(x) - \varepsilon)$ is a neighborhood of x and note that for each $\omega \in N^+$, $\tau(\omega) \leqq \tau(x) + \varepsilon$. Again the point $z \equiv x(\tau(x) - \varepsilon) \in (A(M) \setminus P_\alpha)$, the last set being open. Thus there is a neighborhood N_z of z such that $N_z \subset (A(M) \setminus P_\alpha)$. Then $N^- = N_z(-\tau(x) + \varepsilon)$ is a neighborhood of x and note that for each $\omega \in N^-$ we have $\tau(\omega) \geqq \tau(x) - \varepsilon$. Thus if ω is in the neighborhood $N_\varepsilon = N^+ \cap N^-$ of x , we have

$$\tau(x) - \varepsilon \leqslant \tau(\omega) \leqslant \tau(x) + \varepsilon \quad .$$

This proves continuity of $\tau(x)$ in $A(M)\setminus M$. We now show that $\tau(x) \to -\infty$ as $x \to M$, $x \in A(M)\setminus M$. If this were not true, then there will be a $T > 0$ and a sequence $\{x_n\}$ in $A(M)\setminus M$, such that $x_n \to x \in M$ and $-T \leq \tau(x_n) \leq 0$. Since $\{\tau(x_n)\}$ is a bounded sequence it contains a convergent subsequence. We may therefore assume that $\tau(x_n) \to \tau$, where $-T \leq \tau \leq 0$. Then by the continuity axiom $x_n\, \tau(x_n) \to x\tau$. As M is invariant $x\tau \in M$, on the other hand $x_n\, \tau(x_n) \in \partial P_\alpha$ which is compact. Therefore $x\tau \in \partial P_\alpha$. But $\partial P_\alpha \cap M = \emptyset$, a contradiction. We now define the function $\Phi(x)$ on $A(M)$ as follows

$$\Phi(x) = 0 \qquad \text{for } x \in M, \text{ and}$$

$$\Phi(x) = e^{\tau(x)} \qquad \text{for } x \in A(M) \setminus M.$$

The above observations show that this function is continuous on $A(M)$. It is clearly positive for $x \notin M$, and

2.7.16 $$\Phi(xt) = e^{\tau(xt)} = e^{\tau(x)-t} = \Phi(x)e^{-t} \quad .$$

Lastly to see that this $\Phi(x)$ is uniformly unbounded, recall that $A(M) = \bigcup\{P_\alpha[-n,0] : n = 1,2,3, \dots\}$. Each $P_\alpha[-n,0]$ is compact and positively invariant. Observe that if $x \notin P_\alpha[-n,0]$, then $\tau(x) > n$, so that $\Phi(x) > e^n$. This proves the theorem completely.

2.7.17 *THEOREM*

If $M \subset X$ *is any compact asymptotically stable set, then there exists a continuous uniformly unbounded function* $\Phi(x)$ *on* $A(M)$ *such that*

i) $\Phi(x) = 0$ *for* $x \in M$, *and* $\Phi(x) > 0$ *for* $x \notin M$,

ii) $\Phi(xt) < \Phi(x)$ *for* $x \notin M$ *and* $t > 0$.

Proof. Consider any function $\phi(x)$ defined in a neighborhood N of M and satisfying conditions of Theorem 2.7.1. Choose $\varepsilon, \alpha > 0$ as before and consider P_α. For each $x \in A(M) \setminus P_\alpha$ define $\tau(x)$ as before. This $\tau(x)$ is continuous and $\tau(x) \to 0$ as $x \to P_\alpha$, $x \notin P_\alpha$. Now define

$$\Phi(x) \equiv \phi(x) \qquad \text{for } x \in P_\alpha \text{, and}$$

$$\Phi(x) = \alpha e^{\tau(x)} \qquad \text{for } x \in A(M)\setminus P_\alpha .$$

This $\Phi(x)$ has the desired properties as may easily be verified.

2.7.18 *THEOREM*

Let $M \subset X$ *be compact and let there exist a continuous uniformly unbounded function* $\Phi(x)$ *defined on an open neighborhood* N *of* M *such that*

i) $\Phi(x) = 0$ *for* $x \in M$, *and* $\Phi(x) > 0$ *for* $x \notin M$,

ii) $\Phi(xt) < \Phi(x)$ *for* $x \notin M$, $t > 0$ *and* $x[0,t] \subset N$.

Then M *is asymptotically stable and* $N \subset A(M)$. *If, in addition, any condition guaranteeing the invariance of* N *holds, then* $N = A(M)$.

Proof: The proof follows from the observation that if $\Phi(x)$ is uniformly unbounded on N, then for any $\alpha > 0$, the set $K_\alpha = \{x : \Phi(x) \leqq \alpha\}$ is compact. Then by Theorem 2.7.9, $K_\alpha \subset A(M)$. Now $N = \bigcup \{K_n : n = 1,2, \dots \}$, and since each $K_n \subset A(M)$ we have $N \subset A(M)$. Lastly if N is invariant we must have $N = A(M)$ as N is a neighborhood of M. The remaining details of the proof will be the same as in any proof of sufficiency of Theorem 2.7.1. These we leave to the reader.

For global asymptotic stability we can state the following two theorems as corollaries of the above results.

2.7.19 *THEOREM*

A compact invariant set $M \subset X$ *is globally asymptotically stable if there exists a continuous function* $\Phi(x)$ *defined on* X *such that*

i) $\Phi(x) = 0$ *for* $x \in M$, $\Phi(x) > 0$ *for* $x \notin M$,

ii) $\Phi(xt) = e^{-t} \Phi(x)$.

(Note that any such $\Phi(x)$ *will be necessarily uniformly unbounded on* X*).*

Proof: The sufficiency follows from Theorem 2.7.18 as X is an invariant neighborhood of M. The necessity follows from Theorem 2.7.14.

2.7.20 *THEOREM*

A compact set $M \subset X$ *is globally asymptotically stable if and only if there exists a continuous uniformly unbounded function* $\phi(x)$ *defined on* X *such that*

i) $\Phi(x) = 0$ *for* $x \in M$, $\Phi(x) > 0$ *for* $x \notin M$,

ii) $\Phi(xt) < \Phi(x)$ *for* $x \notin M$ *and* $t > 0$.

Proof: Sufficiency follows from Theorem 2.7.18, the necessity from Theorem 2.7.17.

2.7.21 *Remark*

In dynamical systems defined in locally compact metric spaces, one may define ultimate boundedness of the dynamical system by the property that there is a compact set $K \subset X$ with $\Lambda^+(x) \neq \emptyset$, and $\Lambda^+(x) \subset K$ for each $x \in X$, i.e. whenever there exists a compact global attractor in X . It is shown in Theorem 2.6.15 that if $K \subset X$ is a compact weak attractor, then $D^+(K)$ (the first positive prolongation of K) is a compact positively invariant set which is asymptotically stable and has the same region of attraction as K. Following Ura [7], one can show now that the largest invariant set in $D^+(K)$ is compact and asymptotically stable with the same region of attraction as of K. These observations will allow one to write theorems on ultimate boundedness which are similar to those on global asymptotic stability. We leave these to the reader.

2.7.22 *Remark*

If $M \subset X$ is a compact asymptotically stable set, then following the methods in Bhatia [1] one can obtain a Liapunov function $\phi(x)$ defined in $A(M)$ with the following properties

i) $\phi(x) = 0$ for $x \in M$, $\phi(x) > 0$ for $x \notin M$,

ii) $\phi(xt) \leq e^{-t}\, \phi(x)$ for all $x \in A(M)$ and $t > 0$.

This function, however, need not be uniformly unbounded on $A(M)$. To obtain a function which is uniformly unbounded and has the above two properties, we may use the above function in the construction of $\Phi(x)$ of the proof of Theorem 2.7.17. We thus have the following stronger result for a compact (not necessarily invariant) set M.

2.7.23 *THEOREM*

If $M \subset X$ *is a compact asymptotically stable set with the region of attraction* $A(M)$, *then there exists a continuous uniformly unbounded function* $\Phi(x)$ *on* $A(M)$ *having the following properties*

i) $\Phi(x) = 0$ *for* $x \in M$ *and* $\Phi(x) > 0$ *for* $x \notin M$,

ii) $\Phi(xt) \leq e^{-t}\, \Phi(x)$ *for* $x \in A(M)$ *and* $t \geq 0$.

2.7.24 *Remark*

In Theorem 2.7.18, 2.7.19 and 2.7.20 the proof of sufficiency can be completed without the explicit assumption that $\Phi(x) > 0$ for $x \notin M$, for this follows from the remaining conditions on $\Phi(x)$.

2.7.25 *Notes and References*

This section contains results of Bhatia [6]. Some remarks are in order. Earlier results in this direction, for example those of Zubov [6], Auslander and Seibert [2], and Bhatia [1], used essentially the same methods as used for the well-developed theory in the case of ordinary differential systems. For results on ordinary differential equations see, for example, A. M. Liapunov, I. A. Malkin, Barbashin, Krasovskii, Kurzweil, Vrkoch, K. P. Persidskii, S. K. Persidskii, Zubov, Massera, Antosiewicz, Yoshizawa, W. Hahn. The basic feature of the results in this section is that Liapunov functions are shown to exist on the whole region of attraction as against on a sufficiently small neighborhood in earlier results. The functions, in general, have sufficient properties to allow the derivation of theorems on global asymptotic stability and ultimate boundedness as corollaries. Indeed Auslander, Seibert established formally the long suspected duality between stability and boundedness in locally compact separable metric spaces.

2.8 *Topological properties of* $A_\omega(M), A(M)$ *and* $P_\alpha(M)$.

In this section we shall present some additional properties of attractors, region of attractions and the level lines of the corresponding Liapunov functions. We shall present results for the case of weak attraction and asymptotic stability. The latter results are valid with few obvious changes also for the case of complete instability, i.e., in all cases of strong stability properties. The results that we shall present are extensions and improvements of the ones presented in Section 1.9 and they are based upon the following two lemmas, the first of which is an obvious restatement of the results proved in Theorems 2.7.9 and 2.7.17.

2.8.1 *LEMMA*

Let X *be a locally compact metric space.*

Let $M \subset X$ *be a compact asymptotically stable set.*

Let $v = \Phi(x)$ *be any continuous function defined on some neighborhood* N *of* M *and having the properties*

i) $\Phi(x) = 0$ *for* $x \in M, \Phi(x) > 0$ *for* $x \notin M$;

ii) $\Phi(xt) < \Phi(x)$ *for* $x \notin M, t > 0$, *and* $x[0,t] \subset N$.

(such functions can always be defined on $A(M)$*). Let* $\varepsilon > 0$ *be such that* $S[M,\varepsilon]$ *is a compact subset of* N. *Let* α, $0 < \alpha < m(\varepsilon)$ *where*

2.8.2 $$m(\varepsilon) = \min\{\Phi(x) : x \in H(M,\varepsilon)\}.$$

Then the set

2.8.3 $$P_\alpha = K_\alpha \cap S[M,\varepsilon],$$

where

2.8.4 $$K_\alpha = \{x \in N : \Phi(x) \leq \alpha\},$$

is a compact positively invariant set, with $P_\alpha \subset A(M)$.

2.8.5 *LEMMA*

For each sufficiently small α, *the set* P_α *defined in* 2.8.3 *is a retract of* $A(M)$.

Proof. This is so because we can define a map $h: A(M) \to P_\alpha$ by $h(x) = x$ if $x \in P_\alpha$, and $h(x) = x\tau(x)$ if $x \notin P_\alpha$, where $\tau(x)$ is defined as in the proof of Theorem 2.7.14. Because of the continuity of $\tau(x)$ and of the phase map π, and the fact that $\tau(x) = 0$ for $x \in \partial P_\alpha$, it follows that h is a continuous map of $A(M)$ into P_α which is an identity on P_α. Thus by definition P_α is a retract of $A(M)$, and hence also a retract of every subset of $A(M)$ which contains P_α.

We are now in the position to prove the following important result which is a generalization of Theorem 1.9.6.

2.8.6 *THEOREM*

Let $M \subset E$ *be a compact set which is a weak attractor for a dynamical system* (E, R, π). *Let the region of attraction* $A(M)$ *of* M *be homeomorphic to* E. *Then* M *contains a rest point. In particular, when* $A(M) = E$ *(i.e.,* M *is a global weak attractor), then* M *contains a rest point.*

Proof. By Theorem 2.6.15 $D^+(M)$ is an asymptotically stable compact set with $A(D^+(M)) = A(M)$. Let $\Phi(x)$ be any function for the asymptotically stable set $D^+(M)$ as in Lemma 2.8.1, and consider a set P_α for $\Phi(x)$. Then P_α is compact, positively invariant, and is a retract of $A(M)$. As $A(M)$ is homeomorphic to E we can choose a compact set B, $P_\alpha \subset B \subset A(M)$, where B is homeomorphic to the unit ball in E. Then P_α is a retract of B. Thus P_α has the fixed point property, as B has, by the Brouwer Fixed-Point Theorem. Since P_α is positively invariant, the transition π^τ maps P_α into P_α for each $\tau \geq 0$. Thus for each fixed $\tau > 0$, π^τ has a fixed point in P_α, i.e., corresponding to any $\tau > 0$ there is an $x_\tau \in P_\alpha$ such that $\pi^\tau(x_\tau) = \pi(x_\tau,\tau) = x_\tau$. Thus the orbit $\gamma(x_\tau)$ is closed and has a period τ, moreover $\gamma(x_\tau) \subset P_\alpha$, because $\gamma(x_\tau) = \gamma^+(x_\tau) \subset P_\alpha$. We have thus shown that, corresponding to any sequence $\{\tau_n\}$, $\tau_n > 0$, $\tau_n \to 0$, there is a sequence of closed orbits $\{\gamma_n\}$, $\gamma_n = \gamma(x_{\tau_n})$, with γ_n having a period τ_n. This sequence being in P_α, P_α contains a rest point x^* (say) (lemma 1.9.5). However, $M \subset D^+(M) \subset P_\alpha \subset A(M)$. As M is a weak attractor we have $\Lambda^+(x) \cap M \neq \emptyset$ for each $x \in A(M)$. Thus $\Lambda^+(x^*) \cap M \neq \emptyset$. But $\Lambda^+(x^*) = \{x^*\}$, as x^* is a rest point. Hence $x^* \in M$. The theorem is proved.

For the following corollaries the dynamical system is assumed to be defined on E .

2.8.7 *COROLLARY*

If the dynamical system is ultimately bounded, then it contains a rest point.

This is so, because ultimate boundedness is equivalent to the existence of a compact globally asymptotically stable set (Remark 2.7.21) which by the above theorem contains a rest point.

2.8.8 *COROLLARY*

The region of attraction of a compact minimal weak attractor M *cannot be homeomorphic to* E, *unless* M *is a rest point.*

Note, however, that if a rest point $p \in E$ is weakly attracting, or attracting, then $A(p)$ need not be homeomorphic to E , as the analytic example 1.5.32 (ii) shows. However, if a rest point P is asymptotically stable then its region of attraction is homeomorphic to E . This we shall prove next; its proof depends on the following topological theorem

2.8.9 *THEOREM*

Let $\{U_n\}$ *be a monotone sequence of open* n-*cells in* E, *i.e.,* $U_n \subset U_{n+1}$, $n = 1,2,\ldots$. *Then* $\bigcup\{U_n : n = 1,2,\ldots\}$ *is an open* n-*cell.*

2.8.10 *THEOREM*

If a rest point $p \in E$ *is asymptotically stable, then* $A(p)$ *is homeomorphic to* E.

Proof. Since $A(p)$ is a neighborhood of p, there is an $\varepsilon > 0$ such that the closed ball $S[p,\varepsilon] \subset A(p)$. For each $t \in R$, the transition π^t being a homeomorphism of E onto E, the image $S(p,\varepsilon)t$ of the open ball $S(p,\varepsilon)$ by π^t is an open n-cell. We claim now that for any given t_1 there exists a t_2, $t_2 < t_1$ such that $S(p,\varepsilon)t_1 \subset S(p,\varepsilon)t_2$. This is so because $\overline{S(p,\varepsilon)t_1}$ being a subset of the compact set $S[P,\varepsilon]t_1$ is itself compact. Further, $S[p,\varepsilon]t_1 \subset A(p)$, and $A(p)$ is open. Since p is uniformly attracting (Theorem 1.5.27) there exists a $T > 0$ such that $S[p,\varepsilon]\,(t_1 + t) \subset S(p,\varepsilon)t_1$ for $t \geq T$. In particular, $S[p,\varepsilon](t_1 + T) \subset S(p,\varepsilon)t_1 \subset S[p,\varepsilon]t_1$. Hence $S[p,\varepsilon]t_1 \subset S(p,\varepsilon)(t_1 - T) \subset S[p,\varepsilon](t_1 - T$ Setting $t_2 = t_1 - T$, we get $S[p,\varepsilon]t_1 \subset S(p,\varepsilon)t_2$. The above analysis shows that we can choose a sequence $\{t_n\}$, $t_n \to -\infty$, such that $\{S(p,\varepsilon)t_n\}$ is a monotone sequence of open n-cells. By Theorem 2.8.9 $\bigcup\{S(p,\varepsilon)t_n;\ n = 1,2,\ldots\}$ is an open n-cell. But this last union is $A(p)$, so that $A(p)$ is an open n-cell and hence homeomorphic to E.

2.8.11 *COROLLARY*

If p *is an asymptotically stable rest point, then* $A(p) \setminus \{p\}$ *is homeomorphic to* $E \setminus \{0\}$, *where* 0 *is the origin in* E.

We can now prove the following result.

2.8.12 *THEOREM*

Let $M \subset E$ *be a compact globally asymptotically stable set. Then*

$E \setminus M = C(M)$ *is homeomorphic to* $E \setminus \{0\}$.

Proof. By Theorem 2.8.6, M contains a rest point. We may assume without loss of generality that M contains the origin 0 and 0 is a rest point. Consider now the homeomorphism $h: E \setminus \{0\} \to E \setminus \{0\}$ defined by $h(x) = \frac{x}{||x||^2}$, where $||x||$ is the euclidean norm of x. h maps the given dynamical system into a dynamical system on E, with 0 becoming a negatively asymptotically stable rest point, and $E \setminus M$ is mapped onto $A(0) \setminus \{0\}$, where $A(0)$ is now the region of negative attraction of 0. By the Corollary 2.8.11, $A(0) \setminus \{0\}$ is homeomorphic to $E \setminus \{0\}$. Hence the result follows.

We shall now present one example of application of Theorem 2.8.12.

2.8.13 *Example*

Consider a flow $\mathcal{F}$ in E with only two rest points x and y, $x \neq y$. Theorem 2.8.12 shows that y cannot be asymptotically stable with $A(\{y\}) = E \setminus \{x\}$, since $C(\{x\} \cup \{y\})$ is not homeomorphic to $E \setminus \{0\}$.

2.8.14 *Notes and references*

Most of the results presented in this section are contained in the paper by N. P. Bhatia and G. P. Szegö [1].

An analytic example showing that if $p \in E$ is attracting, then $A(p)$ need not be homeomorphic to E is 1.5.32 (ii) (J. Auslander, N. P. Bhatia and P. Seibert at pg. 58).

Theorem 2.8.9 is due to M. Brown. The results contained in this section and in particular Theorem 2.8.10 and the natural conjecture which generalizes this theorem to sets $M \subset E$ such that $E \setminus M$ is homeomorphic to $E \setminus \{0\}$ are rather useful. In particular they may have a strong influence on the solution of one of the most important still open problems in the stability theory of dynamical systems, viz. the problem of local properties and the related theory of separatrices.

A separatrix, according to S. Lefschetz (1, pg. 223) is, in E^2, "a trajectory (not a critical point) behaving topologically abnormally in comparison with neighboring paths".

A theory of separatrices in E^2 was formally suggested by Markus [5] who gives a definition of separatrix and concludes that the union σ (separating set) of all separatrices of a differential system in E^2 is closed.

Each component of the set $C(\sigma)$ is called by Markus a canonical region. Markus proves that in each canonical region the flow is "parallel" i.e., either parallelizable or homeomorphic to a family of concentric cycles.

Clearly since the flow is parallel in each canonical region it admits there a transversal section. The results presented in this section are helpful in generalizing some of these results to flows in E. For instance, one can show (after a suitable generalization of the concept of separatrix) that the number of canonical regions homeomorphic to balls cannot exceed the number of equilibrium points of the flow. If, in addition, one defines the separating set in such a way that in the corresponding canonical regions the flow has only strong stability properties then the characterization of the separating set above (which may have a very complicated structure) would be enough for the complete global description of the stability properties of the flow.

2.9 *Minimal Sets and Recurrent Motions.*

A rest point and a periodic trajectory are examples of compact minimal sets (for definition see 2.2.4). A rest point and a periodic motion are also Poisson stable. Example 2.5.10 indicates that the closure of a Poisson stable trajectory need not be a minimal set (in the example the closure of every Poisson stable trajectory except the rest point is the whole torus, which is not minimal as it contains a rest point). G.D. Birkhoff discovered an intrinsic property of motions in a compact minimal set, which is usually called the property of recurrence. The aim of this section is to study this concept of recurrence.

We start with some characteristic properties of minimal sets.

2.9.1 *THEOREM*

Every compact invariant set $K \subset X$ *contains a minimal set.*

Proof. Consider the set G of all closed invariant subsets of K. This set G is partially ordered by the inclusion relation $\subset$. Since K is compact it has the finite intersection property [Dugunji I, pg. 223]. Thus every chain has an upper bound. Hence by Zorn's lemma there is a maximal element $M \subset G$. Then M is maximal and the theorem is proved.

2.9.2 *COROLLARY*

For any $x \in X$, *if the motion* π_x *is positively (negatively) Lagrange stable, then* $\Lambda^+(x)$ $(\Lambda^-(x))$ *contains a minimal set.*

An elementary characterization of a minimal set is given by

2.9.3 *THEOREM*

A set $M \subset X$ *is minimal if and only if for each* $x \in M$ *one has* $\overline{\gamma(x)} = M$.

Proof. Let M be minimal, and suppose if possible that there is an $x \in M$ such that $\overline{\gamma(x)} \neq M$. As M is closed and invariant we have indeed $\overline{\gamma(x)} \subset M$. Thus $\overline{\gamma(x)}$ is closed and invariant and a proper subset of M, a contradiction. Hence for each $x \in M$ we have $\overline{\gamma(x)} = M$. Conversely, assume that for each $x \in M$, $\overline{\gamma(x)} = M$. Let if possible M be not minimal. Then there is a non-empty closed and invariant subset N of M, $N \neq M$. Then for any $x \in N$, $\overline{\gamma(x)} \subset N \neq M$, a contradiction. The theorem is proved.

We now introduce the notion of recurrence à la Birkhoff.

2.9.4 *DEFINITION* (recurrence)

For any $x \in X$, *the notion* π_x *is said to be* recurrent *if for each* $\varepsilon > 0$ *there exists a* $T = T(\varepsilon) > 0$, *such that*

$$\gamma(x) \subset S(x[t-T, t+T], \varepsilon)$$

for all $t \in R$.

2.9.5 *Remark.*

It is clear that if a motion π_x is recurrent then every motion π_y with $y \in \gamma(x)$ is also recurrent. Thus we shall also speak of the trajectory $\gamma(x)$ being recurrent.

2.9.6 *Exercise.*

Show that every recurrent motion is Poisson stable.

That the concept of recurrence is basic in the theory of compact minimal sets is seen from the following theorem of Birkhoff [2].

2.9.7 *THEOREM*

Every trajectory in a compact minimal set is recurrent:

Proof. Let M be a compact minimal set. Suppose that there is an $x \in M$ such that the motion π_x is not recurrent. Then there is an $\varepsilon > 0$ and sequences $\{T_n\}$, $\{t_n\}$, $\{\tau_n\}$, with $T_n > 0$, $T_n \to +\infty$, and

$$x\tau_n \notin S(x[t_n - T_n,\ t_n + T_n],\ \varepsilon)\ ,\ n=1,2,\ldots\ .$$

This shows that

$\rho(x\tau_n,\ x(t_n+t)) \geq \varepsilon$ whenever $|t| \leq T_n$, $n=1,2,\ldots$. The sequences $\{xt_n\}$, $\{x\tau_n\}$ are contained in the compact set M and may without loss of generality be assumed to be convergent. So let $xt_n \to y$ and $x\tau_n \to z$. Then $y, z \in \overline{\gamma(x)} = M$. Consider now the motion π_y. Let $T > 0$ be arbitrary but fixed. Then there is a $\delta > 0$ such that

$$\rho(yt,\ \omega t) < \frac{\varepsilon}{3} \text{ whenever } |t| \leq T \text{ and } \rho(y,\omega) < \delta.$$

We can now find an integer n such that $T_n > T$, $\rho(y,\ xt_n) < \delta$ and $\rho(z,\ x\tau_n) < \frac{\varepsilon}{3}$. Then we have

$$\rho(yt,\ x(t_n+t)) < \frac{\varepsilon}{3} \text{ for } |t| \leq T.$$

Moreover, from the choice of $x\tau_n$ we have

$$\rho(x(t_n+t), x\tau_n) \geqq \varepsilon \quad \text{for} \quad |t| \leqq T < T_n.$$

The above inequalities show that

$$\rho(yt, z) \geqq \rho(x(t_n+t), x\tau_n) - \rho(yt, x(t_n+t))$$

$$- \rho(x\tau_n, z) \geqq \varepsilon - \frac{\varepsilon}{3} - \frac{\varepsilon}{3} = \frac{\varepsilon}{3},$$

whenever $|t| \leqq T$. As T was arbitrary, we conclude that $\rho(yt, z) \geqq \frac{\varepsilon}{3}$ for all $t \in R$. Thus $z \notin \overline{\gamma(y)}$, i.e. $z \notin M$ as $M = \overline{\gamma(y)}$. This contradiction proves the theorem.

2.9.8 *THEOREM*

If a trajectory $\gamma(x)$ *is recurrent and* $\overline{\gamma(x)}$ *is compact then* $\overline{\gamma(x)}$ *is also minimal.*

Proof. Set $\overline{\gamma(x)} = M$. Let if possible M be not minimal. Then there exists a non-empty compact invariant subset N of M, $N \neq M$. Clearly $x \notin N$ (otherwise $\overline{\gamma(x)} \subset N$ which is impossible). Now let $\rho(x,N) = \varepsilon$ (> 0). As π_x is recurrent, there is a $T > 0$ such that

2.9.9 $$\gamma(x) \subset S(x[t-T, t+T], \tfrac{\varepsilon}{3}) \quad \text{for all} \quad t \in R.$$

Now choose any $y \in N$. Since $y \in M = \overline{\gamma(x)}$, and $y \notin \gamma(x)$, we have $y \in \Lambda^+(x)$ or $y \in \Lambda^-(x)$. Let $y \in \Lambda^+(x)$. Then there is a sequence $\{t_n\}$, $t_n \to +\infty$, such that $xt_n \to y$. By the axioms of the dynamical system, there is a $\delta > 0$ such that $\rho(yt, zt) < \frac{\varepsilon}{3}$ whenever $\rho(y,z) < \delta$ and $|t| \leq T$ This shows that there is an n with

$$\rho(yt, x(t_n+t)) < \frac{\varepsilon}{3} \quad \text{for} \quad |t| \leq T.$$

From this it follows that

$$\rho(x, x(t_n+t)) \geq \rho(x,yt) - \rho(yt, x(t_n+t))$$

$$\geq \varepsilon - \frac{\varepsilon}{3} = \frac{2\varepsilon}{3} \quad \text{for} \quad |t| \leq T.$$

This however contradicts 2.9.9 . The theorem is proved.

2.9.10 *COROLLARY.*

If the space X *is complete, then the closure* $\overline{\gamma(x)}$ *of any recurrent trajectory* $\gamma(x)$ *is a compact minimal set.*

The proof follows from the observation that the conditions imply compactness of $\overline{\gamma(x)}$, so that the result follows from theorem 2.9.8. The details are left to the reader as an exercise.

Another way of defining a Lagrange stable recurrent motion is provided via the concept of a <u>relatively dense set</u> of numbers(0.2.14).

2.9.11 *THEOREM*

A Lagrange stable motion π_x *is recurrent if and only if for each* $\varepsilon > 0$ *the set*

$$K_\varepsilon = \{t: \rho(x, xt) < \varepsilon\}$$

is relatively dense.

Proof. Let for each $\varepsilon > 0$ the set K_ε be relatively dense. For any $\varepsilon > 0$ there is by definition a $T_\varepsilon = T > 0$ such that

$$K_\varepsilon \cap (t-T, t+T) \neq \emptyset \text{ for all } t \in R.$$

As $\overline{\gamma(x)}$ is compact, to show that the motion π_x is recurrent we need show only that $\overline{\gamma(x)}$ is minimal. Let $\overline{\gamma(x)}$ be not minimal. Then there is a minimal subset M of $\overline{\gamma(x)}$, $M \neq \overline{\gamma(x)}$. Clearly $x \notin M$ (otherwise $\overline{\gamma(x)} \subset M$ which will imply $\overline{\gamma(x)} = M$). Set $\rho(x, M) = 3\varepsilon (> 0)$. Choose any $y \in M$. Then there is a $\delta > 0$ such that $\rho(yt, zt) < \varepsilon$ whenever $\rho(y,z) < \delta$ and $|t| \leq T = T_\varepsilon$. As $y \in M \subset \overline{\gamma(x)}$, and $y \notin \gamma(x)$, we conclude that $y \in \Lambda^+(x)$ or $y \in \Lambda^-(x)$. Let $y \in \Lambda^+(x)$. Then there is a sequence $\{t_n\}$, $t_n \to +\infty$ and $xt_n \to y$. Thus for all sufficiently large n we have $\rho(yt, x(t_n+t)) < \varepsilon$ for $|t| \leq T = T_\varepsilon$. But then for $t \in [t_n-T, t_n+T]$ we have

$$\rho(x, xt) \geq \rho(x, M) - \rho(xt, M)$$

$$\geq 3\varepsilon - \varepsilon = 2\varepsilon.$$

This shows that

$$K_\varepsilon \cap [t_n - T_\varepsilon, t_n + T_\varepsilon] = \emptyset,$$

which is a contradiction. This shows that $\overline{\gamma(x)}$ is minimal and hence the motion π_x is recurrent. The converse holds trivially. The theorem is proved.

2.9.12 *THEOREM*

There exist non-compact minimal sets which contain more than one trajectory.

Proof. Consider a dynamical system defined in a euclidean 3-space, in which the torus T of example 2.5.10 is embedded with the rest point on the torus coinciding with the origin of the euclidean space. We now consider the transformation $y = \frac{x}{||x||}$, $x \neq 0$, which transforms the given euclidean space into a euclidean space. The set $T \setminus \{0\}$ is now transformed into a closed minimal set which is not compact, since it is not bounded, as is evident from the considerations in Example 2.5.10.

Notice that in the example of the unbounded minimal set given above the motions are not recurrent, showing that Theorem 2.9.7 is not true if the minimal set is not compact.

2.9.13 *Notes and References*

G. D. Birkhoff defined the notions of a compact minimal set and of recurrent motions and showed the deep connection between them. The presentation here is adapted from Nemytskii and Stepanov's book. The literature on non-compact minimal sets is very scanty. The example in Theorem 2.9.12 is included to give an idea that these sets do not have many known interesting properties.

2.10 *Stability of a Motion and Almost Periodic Motions.*

In this section we shall assume throughout that the metric space X is complete.

The concept of almost periodicity is intermediate between that of periodicity and recurrence, and the concept of stability of motion plays a central role in its study. We therefore first introduce the concept of stability of a motion.

2.10.1 *DEFINITION*

A motion π_x *is said to be* <u>*positively (Liapunov) stable in a subset*</u> N *of* X *, if for any* $\varepsilon > 0$ *, there is a* $\delta > 0$ *such that* $y \in N \cap S(x,\delta)$ *implies* $\rho(xt, yt) < \varepsilon$ *for* $t \in R^+$.

Any motion π_x *is called* <u>*negatively stable*</u>*, or* <u>*stable in both directions*</u> *in a subset* N *of* X *, if the above condition is satisfied with* $t \in R^+$ *replaced by* $t \in R^-$ *or* $t \in R$ *respectively.*

If in the above definition N *is a neighborhood of* x *, then the qualifier "in the subset* N *of* X *" will be deleted. Thus a motion* π_x *is positively stable if given* $\varepsilon > 0$ *, there is a* $\delta > 0$ *such that* $y \in S(x, \delta)$ *implies* $\rho(xt, yt) < \varepsilon$ *for* $t \in R^+$.

2.10.2 *Exercise*

Show that a motion π_x is positively stable if and only if every motion π_{xt} , $t \in R$, is positively stable.

2.10.3 *DEFINITION*

If $A \subset B \subset X$ *, then the motions through* A *(i.e. motions* π_x

with $x \in A$) *will be called* ***uniformly positively stable, uniformly negatively stable,*** *or* ***uniformly stable in both directions*** *in* B , *if given any* $\varepsilon > 0$, *there is a* $\delta > 0$ *such that* $\rho(xt, yt) < \varepsilon$ *for* $t \in R^+$, $t \in R^-$, *or* $t \in R$ *respectively whenever* $x \in A$, $y \in B$, *and* $\rho(x, y) < \delta$.

2.10.4 *Exercise.*

Show that if A is a compact subset of B , then the motions through A are uniformly positively stable in B , whenever each motion through A is positively stable in B .

We now introduce the concept of almost periodicity.

2.10.5 *DEFINITION*

A motion π_x *is said to be* ***almost periodic*** *if for every* $\varepsilon > 0$ *there exists a relatively dense subset of numbers* $\{\tau_n\}$ *called displacements such that*

$$\rho(xt, x(t + \tau_n)) < \varepsilon$$

for all $t \in R$ *and each* τ_n .

It is obvious that periodic motions and rest points are special cases of almost periodic motions. That every almost periodic motion is recurrent follows from Theorems 2.9.10-12 and we leave this as an exercise. Later in this section we shall consider examples to show that not every recurrent motion is almost periodic, and that an almost periodic motion need not be periodic.

The following theorems show how stability is deeply connected with almost periodic motions. First observe the following lemma.

2.10.6 *LEMMA.*

If a motion π_x *is almost periodic, then every motion* π_y *with*

$y \in \gamma(x)$ *is almost periodic with the same set of displacements* $\{\tau_n\}$ *corresponding to a given* $\varepsilon > 0$.

Proof. Indeed for any $\varepsilon > 0$ there is a set of displacements $\{\tau_n\}$ such that

$\rho(xt, x(t + \tau_n)) < \varepsilon$ for $t \in R$, and each τ_n. If $y \in \gamma(x)$, then there is a $\tau \in R$ such that $y = x\tau$, or $x = y(-\tau)$. The above inequality together with the homomorphism axiom then gives

$\rho(y(t - \tau), y(t - \tau + \tau_n)) < \varepsilon$ for $t \in R$. Setting $t - \tau = s$, we see that

$\rho(ys, y(s + \tau_n)) < \varepsilon$ for $s \in R$ and each τ_n, as τ is fixed. This proves the lemma.

2.10.7 *THEOREM*

Let the motion π_x *be almost periodic and let* $\overline{\gamma(x)}$ *be compact. Then* (i) *every motion* π_y *with* $y \in \overline{\gamma(x)}$ *is almost periodic with the same set of displacements* $\{\tau_n\}$ *for any given* $\varepsilon > 0$, *but with the strict inequality* $<$ *replaced by* $\leq$; (ii) *the motion* π_x *is stable in both directions in* $\overline{\gamma(x)}$.

Proof.

i) For any $y \in \overline{\gamma(x)}$, there is a sequence $\{x_n\} \subset \gamma(x)$ such that $x_n \to y$. By Lemma 2.10.6 , for any $\varepsilon > 0$ there is a set of displacements $\{\tau_n\}$ such that: $\rho(x_n t, x_n(t + \tau_m)) < \varepsilon$ for all $t \in R$, $x_n \in \{x_n\}$, and $\tau_m \in \{\tau_n\}$. Now keeping $t \in R$, and $\tau_m \in \{\tau_n\}$ fixed but arbitrary and proceeding to the limit we get $\rho(yt, y(t + \tau_m)) \leq \varepsilon$ for all $t \in R$ and $\tau_m \in \{\tau_n\}$. This completes the proof of (i) .

ii) Given $\varepsilon > 0$, let $\{\tau_n\}$ be a set of displacements corresponding to $\frac{\varepsilon}{3}$ for the almost periodic motion π_x, and let $T > 0$ be such that $\{\tau_n\} \cap [t - T, t + T] \neq \emptyset$ for $t \in R$. Then by part (i) of the theorem $\rho(yt, y(t + \tau_n)) \leq \frac{\varepsilon}{3}$ for all $y \in \overline{\gamma(x)}$, $t \in R$, and each τ_n. By the continuity axiom, for $\frac{\varepsilon}{3} > 0$ and $T > 0$ as above there is a $\delta > 0$ such that $\rho(y, z) < \delta$ implies $\rho(yt, zt) < \frac{\varepsilon}{3}$ for all $|t| \leq T$, whenever $\{y,z\} \subset \overline{\gamma(x)}$ as this last set is compact. Now for any $y \in \overline{\gamma(x)}$ and $\rho(x, y) < \delta$, we have for any $t \in R$

$$\rho(xt,yt) \leq \rho(xt, x(t + \tau_n)) + \rho(x(t + \tau_n), y(t + \tau_n))$$

$$+ \rho(y(t + \tau_n), y\,t) < \frac{\varepsilon}{3} + \frac{\varepsilon}{3} + \frac{\varepsilon}{3} = \varepsilon ,$$

because for any $t \in R$ we can choose τ_n such that $|t + \tau_n| \leq T$. This proves the theorem completely.

2.10.8 *COROLLARY*

If M *is a compact minimal set, and if one motion in* M *is almost periodic, then every motion in* M *is almost periodic.*

2.10.9 *COROLLARY*

If M *is a compact minimal set of almost periodic motions, then the motions through* M *are uniformly stable in both directions in* M .

The above corollary follows from Theorem 2.10.7 Part (ii) and Exercise 2.10.4 .

We now investigate when a recurrent motion is almost periodic.

2.10.10 *THEOREM*

If a motion π_x *is recurrent and stable in both directions in* $\gamma(x)$, *then it is almost periodic.*

Proof. We have indeed that given $\varepsilon > 0$ there is a $\delta > 0$ such that $\rho(xt, yt) < \varepsilon$ for all $t \in R$, whenever $\{x,y\} \subset \gamma(x)$ and $\rho(x,y) < \delta$. Further, by recurrence of π_x (Theorem 2.9.12), there is a relatively dense set $\{\tau_n\}$ of displacements such that

$$\rho(x, x\tau_n) < \delta \quad \text{for each} \quad \tau_n .$$

From the above two results we conclude

$$\rho(xt, x(t + \tau_n)) < \varepsilon \quad \text{for} \quad t \in R$$

and each τ_n. The theorem is proved.

A stronger result is the following:

2.10.11 *THEOREM.*

If a motion π_x *is recurrent and positively stable in* $\gamma(x)$, *then it is almost periodic.*

Proof. (a) By positive stability of π_x in $\gamma(x)$, we have given $\varepsilon > 0$ there is a $\delta > 0$ such that $\rho(x, x\tau) < \delta$ implies $\rho(xt, x(t + \tau)) < \frac{\varepsilon}{2}$ for all $t \in R^+$. (b) By recurrence of π_x, there is a relatively dense set $\{\tau_n\}$ such that $\rho(x, x\tau_n) < \frac{\delta}{2}$ for each τ_n.

(c) By the continuity axiom, for any τ_n there is a $\sigma > 0$ such that $\rho(x, y) < \sigma$ implies $\rho(x\tau_n, y\tau_n) < \frac{\delta}{2}$. Now let $t \in R$ and τ_n be arbitrary but fixed. Then by recurrence of π_x, there is a $\tau < t$ such that $\rho(x, x\tau) < \min\ (\sigma, \delta)$. Then by (b) $\rho(x\,\tau_n, x(\tau + \tau_n)) < \frac{\delta}{2}$, so that

$$\rho(x, x(\tau + \tau_n)) \leq \rho(x, x\tau_n) + \rho(x\tau_n, x(\tau + \tau_n)) < \frac{\delta}{2} + \frac{\delta}{2} = \delta .$$

Hence by (a), since $t - \tau > 0$, we get $\rho(x(t - \tau), x(t + \tau_n)) < \frac{\varepsilon}{2}$.

Further, $\rho(xt, x(t - \tau)) = \rho(x\tau(t - \tau), x(t - \tau)) < \frac{\varepsilon}{2}$ by (a), as $\rho(x, x\tau) < \min(\sigma, \delta) \leq \delta$, and $t - \tau > 0$. Thus we get

$$\rho(xt, x(t + \tau_n)) \leq \rho(xt, x(t - \tau)) + \rho(x(t - \tau), x(t + \tau_n)) < \frac{\varepsilon}{2} + \frac{\varepsilon}{2} = \varepsilon .$$

This shows that π_x is almost periodic and the theorem is proved.

2.10.12 *THEOREM*

If the motions in $\gamma(x)$ *are uniformly positively stable in* $\gamma(x)$ *and are negatively Lagrange stable, then they are almost periodic.*

Proof. It is sufficient to prove that the motion π_x is recurrent, as the rest follows from the last theorem.

By negative Lagrange stability of π_x, we conclude that $\Lambda^-(x)$ is compact, and indeed $\Lambda^-(x) \subset \overline{\gamma(x)}$. Since $\Lambda^-(x)$ is also invariant, there is a compact minimal set M, $M \subset \Lambda^-(x)$. If π_x is not recurrent, then $M \neq \overline{\gamma(x)}$, and in particular $x \notin M$. Let $\rho(x,M) = \alpha > 0$. We will show, that every motion π_y, $y \in \Lambda^-(x)$, is positively stable in $\gamma(x)$. To this end, given $\varepsilon > 0$, there is a $\delta = \delta(\varepsilon) > 0$ (by uniform positive stability in $\gamma(x)$ of motions in $\gamma(x)$) such that $\{x_n, x_m\} \subset \gamma(x)$ and $\rho(x_n, x_m) < \delta$ imply $\rho(x_n t, x_m t) < \varepsilon$ for $t \geq 0$. Now for $y \in \Lambda^-(x)$, there is a sequence $\{t_n\}$ in R^-, $t_n \to -\infty$, such that $xt_n \to y$. There is then an integer N such that $n \geq N$, $m \geq N$ imply $\rho(xt_n, xt_m) < \delta$ and $\rho(xt_n, y) < \delta$, and consequently $\rho(xt_n(t), xt_m(t)) < \varepsilon$ for $t \geq 0$. Keeping, in this last inequality, $t \in R$ and $n \geq N$ arbitrary but fixed and letting $m \to \infty$, we get $\rho(yt, xt_n(t)) \leq \varepsilon$ for $t \geq 0$ whenever $\rho(y, xt_n) < \delta$. Choosing now $\varepsilon = \frac{\alpha}{2}$ and $t = -t_n$, we see that $\rho(y(-t_n), x) \leq \frac{\alpha}{2}$. Since $y(-t_n) \in M$, this contradicts the assumption that $\rho(x, M) = \alpha$. The theorem is proved.

The remaining portion of this section will be devoted to finding conditions under which a limit set $\Lambda^+(x)$ is compact and minimal and, further, when such a set consists of almost periodic notions only. For this the following definition is useful.

2.10.13 *DEFINITION*

A semi-trajectory $\gamma^+(x)$ *is said to* *uniformly approximate its limit set* $\Lambda^+(x)$, *if given any* $\varepsilon > 0$, *there is a* $T = T(\varepsilon) > 0$ *such that* $\Lambda^+(x) \subset S(x[t, t + T], \varepsilon)$ *for each* $t \in R^+$.

2.10.14 *THEOREM*

Let the motion π_x *be positively Lagrange stable. Then the limit set* $\Lambda^+(x)$ *is minimal if and only if the semi-trajectory* $\gamma^+(x)$ *uniformly approximates* $\Lambda^+(x)$.

Proof. The set $\Lambda^+(x)$ is non-empty, compact, and invariant. Now let $\gamma^+(x)$ uniformly approximate $\Lambda^+(x)$. If $\Lambda^+(x)$ is not minimal, then there are points $y, z \in \Lambda^+(x)$ such that $z \notin \overline{\gamma(y)}$ (otherwise $\overline{\gamma(y)} = \Lambda^+(x)$ for each $y \in \Lambda^+(x)$, and $\Lambda^+(x)$ is minimal; Theorem 2.9.3). Let $\rho(z, \overline{\gamma(y)}) = \varepsilon > 0$. By uniform approximation there is a $T > 0$ such that $S(x[t, t + T], \frac{\varepsilon}{2}) \supset \Lambda^+(x)$ for $t \geqq 0$. Further, there is a $\delta > 0$ such that $\rho(y,w) < \delta$ implies $\rho(yt, wt) < \frac{\varepsilon}{2}$ for $|t| \leqq T$. Since $y \in \Lambda^+(x)$, there is a point $x_1 \in \gamma^+(x)$ such that $\rho(x_1, y) < \delta$. And because $S(x[t, t + T], \frac{\varepsilon}{2}) \supset \Lambda^+(x)$, we have in particular $S(x_1 [0, T], \frac{\varepsilon}{2}) \supset \Lambda^+(x)$. Thus there is a point $x_2 \in x_1[0, T]$ such that $\rho(z, x_2) < \frac{\varepsilon}{2}$. If then $x_2 = x_1 \tau$, where $0 \leqq \tau \leqq T$, then $\rho(x_1 \tau, y\tau) < \frac{\varepsilon}{2}$, so that $\rho(z, y\tau) \leqq \rho(z, x_2) + \rho(x_2, y\tau) < \frac{\varepsilon}{2} + \frac{\varepsilon}{2} = \varepsilon$, as $x_2 = x_1\tau$. This is a contradiction, as $\rho(z, \overline{\gamma(y)}) = \varepsilon$. Thus $\Lambda^+(x)$ is minimal. Now let $\Lambda^+(x)$ be minimal, so that for any $y \in \Lambda^+(x)$, $\overline{\gamma(y)} = \Lambda^+(x)$. Now assume, if possible, that $\gamma^+(x)$ does not uniformly approximate $\Lambda^+(x)$. Then there is an $\varepsilon > 0$, a sequence of intervals $\{(t_n, \tau_n)\}$, and a sequence $\{y_n\} \subset \Lambda^+(x)$ such that $t_n \to +\infty$, $(\tau_n - t_n) \to +\infty$, $y_n \to y$ $(\in \Lambda^+(x))$, and $y_n \notin S(x[t_n, \tau_n], \varepsilon)$. We may also assume that $\rho(y_n, y) < \frac{\varepsilon}{3}$ for all n. Then for arbitrary n

$$\rho(y, x[t_n, \tau_n] \geqq \rho(y_n, x[t_n, \tau_n]) - \rho(y_n, y) > \varepsilon - \frac{\varepsilon}{3} = \frac{2\varepsilon}{3}.$$

Consider now the sequence of points $\{x_n\}$, where

$$x_n = x(\frac{t_n + \tau_n}{2}) = x\, t'_n.$$

Clearly $t'_n \to +\infty$. Since $\overline{\gamma^+(x)}$ is compact, we may assume that $x\, t'_n \to z$ $(\in \Lambda^+(x))$. Since $\Lambda^+(x)$ is minimal, $\overline{\gamma(z)} = \Lambda^+(x)$, so that there is a $T \in R$ such that $\rho(zT, y) < \frac{\varepsilon}{3}$. By the continuity axiom we can choose a $\sigma > 0$ such that $\rho(zT, wT) < \frac{\varepsilon}{3}$ whenever $\rho(z, w) < \sigma$. Now choose N large enough such that $\rho(z, x_N) < \sigma$, and $\frac{\tau_N - t_N}{2} > |T|$. Then

$$x_N T = x(t'_N + T) \in x[t_n, \tau_n],$$

and hence $\rho(y, x_N T) > \frac{2\varepsilon}{3}$. On the other hand $\rho(x_N T, zT) < \frac{\varepsilon}{3}$, so that

$$\rho(y, x_N T) \leqq \rho(x_N T, zT) + \rho(zT, y) < \frac{\varepsilon}{3} + \frac{\varepsilon}{3} = \frac{2\varepsilon}{3}.$$

This contradiction proves the result.

The following theorem gives a sufficient condition for a positive limit set $\Lambda^+(x)$ to be a minimal set of almost periodic motions. No necessary and sufficient condition is known as yet.

2.10.15 *THEOREM.*

Let the motion π_x *be positively Lagrange stable, and let the motions in* $\gamma^+(x)$ *be uniformly positively stable in* $\gamma^+(x)$. *If moreover* $\gamma^+(x)$ *uniformly approximates* $\Lambda^+(x)$, *then* $\Lambda^+(x)$ *is a minimal set of almost periodic motions.*

Proof. By Theorem 2.10.14 , $\Lambda^+(x)$ is a compact minimal set. In view of Theorem 2.10.11 we need only prove that every motion through $\Lambda^+(x)$ is positively stable in $\gamma^+(x)$.

By uniform positive stability of motions in $\gamma^+(x)$, we have given $\varepsilon > 0$, there is a $\delta > 0$ such that $\rho(xt_1, xt_2) < \delta$ implies $\rho(x(t_1 + t), x(t_2 + t)) < \frac{\varepsilon}{3}$ for $t \geqq 0$. Let $\{y,z\} \subset \Lambda^+(x)$ and $\rho(y, z) < \frac{\delta}{3}$. Let $\tau > 0$ be arbitrary. We wish to estimate $\rho(y\tau, z\tau)$. By the continuity axiom there is a $\sigma > 0$ such that $\rho(y, w) < \sigma$, $\rho(z, u) < \sigma$ imply $\rho(y\tau,w\tau) < \frac{\varepsilon}{3}$ and $\rho(z\tau,u\tau) < \frac{\varepsilon}{3}$. If $\zeta = \min[\sigma, \frac{\delta}{3}]$. There are $t_1 > 0$ and $t_2 > 0$ such that $\rho(xt_1, y) < \zeta$ and $\rho(xt_2,y) < \zeta$. Thus

$$\rho(xt_1, xt_2) \leqq \rho(xt_1, y) + \rho(y, z) + \rho(z, xt_2) < \zeta + \frac{\delta}{3} + \zeta \leqq \delta.$$

Consequently $\rho(x(t_1 + \tau), x(t_2 + \tau)) < \frac{\varepsilon}{3}$, and $\rho(y\tau, x(t_1 + \tau)) < \frac{\varepsilon}{3}$ and also $\rho(z\tau, x(t_2 + \tau)) < \frac{\varepsilon}{3}$. The last three inequalities yield

$$\rho(y\tau, z\tau) \leqq \rho(y\tau, x(t_1 + \tau)) + \rho(x(t_1 + \tau), x(t_2 + \tau)) + \rho(x(t_2\ \tau), z\tau)$$

$$< \frac{\varepsilon}{3} + \frac{\varepsilon}{3} + \frac{\varepsilon}{3} = \varepsilon.$$

This shows in fact that the motions through $\Lambda^+(x)$ are uniformly positively stable in $\Lambda^+(x)$. The theorem is proved.

We now give a simple example of an almost periodic motion which is neither a rest point nor a periodic motion.

2.10.16. Example.

Consider a dynamical system defined on a torus by differential equations of the type 2.5.10 , specifically

$$\frac{d\,\phi}{dt} = 1, \frac{d\,\theta}{dt} = \alpha,$$

where α is irrational. For any point P on the torus $\overline{\gamma(P)}$ = the torus, and since α is irrational, no trajectory is periodic. The torus thus is a minimal set of recurrent motions. To see that the motions are almost periodic, we note first that if $P_1 = (\phi_1, \theta_1)$, $P_2 = (\phi_2, \theta_2)$, then $\rho(P_1, P_2) = \sqrt{(\phi_1 - \phi_2)^2 + (\theta_1 - \theta_2)^2}$, where the values of $\phi_1 - \phi_2$ and $\theta_1 - \theta_2$ are taken as the smallest in absolute value of the differences (mod 1). Now any motion on the torus is given by $\phi = \phi_0 + t$, $\theta = \theta_0 + \alpha t$. Then for the motions through P_1 and P_2 we have $\rho(P_1t, P_2t) = \sqrt{(\phi_1 - \phi_2)^2 + (\theta_1 - \theta_2)^2} = \rho(P_1, P_2)$. Thus the motions are uniformly stable in both directions in the torus. Thus by Theorem 2.10.10 the torus is a minimal set of almost periodic motions.

Examples of motions which are recurrent but not almost periodic are more difficult to construct. The first example was given by Poincaré in which he defined a dynamical system on a torus with a minimal subset which is not locally connected. For the details we refer the reader to the book of Nemytskii and Stepanov.

2.10.17 *Notes and References*

This section brings to a completion the classification of compact minimal sets, viz, a rest point, a periodic trajectory, the closure of an almost periodic trajectory, and the closure of a recurrent trajectory. The relationship between almost periodicity and stability is clarified.

The notion of an almost periodic function is due to H. Bohr and Theorem 2.10.7 is due to S. Bochner.

A. A. Markov [3] showed the relationship between stability of motion and almost periodicity (Theorem 2.10.11, 2.10.12).

Definition 2.10.13 and the following material is due to V. Nemytskii. In this connection one may also see the paper of Deysach and Sell on the existence of almost periodic motions.

2.11 *Parallelizable Dynamical Systems.*

So far we have been considering properties involving positively (or negatively) Lagrange stable trajectories. In this section we shall be concerned with dynamical systems none of whose trajectories are either positively or negatively Lagrange stable.

2.11.1 DEFINITION

For any $x \in X$, *the motion* π_x *is called* *__Lagrange unstable__* *if it is neither positively, nor negatively Lagrange stable, i.e. if both* $\overline{\gamma^+(x)}$ *and* $\overline{\gamma^-(x)}$ *are non-compact.*

2.11.2 DEFINITION

A dynamical system (X,R,π) *will be called Lagrange unstable if every motion* π_x *is Lagrange unstable.*

2.11.3 DEFINITION

A point $x \in X$ *will be called a* *__wandering__* *point if* $x \notin J^+(x)$ *(see Section 2.3). It is called* *__non-wandering__* *if* $x \in J^+(x)$.

2.11.4 DEFINITION

The dynamical system (X,R,π) *will be called* *__wandering__* *if every point* $x \in X$ *is wandering. (Such a system is usually called* *__completely unstable__*.*).*

2.11.5 DEFINITION

The dynamical system (X,R,π) *will be called* *__dispersive__* *if for any* $\{x,y\} \subset X$, *there exist neighborhoods* U_x *and* U_y *and a constant* $T > 0$ *such that* $U_x \cap U_y t = \emptyset$ *for all* t, $|t| \geq T$.

2.11.6 *DEFINITION*

A dynamical system (X,R,π) *is called* <u>*parallelizable*</u> *if there exist a set* $S \subset X$ *and a homeomorphism* $h:X \to S \times R$ *such that* $SR = X$ *and* $h(xt) = (x,t)$ *for every* $x \in S$ *and* $t \in R$.

2.11.7 *LEMMA*

For any $x \in X$ *we have* $x \in J^+(x)$ *if and only if* $x \in J^-(x)$.

Proof. If $x \in J^+(x)$, then there are sequences $\{x_n\}$ in X and $\{t_n\}$ in R^+ such that $x_n \to x$, $t_n \to +\infty$, and $x_n t_n \to x$. Setting $x_n t_n = y_n$ and, $\tau_n = -t_n$, we see that $x_n = y_n \tau_n$. Thus indeed we have sequences $\{y_n\}$ and $\{\tau_n\}$ such that $y_n \to x$, $\tau_n \to -\infty$, and $y_n \tau_n \to x$. Thus $x \in J^-(x)$. The converse is now obvious and the lemma is proved.

2.11.8 *Exercise*

Prove that a point $x \in X$ is a wandering point if and only if there is a neighborhood U of x and a $T > 0$ such that $U \cap Ut = \emptyset$ for all t, $|t| \geqq T$.

It is now easy to see that a parallelizable dynamical system is dispersive, a dispersive one is wandering, and a wandering one is Lagrange unstable. The converses do not hold as the following examples will show.

2.11.9 *Example*

Consider a dynamical system in the euclidean (x_1, x_2)-plane. whose phase portrait is as in Figure 2.11.10. The unit circle contains a rest point p and trajectory γ such that for each point $q \in \gamma$, we have $\Lambda^+(q) = \Lambda^-(q) = \{p\}$. All trajectories in the interior of the unit circle have the same property as γ. All trajectories in the exterior of the unit circle spiral to the unit circle as $t \to +\infty$, so that for each point q in the exterior of the unit circle $\{p\} \cup \gamma$ we have $\Lambda^+(q) = \{p\} \cup \gamma$, and $\Lambda^-(q) = \emptyset$. Notice that if we consider the dynamical system obtained from this one by deleting the rest point p (the dynamical system is thus defined on $R^2 \setminus \{p\}$) then this system is Lagrange unstable, but it is not wandering, because for each $q \in \gamma$ we have $J^+(q) = \gamma$, i.e. $q \in J^+(q)$.

2.11.10. *Figure*

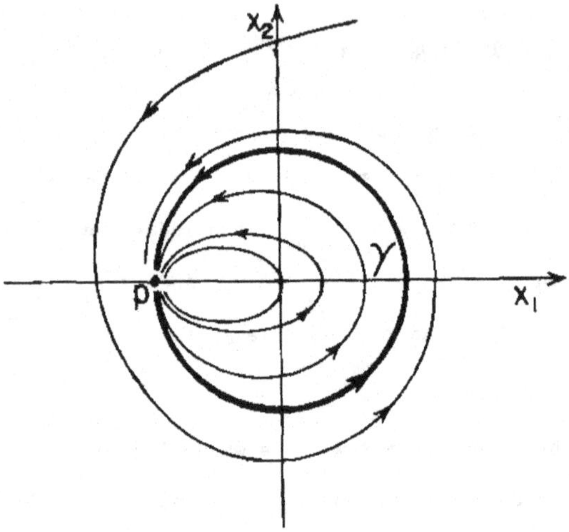

2.11.11 *Remark*

For dynamical systems defined by differential equations in the euclidean plane R^2 the concept of Lagrange instability and the concept of wandering are equivalent. This may easily be proved using the Poincarè-Bendixson-theory of planar systems.

2.11.12 *Example*

In Example 1.4.7ii we have a dynamical system defined in the plane which is wandering but not dispersive. This follows by noticing that for each point $p \in \gamma_{-1}$, $J^+(p) = \gamma_0$, and for all other points p, $J^+(p) = \emptyset$.

2.11.13 *Example*

Consider a dynamical system defined in R^2 by the differential equations

$$\frac{dx_1}{dt} = f(x_1, x_2), \qquad \frac{dx_2}{dt} = 0,$$

where $f(x_1, x_2)$ is continuous, and moreover $f(x_1, x_2) = 0$ whenever the point (x_1, x_2) is of the form $(n, \frac{1}{n})$ with n a positive integer. For simplicity we assume that $f(x_1, x_2) > 0$ for all other points. The phase portrait is as shown in figure 2.11.14. Let us now consider the dynamical system obtained from the above one by deleting the sets

$$I_n = \{(x_1, x_2): \; x_1 \leqq n, \; x_2 = \frac{1}{n}\}, \quad n = 1, 2, 3, \ldots,$$

from the plane R^2. This system is dispersive, but is not parallelizable. This may easily be seen, and it will indeed become clear as we develop the theory further.

2.11.14 *Figure.*

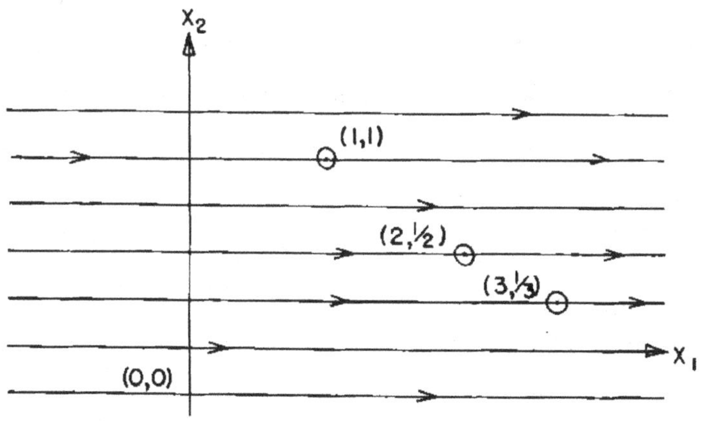

We now develop a criterion for dispersive flows.

2.11.15 *THEOREM*

A dynamical system (X, R, π) *is dispersive if and only if for each* $x \in X$, $J^+(x) = \emptyset$

Proof. Let (X, R, π) be dispersive. Let if possible $x \in X$ and $J^+(x) \neq \emptyset$. Then if $y \in J^+(x)$, there are sequences $\{x_n\}$, $\{t_n\}$, $x_n \to x$, $t_n \to +\infty$, and $x_n t_n \to y$. This shows that for any neighborhoods U_x, U_y of x and y respectively $U_x t_n \cap U_y \neq \emptyset$ as the element $x_n t_n = y_n$ is contained in this intersection. Since $t_n \to +\infty$, this contradicts the definition of a dispersive flow. Hence $J^+(x) = \emptyset$ for each $x \in X$. Conversely, let $J^+(x) = \emptyset$ for each $x \in X$. We first show that in this case $J^-(x) = \emptyset$ for each $x \in X$. Otherwise, if for some $x, J^-(x) \neq \emptyset$, then there is a y, $y \in J^-(x)$. This implies that $x \in J^+(y)$, contradicting the assumption that $J^+(y) = \emptyset$ for each $y \in X$. Now if $J^+(x) = \emptyset$ for each $x \in X$, we claim that for $\{x, y\} \subset X$ there are neighborhoods U_x of x and U_y of y and a $T \geq 0$ such that $U_x t \cap U_y = \emptyset$ for all $t \geq T$. For if not, then there will be sequences $\{x_n\}$, $\{y_n\}$, $\{t_n\}$, $x_n \to x$, $y_n = x_n t_n$, $y_n \to y$, and $t_n \to +\infty$, so that $y \in J^+(x)$. Similarly, since $J^-(x) = \emptyset$ for each $x \in X$, for any $\{x, y\} \subset X$, there are neighborhoods U_x' of x and U_y' of y and a $T' \geq 0$ such that $U_x' t \cap U_y' = \emptyset$ for $t \leq -T'$. Setting now $U_x^* = U_x \cap U_x'$, $U_y^* = U_y' \cap U_y$, and $T^* = \max(T, T')$, we see that $U_x^* t \cap U_y^* = \emptyset$ for $|t| \geq T^*$, i.e., (X,R,π) is dispersive.

2.11.16 *Remark*

Using the above theorem the dynamical system described in example 2.11.13 is clearly seen to be dispersive.

We now give another criterion for dispersive flows, which is sometimes more useful than the one given above.

2.11.17 *THEOREM*

The dynamical system (X, R, π) *is dispersive if and only if for each* $x \in X$, $D^+(x) = \gamma^+(x)$ *and there are no rest points or periodic trajectories.*

Proof. If (X, R, π) is dispersive, then $J^+(x) = \emptyset$ for each $x \in X$. Consequently $D^+(x) = \gamma^+(x) \cup J^+(x) \equiv \gamma^+(x)$ for each $x \in X$, and there are no rest points or periodic orbits. For if x is a rest point or $\gamma(x)$ is periodic then $\gamma(x) \equiv \Lambda^+(x) \subset J^+(x)$. Conversely, if $D^+(x) = \gamma^+(x)$ and there are no rest points or periodic orbits, then $J^+(x) = \emptyset$. For indeed $D^+(x) \equiv \gamma^+(x) \cup J^+(x) = \gamma^+(x)$ implies that $J^+(x) \subset \gamma^+(x)$. $J^+(x)$ being closed and invariant, we conclude if $J^+(x)$ is not empty, that $\gamma(x) \subset J^+(x) \subset \gamma^+(x)$, i.e., $\gamma(x) = \gamma^+(x)$. This shows that if $\tau < 0$ is arbitrary, then there is a $\tau' \geq 0$ such that $x\tau = x\tau'$, i.e., $x = x(\tau' - \tau)$. Since $\tau' - \tau > 0$, the last equality shows that the trajectory $\gamma(x)$ is closed and has a period $\tau' - \tau$. Since we assumed that there are no rest points or periodic orbits, we have arrived at a contradiction. Thus $J^+(x) = \emptyset$ for each $x \in X$, and the dynamical system is dispersive. The theorem is proved.

We now develop a criterion for parallelizable dynamical systems. For this purpose the following definition is needed.

2.11.18 *Definition*

A set $S \subset X$ *is called a* <u>*section*</u> *of* (X, R, π) *if for each* $x \in X$ *there is a unique* $\tau(x)$ *such that* $x\tau(x) \in S$.

Not every dynamical system has a section. Indeed any (X, R, π) has a section if and only if it has no rest points or periodic trajectories.

The function $\tau(x)$ will be basic in what follows. In general

$\tau(x)$ is not continuous, but the existence of a section S with continuous $\tau(x)$ implies certain properties of the dynamical system which we sum up in the following lemma.

2.11.19 *LEMMA*

If S *is a section of the dynamical system* (X, R, π) *with* $\tau(x)$ *continuous on* X *then*

i) S *is closed in* X ,

ii) S *is connected, arcwise connected, simply connected if and only if* X *is respectively connected, arcwise connected, simply connected,*

iii) *If* $K \subset S$ *is closed in* S , *then* Kt *is closed in* X *for every* $t \in R$,

iv) *If* $K \subset S$ *is open in* S , *then* KI , *where* I *is any open interval in* R , *is open in* X .

Proof.

i) If $\{x_n\}$ in S , and $x_n \to x \in X$, then $\tau(x_n) \to \tau(x)$ by continuity. Since $\tau(x_n) = 0$ for each n , we get $\tau(x) = 0$. Thus $x\tau(x) = x\,0 = x \in S$ by definition of $\tau(x)$ and S . Hence S is closed in X .

ii) We shall prove only the first part. The interested reader can supply the proofs of the remaining parts. Let S be not connected. Then there are disjoint closed sets S_1, S_2 such that $S_1 \cup S_2 = S$. As $X = SR$, we have $X = S_1 R \cup S_2 R$. Note that S_1R and S_2R are disjoint. We prove that they are closed. Consider S_1R , and let $\{x_n\}$ in S_1R , $x_n \to x$. Then $\tau(x_n) \to \tau(x)$, and by continuity of the phase map $\pi : x_n\tau(x_n) \to x\,\tau(x)$. Since $\{x_n\tau(x_n)\}$ in S_1 and S_1 is closed, we conclude that $x\tau(x) \in S_1$. Then

$x = x\tau(x)(-\tau(x)) \in x\tau(x)R \subset S_1R$. Thus S_1R is closed. Similarly we can prove that S_2R is closed. Thus X being the union of two disjoint non-empty closed sets is not connected. We conclude that if X is connected, so must S be . The converse follows similarly.

iii) The proof follows by observing that if K is closed in S , then it is closed in X .

iv) The simple proof is left to the reader.

The following theorem now gives a criterion for parallelizable dynamical systems.

2.11.20 *THEOREM*

A dynamical system (X, R, π) *is parallelizable if and only if it has a section* S *with* $\tau(x)$ *continuous on* X .

Proof. Sufficiency. Indeed SR = X. Define $h: X \to S \times R$ by $h(x) = (x\tau(x), -\tau(x))$. Then h is 1-1 and continuous by the continuity of $\tau(x)$ and the phase map π. The inverse $h^{-1}: S \times R \to X$ is given by $h^{-1}(x, t) = xt$ and is clearly continuous. This h is thus a homeomorphism of X onto $S \times R$, i.e., (X, R, π) is parallelizable. To see necessity, we note that if the dynamical system is parallelizable, then the set S in its definition is a section of X. Further for any $x \in X$ set $\tau(x) = -t$ where t is given by $h(x) = (x(-t), t)$. Then continuity of $\tau(x)$ follows from that of h. The proof is completed.

2.11.21 *Remark*

The above theorem shows that the dynamical system of example 2.11.13 is not parallelizable. Notice however that the phase space in this example is not locally compact.

The following is the most important theorem in this section.

2.11.22 *THEOREM*

A dynamical system (X, R, π) *on a locally compact separable metric space* X *is parallelizable if and only if it is dispersive.*

The proof of this theorem depends on properties of certain sections which we now describe.

2.11.23 *DEFINITION*

An open set U *in* X *will be called a* *tube* *if there exists a* $\tau > 0$ *and a subset* $S \subset U$ *such that*

i) $SI_\tau \subset U$, *and*

ii) *for each* $x \in U$ *there is a unique* $\tau(x)$, $|\tau(x)| < \tau$, *such that* $x\tau(x) \in S$. *Here* $I_\tau \equiv (-\tau, \tau)$.

It is clear that if (i) and (ii) hold, then $U = SI_\tau$. U is also called a τ-tube with section S, and S a (τ- U)-section of the tube U.

If $I_\tau = R$, then U is an ∞-tube, and S an $(\infty - U)$-section. In this last case indeed $U = SR$. Note also that the function $\tau(x)$ which maps U into I_τ is 1-1 along each trajectory in U.

2.11.24 *LEMMA*

Let U *be a* τ*-tube with section* S. *If* $K \subset S$ *is compact, then the function* $\tau(x)$ *is continuous on* KI_s *for any* s, $0 < s < \tau$.

Proof. To show : if $\{x_n\}$ in KI_s and $x_n \to x \in KI_s$, then $\tau(x_n) \to \tau(x)$. Note that the sequence $\{x_n\tau(x_n)\}$ is in K, and we may assume that it is convergent as K is compact. Further $\{\tau(x_n)\}$ is in I_s and hence bounded so that we may also assume that $\{\tau(x_n)\}$ is convergent. Thus let $x_n\tau(x_n) \to x^* \in K$, and $\tau(x_n) \to \tau^* \in \bar{I}_{\tau'}$. Since $x_n \to x$, we have

$x^* = x\tau^*$. Since $|\tau^*| \leq \tau' < \tau$, we have $\tau^* = \tau(x)$ by uniqueness. The lemma is proved.

The next theorem shows that if $x \in X$ is not a rest point, then there is a tube containing x.

2.11.25 *THEOREM*

If $x \in X$ *is not a rest point, then there exists a tube containing* x.

Proof. Since x is not a rest point, there is a $T_o > 0$ such that $\rho(x, xT_o) > 0$. Consider the function

$$\psi(y, t) = \int_t^{t + T_o} \rho(x, y\tau)\, d\tau.$$

It follows that

$$\psi(y, t_1 + t_2) = \int_{t_1 + t_2}^{t_1 + t_2 + T_o} \rho(x, y\tau)\, d\tau$$

$$= \int_{t_2}^{t_2 + T_o} \rho(x, y(\tau + t_1))\, d\tau$$

$$= \int_{t_2}^{t_2 + T_o} \rho(x, yt_1(\tau))\, d\tau$$

$$= \psi(yt_1, t_2)\ .$$

Further the function $\psi(y, t)$ is continuous in (y, t) and has the partial derivative

$$\psi_t(y, t) = \rho(x, y(t + T_0)) - \rho(x, yt).$$

Since

$$\psi_t(x, 0) = \rho(x, xT_0) > 0,$$

there is an $\varepsilon > 0$ such that $\psi_t(y, 0) > 0$ for $y \in S(x, \varepsilon)$. Define $\tau_0 > 0$ such that $x[-3\tau_0, 3\tau_0] \subset S(x, \varepsilon)$. Then, in particular, $\psi(x, \tau_0) > \psi(x, 0) > \psi(x, -\tau_0)$. Now choose $\zeta > 0$ such that

$$(S[x\tau_0, \zeta] \cup S[x(-\tau_0), \zeta]) \subset S(x, \varepsilon),$$

and such that for $y \in S(x\tau_0, \zeta)$ we have $\psi(y, 0) > \psi(x, 0)$, and for $y \in S(x(-\tau_0), \zeta)$ we have $\psi(y, 0) < \psi(x, 0)$. Finally determine $\delta > 0$ such that

$$S[x, \delta]\, \tau_0 \subset S(x\tau_0, \zeta), \qquad S[x, \delta]\, (-\tau_0) \subset S(x(-\tau_0), \zeta),$$

and

$$S[x, \delta]\, [-3\tau_0, 3\tau_0] \subset S(x, \varepsilon).$$

We will show that if $y \in S[x, \delta]$, then there is exactly one $\tau(y)$, $|\tau(y)| < \tau_0$ such that $\psi(y, \tau(y)) = \psi(x, 0)$. This follows from the fact that $\psi(y, t) = \psi(yt, 0)$ is an increasing function of t, and $\psi(y, \tau_0) > \psi(x, 0) > \psi(y, -\tau_0)$.

Consider now the open set $U = S(x, \delta)\, I_{\tau_0}$, and set

$S = \{y \in U: \psi(y, 0) = \psi(x, 0)\}$. We claim that S is a $(2\tau_0 - U)$-section. For this we need prove that if $y \in U$, then there is a unique $\tau(y)$, $|\tau(y)| < 2\tau_0$ such that $y\tau(y) \in S$. Indeed for any $y \in U$, there is a t', $|t'| < \tau_0$, such that $y' = yt' \in S(x, \delta)$, and for $y' \in S(x, \delta)$ there is a t'', $|t''| < \tau_0$, such that $y't'' \in S$. Thus $y(t' + t'') = y\tau(y) \in S$, where $\tau(y) = t' + t''$, and $|\tau(y)| \leq |t'| + |t''| < 2\tau_0$. Now let if possible there be two numbers, $\tau'(y)$, $\tau''(y)$, $|\tau'(y)| < 2\tau_0$, $|\tau''(y)| < 2\tau_0$, such that $y\,\tau'(y) \in S$ and $y\,\tau''(y) \in S$, and let $y' = yt' \in S(x, \delta)$, where $|t'| \leq \tau_0$. Then $\psi(y', \tau'(y) - t') = \psi(y, \tau'(y)) = \psi(y\tau'(y), 0)$, and $\psi(y', \tau''(y) - t') = \psi(y, \tau''(y)) = \psi(y\tau''(y), 0)$, so that $\psi(y', \tau'(y) - t') = \psi(y', \tau''(y) - t') = \psi(x, 0)$. Now $|\tau'(y) - t'| \leq 3\tau_0$, and $|\tau''(y) - t'| \leq 3\tau_0$, and $\psi_t(y', t) > 0$ for $|t| \leq 3\tau_0$, i.e. $\psi(y', t)$ is strictly increasing for $|t| \leq 3\tau_0$. Hence $\tau'(y) - t' = \tau''(y) - t'$, or $\tau'(y) = \tau''(y)$. The theorem is proved.

2.11.26 *Remark*

If X is locally compact, then we can restrict $\delta > 0$ in the above proof to ensure that $S[x, \delta]$ is compact. Thus the $(2\tau_0 - U)$-section S constructed in the above proof will also be locally compact. By Lemma 2.11.25 we may further assume the function $\tau(x)$ corresponding to the section S to be continuous on U.

In fact the following more general theorem can now be proved.

2.11.27 *THEOREM*

Let $x \in X$ *be not a rest point. Let* $\tau > 0$ *be given, restricted only by* $\tau < \frac{w}{4}$ *if the motion* π_x *is periodic with least period* w. *Then there exists a tube* U *containing* x *with a* $(\tau - U)$*-section* S. *Further, if* X *is locally compact, then the function* $\tau(x)$ *corresponding to the section* X *can be assumed continuous on* U.

The proof of this theorem is left to the reader.

For wandering points $x \in X$ one can prove:

2.11.28 *THEOREM*

If $x \in X$ *is a wandering point, i.e.,* $x \notin J^+(x)$ *, and moreover* X *is locally compact, then there exists a tube* U *containing* x *, with an* $(\infty - U)$*-section* S *, and with* $\tau(x)$ *continuous on* U *.*

Proof. Indeed there is a tube W containing x, with a $(\tau - W)$-section S, and $\tau(x)$ continuous on W. Since x is wandering, we claim that there is a $\delta > 0 \ni S(x,\delta) \cap S = S^*$ is an $(\infty - U)$-section of the open set $U = S^* R$, which is a ∞ - tube containing x. To see this notice that there is a $\delta > 0$, such that every trajectory $\gamma(y)$ with $y \in S^*$, intersects S^* only at the point y. For otherwise, there will be a sequence $\{y_n\}$ in S, $y_n \to x$, and a sequence $\{t_n\}$ in R, $t_n \to +\infty$ (or $t_n \to -\infty$), such that $y_n t_n \to x$, i.e., either $x \in J^+(x)$, or $x \in J^-(x)$, both of which are ruled out by the assumption that x is wandering. We have thus shown that there is a $\delta > 0$ such that $S^* = S(x, \delta) \cap S$ is an $(\infty - U)$-section of $U = S^* R$. U is further open, and continuity of $\tau(x)$ on U follows from its continuity on $W \cap U$, and continuity of the phase map π. This we leave to the reader to verify. The theorem is proved.

For further development we need the following definition.

2.11.29 *DEFINITION*

Given an open ∞*-tube* U *with a section* S *and* $\tau(x)$ *continuous on* U *, let there be given sets* N *,* K *,* $N \subset K \subset S$ *, where* N *is open in* S *and* K *is compact, we shall call* KR *the compactly based tube over* K *. Then indeed* $\tau(x)$ *restricted to* KR *is continuous on* KR *.*

2.11.30 *Remark.*

A compactly based tube need not be closed in X. As an example, one may consider a dynamical system defined in the euclidean plane R^2, as shown in figure 2.11.31. The x_2-axis consists entirely of rest points, all other trajectories are parallel to the x_1-axis, with each having a rest point on the x_2-axis as the only point in its positive limit set, whereas the negative limit sets are empty. Here, for example the set $\{(x_1, x_2): 0 \leq x_2 \leq 1, \ x_1 > 0\}$ is a compactly based tube, which is not closed in X.

2.11.31 *Figure*

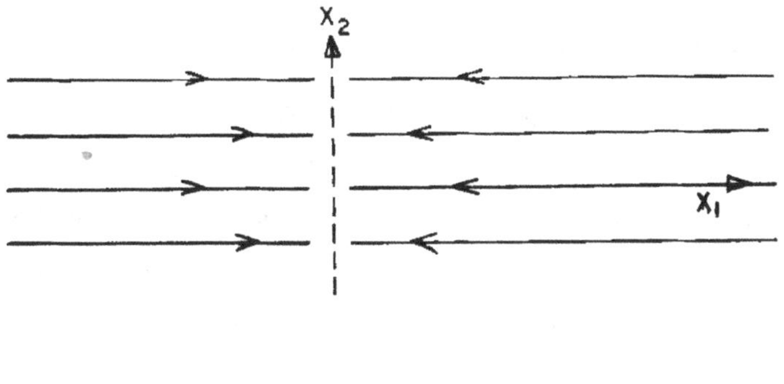

We can now prove the following.

2.11.32 *THEOREM*

If X *is locally compact and separable, and if every* $x \in X$ *is a wandering point, then there exists a countable covering* $\{K_nR\}$ *of* X, *by compactly based tubes* K_nR *each with* $\tau_n(x)$ *continuous on* K_nR.

Proof. The proof is immediate, when we notice that by using theorem 2.11.28, one can find a compactly based tube containing a wandering point of X. The rest follows by the assumption of separability of X.

We gave an example above to show that a compactly based tube need not be closed in X. One may wonder if for a wandering dynamical system (X, R, π), a compactly based tube is not closed. Here is a counter-example.

2.11.33 *Example*

Consider again example 1.4.7ii, referred to in example 2.11.12. Any compactly based tube containing a point $p \in \gamma_{-1}$ is not closed, because its closure will contain γ_0 which is not in such a tube. This is an example of a wandering dynamical system which is not dispersive. In the case that (X, R, π) is dispersive one obtains.

2.11.34 *LEMMA*

A compactly based ∞*-tube* U *with section* K *of a dispersive dynamical system* (X, R, π) *is closed in* X.

Proof. $U = KR$ and if $\{x_n\}$ is a sequence in KR, then there are sequences $\{y_n\}$ in K and $\{\tau_n\}$ in R such that $x_n = y_n \tau_n$. We may assume that $y_n \to y \in K$ as K is compact. If now $x_n \to x$, we claim that the sequence $\{\tau_n\}$ is bounded, so that $x \in y R \subset KR$. For otherwise if $\{\tau_n\}$ contains an unbounded sebsequence $\{\tau_{n_k}\}$, say $\tau_{n_k} \to +\infty$, then clearly $x \in J^+(y)$, which is absurd, as $J^+(y) = \emptyset$ for each $y \in X$ by Theorem 2.11.15. The lemma is proved.

We now prove the last lemma required to prove Theorem 2.11.22.

2.11.35 *LEMMA*

Let U_1, U_2 *be two compactly-based tubes of a dispersive dynamical system with sections* K_1, K_2 *and continuous functions* $\tau_1(x)$ *and* $\tau_2(x)$ *respectively. If* $U_1 \cap U_2 = \emptyset$ *then* $U = U_1 \cup U_2$ *is a compactly based tube with a section* $K \supset K_1$ *and a continuous function* $\tau(x)$. *Moreover, if the time distance between* K_1 *and* K_2 *along orbits in*

$U_1 \cap U_2$ is less than $\tau (> 0)$, the time distance between K *and* K_2 *along orbits in* U *is also less than* τ .

Proof. U_1 and U_2 are invariant and closed. Therefore $U_1 \cap U_2$ is invariant and closed. Further, $K_2 \cap U_1$

2.11.36 *Figure*

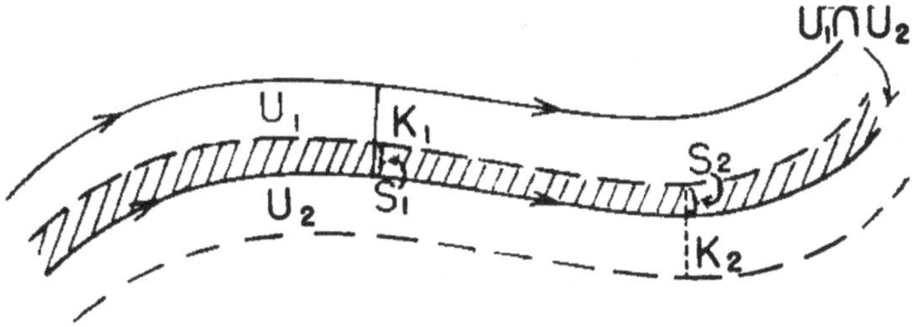

is compact and non-empty. Set $S_2 = K_2 \cap U_1$ and $S_1 = K_1 \cap U_1$. Any orbit in $U_1 \cap U_2$ intersects K_2 and hence S_2 in exactly one point, and also intersects K_1 and hence S_1 in exactly one point. Thus for any $x \in U_1 \cap U_2$ we have $\tau_1(x) = \tau_2(x) + \tau_1(x\, \tau_2(x))$. This is so because $x\tau_1(x) = x\tau_2(x)(\tau_1(x\, \tau_2(x))) = x(\tau_2(x) + \tau_1(x\, \tau_2(x)))$, and there are no rest points or periodic orbits. The function $\tau_1(x)$ is continuous on S_2 (which is compact), and by Tietze's theorem it can be extended to a continuous function $\tau(x)$ defined on K_2, where $\tau(x) \equiv \tau_1(x)$ for $x \in S_2$. Further if $|\tau_1(x)| < \tau$ for $x \in S_2$, we can have $|\tau(x)| < \tau$ on K_2. Notice now that $\{x\, \tau(x) : x \in S_2\} = S_1$, and $\tau(x)$ being continuous $\{x\, \tau(x) : x \in K_2\}$ is compact as K_2 is compact. We set now $K = K_1 \cup \{x\, \tau(x) : x \in K_2\}$, and define $\tau^*(x)$ on $KR = K_1R \cup K_2R$ as follows: $\tau^*(x) = \tau_1(x)$ for $x \in K_1R$, and $= \tau_2(x) + \tau(x\, \tau_2(x))$ if $x \in K_2R$. $\tau^*(x)$ is continuous on KR and we need only verify that if $x \in U_1 \cap U_2$, then $\tau_1(x) = \tau_2(x) + \tau(x\, \tau_2(x))$, which has already been proved. The lemma is proved.

Proof of Theorem 2.11.22. Only the sufficiency part needs proof. It is sufficient to prove that X has a section S with $\tau(x)$ continuous on X. By Theorem 2.11.32 there is a countable covering $\{U_n\}$ of X by compactly based tubes U_n with sections K_n and continuous functions $\tau_n(x)$. We replace this covering by a like covering $\{U^n\}$ of compactly based tubes which we construct as follows. Set $K_1 = K^1$, and $U_1 = U^1$. Beginning with U^1 and U_2 we use lemma 2.11.35 to enlarge K^1 to a compact set K^2, thus obtaining the compactly based tube $U^2 = U^1 \cup U_2$ with $\tau^2(x)$ continuous on U^2. This leaves K^1 unaltered. Having found U^n, we take it together with U_{n+1} and construct similarly U^{n+1} with $K^{n+1} \supset K^n$, and $\tau^{n+1}(x)$ continuous on U^{n+1}. Now set $S = \bigcup K^n$, then $X = SR$, and the function $\tau(x)$ defined by $\tau(x) = \tau^n(x)$ for $x \in U^n$ is continuous on X with the property that $x\tau(x) \in S$; moreover, $\tau(x)$ is unique for each $x \in X$. Thus X has a section S with continuous $\tau(x)$ defined on X. The system (X, R, π) is thus parallelizable and the theorem is proved.

We shall now make some applications of the theory of sections to asymptotic stability of compact sets of a dynamical system defined in locally compact metric space X.

Theorem 2.7.11 shows that if $M \subset X$ is compact and asymptotically stable, then its region of attraction A(M) contains a countable dense subset in it. Thus if A(M) is the region of attraction of a compact asymptotically stable set M, then $A(M) \setminus M = A^*(M)$ is a locally compact separable metric subspace of X. We will show more (see Exercise 2.6.24).

2.11.37 *THEOREM*

If M *is a compact asymptotically stable set of* X, *then for each* $x \in A(M)$, $J^+(x) \subset M$.

Proof. If possible let $x \in A(M)$ and $J^+(x) \not\subset M$. Since $J^+(x) \supset \Lambda^+(x)$ and $\Lambda^+(x) \neq \emptyset$, we have $J^+(x) \neq \emptyset$. Thus there is a $y \in J^+(x)$, $y \notin M$. Consequently by lemma 2.3.12 for any $w \in \Lambda^+(x) \subset M$, we have $J^+(x) \subset D^+(w) \subset D^+(M)$. But $D^+(M) = M$ by Theorem 2.6.6 , as M is stable. Therefore $J^+(x) \subset M$ for each $x \in A(M)$.

2.11.38 *COROLLARY*

If M *is a compact, invariant, asymptotically stable set of a dynamical system* (X, R, π), *then the dynamical system induced by the given one on the invariant set* $A^*(M) = A(M) \setminus M$ *is dispersive. If* X *is locally compact, then by the above observation and Theorem* 2.11.22 *it is parallelizable.*

2.11.39 *Remark*

It is clear now that if M is a compact asymtotically stable set with region of attraction $A(M)$ and M^* is the largest invariant set in M, then the dynamical system induced on $A(M) \setminus M^*$ is dispersive. If X is locally compact, then it is parallelizable. Now let $\Phi(x)$ be any function defined in a neighborhood N of M and satisfying conditions of Lemma 2.8.1. If $\varepsilon > 0$ is chosen such that $S[M,\varepsilon] \subset N$, and $S[M, \varepsilon]$ is compact, then let $m_0 = \min\{\Phi(x) : x \in H(M,\varepsilon)\}$. For any α,

$$0 < \alpha < m_0, \quad \text{set} \quad P_\alpha = \{x \in S[M,\varepsilon] : \Phi(x) \le \alpha\}.$$

Then P_α is compact invariant subset of $A(M)$ as is clear from the proof of theorem 2.7.9. Further , ∂P_α is a section of the parallelizable induced dynamical system on $A(M) \setminus M^*$ because, as shown in the proof of Theorem 2.7.14, there is a continuous function t_x defined on $A(M) \setminus M^*$ such that

$$xt_x \in \partial P_\alpha \quad \text{for each} \quad x \in A^*(M) = A(M) \setminus M^*.$$

Conversly, the observation that $A(M) \setminus M^*$ is parallelizable may be used to demonstrate the existence of functions satisfying various theorems in section 2.7, by using any section of $A(M) \setminus M^*$ and corresponding continuous function $\tau(x)$.

2.11.40 *Notes and References*

This section is based on the work of Nemytskii, Bebutov, Stepanov, Whitney, Barbashin and Dugundji and Antosiewicz.

High point of this section is Theorem 2.11.32 which appears in Nemytskii [10]. Antosiewicz and Dugundji give an alternative proof.

We have not introduced the notion of a saddle point at infinity which is used by Nemytskii in his exposition.

The introduction of the prolongational limit set to describe some classical notions like the wandering point and dispersive flows are of recent origin. See J. Auslander [3], Bhatia [4].

2.12 *Stability and asymptotic stability of closed sets.*

We shall first discuss stability of closed sets.

2.12.1 *DEFINITION*

A closed set $M \subset X$ *will be said to be*

i) *stable, if for each* $\varepsilon > 0$ *and* $x \in M$, *there is a* $\delta = \delta(x,\varepsilon) > 0$ *such that*

$$S(x,\delta)R^+ \subset S(M,\varepsilon),$$

ii) *equi-stable, if for each* $x \notin M$, *there is a* $\delta = \delta(x) > 0$ *such that* $x \notin \overline{S(M,\delta)R^+}$, *and*

iii) *uniformly stable, if for each* $\varepsilon > 0$, *there is a* $\delta = \delta(\varepsilon) > 0$ *such that*

$$S(M,\delta)R^+ \subset S(M,\varepsilon).$$

2.12.2 *Proposition*

If X *is locally compact and* M *is compact, then* M *is uniformly stable whenever it is either equi-stable or stable (or both).*

Proof. (i) If M is stable, then for a given $\varepsilon > 0$, let $\delta_x > 0$ be a number corresponding to $x \in M$ such that $S(x,\delta_x)R^+ \subset S(M,\varepsilon)$. Since $\cup\{S(x,\delta_x) : x \in M\}$ is an open cover of M, there is a finite open cover, say $\cup\{S(x_i,\delta_{x_i}) : i = 1,2,.,n; x_i \in M\}$ of M. But then there is a $\delta > 0$ such that $S(M,\delta) \subset \cup\{S(x_i,\delta_{x_i}) : i = 1,2,\ldots,n\}$. Notice now that $S(M,\delta)R^+ \subset [\cup\{S(x_i,\delta_{x_i}) : i = 1,2,.,n\}]R^+ \subset S(M,\varepsilon)$. Thus M is uniformly stable.
(ii) Let M be equi-stable. Since M is compact and X is locally compact, there is a $\delta > 0$ such that $S[M,\delta]$, and hence also $H(M,\delta)$ are compact.

Then for each $x \notin H(M,\delta)$, there is a $\delta_x > 0$ such that $x \notin \overline{S(M,\delta_x)R^+}$. But then $x \in C(S_x)$, where $S_x \equiv \overline{S(M,\delta_x)R^+}$. Since each $C(S_x)$ is open, and $H(M,\delta) \subset \bigcup\{C(S_x) : x \in H(M,\delta)\}$, we have an open cover of the compact set $H(M,\delta)$. Thus there are points $x_1, x_2, \ldots, x_n$ in $H(M,\delta)$ such that $H(M,\delta) \subset \bigcup\{C(S_{x_i}) : i = 1,2,\ldots,n\}$. Since $\bigcup\{C(S_{x_i}) : i = 1,2,\ldots,n\} = C(\bigcap\{S_{x_i} : i = 1,2,\ldots,n\})$ we have $\bigcap\{S_{x_i} : i = 1,2,\ldots,n\} \subset S(M,\varepsilon)$. If now $\delta = \min\{\delta_{x_1}, \delta_{x_2}, \ldots, \delta_{x_n}\}$, then $S(M,\delta)R^+ \subset \bigcap\{S_{x_i} : i = 1,2,\ldots,n\} \subset S(M,\varepsilon)$. Thus M is uniformly stable.

2.12.3 *Remark*

Note that part (i) of the above theorem did not use the fact that X is locally compact. Further, uniform stability implies both stability and equi-stability, but it cannot be asserted that a closed set which is both stable and equi-stable is uniformly stable.

2.12.4 *THEOREM*

There exist closed sets which are both stable and equi-stable but are not uniformly stable.

We leave the proof to the reader.

2.12.5 *Proposition*

If a closed set is either stable, or equi-stable, then it is positively invariant.

The proof is simple and is left as an exercise.

We now indicate the connection between various kinds of stability and Liapunov Functions.

2.12.6. *THEOREM*

A closed set M *is stable if and only if there exists a function* $v = \phi(x)$ *defined on* X *with the following properties:*

i) For every $\varepsilon > 0$, *there is a* $\delta > 0$ *such that* $\phi(x) \geq \delta$ *whenever* $\rho(x,M) \geq \varepsilon$, *and for any sequence* $\{x_n\}$, $\phi(x_n) \to 0$ *whenever* $x_n \to x \in M$.

ii) $\phi(xt) \leq \phi(x)$ *for all* $x \in X$, $t \geq 0$.

Proof. (a) *Sufficiency.* Given $\varepsilon > 0$, set $m_0 = \inf\{\phi(x):\rho(x,M) \geq \varepsilon\}$. By (i) $m_0 > 0$. Then for $x \in M$ find $\delta > 0$ such that $\phi(y) < m_0$ for $y \in S(x,\delta)$. This is also possible by (i). We claim that $S(x,\delta)R^+ \subset S(M,\varepsilon)$. For otherwise there is $y \in S(x,\delta)$, and $t \geq 0$ such that $\rho(yt,M) = \varepsilon$. But then $\phi(yt) \leq \phi(y) < m_0$ on one hand by (ii), and also $\phi(yt) \geq \inf\{\phi(x):\rho(x,M) \geq \varepsilon\}$, as $\rho(yt,M) = \varepsilon$, i.e., $\phi(yt) \geq m_0$. This contradiction proves the result.

(b) Necessity. Let M be stable. Define

$$\phi(x) = \sup\{ \frac{\rho(xt,M)}{1+\rho(xt,M)}: \quad t \geq 0\}.$$

This $\phi(x)$ is defined on X, and has all the properties required in the theorem. The varification is left to the reader. The theorem is proved.

2.12.7 *Remark*

Condition (i) in the above theorem is equivalent to the requirement that if $\{x_n\}$ is any sequence such that $x_n \to x \in M$, then $\phi(x_n) \to 0$, and there is a continuous strictly increasing function $\alpha(\mu)$, defined for $\mu \geq 0$, such that $\phi(x) \geq \alpha(\rho(x,M))$.

2.12.8 *THEOREM*

A closed set $M \subset X$ *is equi-stable if and only if there is a function* $\phi(x)$ *defined on* X *such that*

i) $\phi(x) = 0$ *for* $x \in M$, $\phi(x) > 0$ *for* $x \notin M$,

ii) for every $\varepsilon > 0$ *there is a* $\delta > 0$ *such that* $\phi(x) \leq \varepsilon$ *if* $\rho(x,M) \leq \delta$, *and*

iii) $\phi(xt) \leq \phi(x)$ *for* $x \in X$, $t \geq 0$.

Proof. (a) Sufficiency. Let $x \notin M$. Set $\rho(x,M) = \varepsilon$. Then by (ii) there is a $\delta > 0$ such that $\phi(x) \leq \frac{\varepsilon}{2}$ for $\rho(x,M) \leq \delta$. Then indeed $\overline{S(M,\delta)R^+} \subset \overline{S(M,\frac{\varepsilon}{2})}$, by (iii). Hence $x \notin \overline{S(M,\delta)R^+}$.

(b) Necessity. Set for each $x \notin M$,

$$\phi(x) = \sup\{\delta > 0 : x \notin \overline{S(M,\delta)R^+}\} ,$$

and $\phi(x) = 0$ for $x \in M$. This $\phi(x)$ has all the desired properties, whose verification is left to the reader. Note that $\phi(x) \leq \rho(x,M)$.

2.12.9 *Remark*

Condition (ii) in the above theorem is equivalent to the existence of a continuous strictly increasing function $\alpha(r)$, $\alpha(0) = 0$, such that $\phi(x) \leq \alpha(\rho(x,M))$.

2.12.10 *THEOREM*

A closed set M *is uniformly stable if and only if there is a function* $\phi(x)$ *defined on* X *such that*

i) for every $\varepsilon > 0$ *there is a* $\delta > 0$ *such that* $\phi(x) \geq \delta$ *whenever* $\rho(x,M) \geq \varepsilon$,

ii) for every $\varepsilon > 0$ *there is a* $\delta > 0$ *such that* $\phi(x) \leq \varepsilon$ *whenever* $\rho(x,M) \leq \delta$,

iii) $\phi(xt) \leq \phi(x)$ *for* $x \in X$, $t \geq 0$.

Proof. We leave the details to the reader, but remark that in the proof of necessity one may choose either of the functions given in the necessity proofs of the two theorems 2.12.6, and 2.12.8.

2.12.11. *Remark.*

Our theorems above differ from the usual theorems on stability in that the existence of the functions is shown in all of X rather than in a small neighborhood of M. Theorems 2.12.6 and 2.12.8 are new. Indeed for sufficiency the unctions need be defined on just a neighborhood of M.

We shall now discuss asymptotic stability of closed sets and its relation ith the Lyapunov Functions.

12.12 *DEFINITION*

A closed set $M \subset X$ *will be said to be*

i) a semi-weak attractor, if for each $x \in M$, *there is a* $\delta_x > 0$, *and for each* $y \in S(x,\delta_x)$ *there is a sequence* $\{t_n\}$ *in* R, $t_n \to +\infty$, *such that* $\rho(yt_n,M) \to 0$,

ii) a semi-attractor, if for each $x \in M$, *there is a* $\delta_x > 0$, *such that for each* $y \in S(x,\delta_x), \rho(yt,M) \to 0$ *as* $t \to +\infty$,

iii) a weak attractor, if there is a $\delta > 0$ *and for each* $y \in S(M,\delta)$, *there is a sequence* $\{t_n\}$ *in* R, $t_n \to +\infty$, *such that* $\rho(yt_n,M) \to 0$,

iv) an attractor, if there is a $\delta > 0$ *such that for each* $y \in S(M,\delta)$, $\rho(yt,M) \to 0$ *as* $t \to +\infty$,

v) a uniform attractor, if there is an $\alpha > 0$, *and for each* $\varepsilon > 0$ *there is a* $T = T(\varepsilon) > 0$, *such that* $x[T,+\infty) \subset S(M,\varepsilon)$ *for each* $x \in S[M,\alpha]$,

vi) an equi-attractor, if it is an attractor, and if there is an $\alpha > 0$ *such that for each* ε, $0 < \varepsilon < \lambda$, *and* $T > 0$, *there exists a* $\delta > 0$ *with the property that* $x[0,T] \cap S(M,\delta) = \emptyset$ *whenever* $\varepsilon \leq \rho(x,M) \leq \lambda$,

vii) semi-asymptotically stable, if it is stable and a semi-attractor,

viii) asymptotically stable, if it is uniformly stable and is an attractor,

ix) uniformly asymptotically stable, if it is uniformly stable and a uniform attractor.

12.13 *DEFINITION*

i) For any set $M \subset X$, *the set* $A_\omega(M) = \{y \in X:$ *there is a sequence* $\{t_n\}$ *in* R, $t_n \to +\infty$, *such that* $\rho(yt_n,M) \to 0\}$ *is called the* <u>*region of weak attraction*</u>

of M, *and*

ii) *the set* $A(M) = \{y \in X: \rho(yt, M) \to 0$ *as* $t \to +\infty\}$ *is called the* *region of attraction of* M.

2.12.14 *Proposition*

If M *is an attractor then*, $A_\omega(M) \equiv A(M)$.

The proof is trivial and is left as an exercise.

2.12.15 *Proposition*

If M *is a weak attractor (attractor), then* $A_\omega(M)$ $((A(M))$ *is an open invariant set which contains* $S(M,\delta)$ *for some* $\delta > 0$.

The proof is simple and is left as an exercise.

2.12.16 *THEOREM*

If a compact set M *is a semi-weak attractor (semi-attractor), then it is weak attractor (attractor).*

Proof is similar to that of Proposition 2.12.2.

We now discuss the existence of Liapunov functions for various kinds of asymptotic stability.

2.12.17 *THEOREM*

A closed set M *is semi-asymptotically stable, if and only if there exists a function* $\phi(x)$ *defined on* X *which has the following properties:*

i) *For each* $y \in M$, $\phi(x)$ *is continuous in some neighborhood* $S(y,\delta_y)$ *of* y,

ii) $\phi(x) = 0$ *for* $x \in M$, $\phi(x) > 0$ *for* $x \notin M$,

iii) *there is a strictly increasing function* $\alpha(\mu), \alpha(0) = 0$, *defined for* $\mu \geq 0$, such that

$$\phi(x) \geq \alpha(\rho(x,M)),$$

iv) $\phi(xt) \leq \phi(x)$ *for all* $x \in X$, $t \geq 0$, *and for each* $y \in M$, *there is a* $\delta_y > 0$ *such that if* $x \notin M$, $x \in S(y,\delta_y)$, *then* $\phi(xt) < \phi(x)$ *for* $t > 0$ *and* $\phi(xt) \to 0$ *as* $t \to +\infty$.

Proof. (a) Sufficiency. Stability follows from Theorem 2.12.6. The semi-attractor property follows from (iii) and (iv).

(b) Necessity. Consider the function

$$\phi(x) = \sup\{\frac{\rho(xt,M)}{1+\rho(xt,M)} : t \geq 0\}.$$

This has all the properties (i) to (iv) except that it may not be strictly decreasing along trajectories originating in any neighborhood of points of M. Before proving this we complete our construction. We define

$$\omega(x) = \int_0^\infty \phi(x\tau)e^{-\tau}d\tau \quad .$$

This $\omega(x)$ has all the properties (i) to (iv) except possibly (iii). The construction is now completed by setting

$$\Phi(x) = \phi(x) + \omega(x).$$

To see for example that for each $y \in M$, there is a $\delta_y > 0$ such that $\phi(x)$ is continuous in $S(y,\delta_y)$, we need prove that $\phi(x)$ is continuous in an open set containing M. If A(M) is the region of attraction of M, then A(M) is invariant, and indeed $I(A(M))$ is also invariant and open and contains M. We now define the set $W_\varepsilon = \{x \in I(A(M)): \gamma^+(x) \subset S(M,\varepsilon)\}$. This W_ε is open, positively invariant, and contains an open set containing M, and has the important property that, if $x \in I(A(M))$, then there is a $T > 0$ such that $xT \in W_\varepsilon$. Now let $x \in I(A(M))$ and let $\rho(x,M) = \lambda$. There is a $T > 0$ such that $xT \in W_{\lambda/4}$. Since $W_{\lambda/4}$ is open we can find a neighborhood $S(xT,\sigma) \subset W_{\lambda/4}$. Then $S(xT,\sigma)(-T) = N$ is a neighborhood of x, and indeed $N \subset I(A(M))$. We can thus choose an $\eta > 0$ such that $\eta < \lambda/4$, and $S(x,\eta) \subset N$. Then if $y \in S(x,\eta)$,

$$\phi(x) - \phi(y) = \sup\{\frac{\rho(xt,M)}{1+\rho(xt,M)} : t \geq 0\} - \sup\{\frac{\rho(yt,M)}{1+\rho(yt,M)} : t \geq 0\}$$

$$= \sup\{\frac{\rho(xt,M)}{1+\rho(xt,M)} : 0 \leq t \leq T\} - \sup\{\frac{\rho(yt,M)}{1+\rho(yt,M)} : 0 \leq t \leq T\},$$

and so

$$|\phi(x) - \phi(y)| \leq \sup\{|\frac{\rho(xt,M)}{1+\rho(xt,M)} - \frac{\rho(yt,M)}{1+\rho(yt,M)}| : 0 \leq t \leq T\}$$

$$= \sup\{|\frac{\rho(xt,M) - \rho(yt,M)}{(1+\rho(xt,M))(1+\rho(yt,M))}| : 0 \leq t \leq T\}$$

$$\leq \sup\{|\rho(xt,M) - \rho(yt,M)| : 0 \leq t \leq T\}.$$

$$\leq \sup\{\rho(xt,yt) : 0 \leq t \leq T\}.$$

By the continuity axiom the right hand side tends to zero as $y \to x$, hence $\phi(x)$ is continuous in $I(A(M))$. The rest of the observations on $\phi(x), \omega(x)$ are easy to verify and are left as an exercise.

2.12.18 *THEOREM*

Let M *be a closed set. Then* M *is asymptotically stable if and only if there is a function* $\phi(x)$ *defined in* X *with the following properties:*

i) $\phi(x)$ *is continuous in some neighborhood of* M *which contains the set* $S(M,\delta)$ *for some* $\delta > 0$,

ii) $\phi(x) = 0$ *for* $x \in M$, $\phi(x) > 0$ *for* $x \notin M$,

iii) there exist strictly increasing functions $\alpha(\mu), \beta(\mu), \alpha(0) = \beta(0) = 0$, *defined for* $\mu \geq 0$, *such that*

$$\alpha(\rho(x,M)) \leq \phi(x) \leq \beta(\rho(x,M)),$$

iv) $\phi(xt) \leq \phi(x)$ *for all* $x \in X, t > 0$, *and there is a* $\delta > 0$ *such that if* $x \in S(M,\delta), x \notin M$, *then* $\phi(xt) < \phi(x)$ *for* $t > 0$, *and* $\phi(xt) \to 0$ *as* $t \to +\infty$.

The proof follows exactly the same lines as that of the previous theorem and is left as an exercise. We note, however, that in the proof of necessity, since $A(M)$ is open and invariant, and $A(M) \supset S(M,\delta)$ for some $\delta > 0$, the functions $\phi(x)$ and $\omega(x)$ can be taken as being defined and continuous on $A(M)$. In the present case $\phi(x)$ will have the property (iii), whereas $\omega(x)$ may not satisfy the left inequality in (iii) although it will satisfy the right inequality. Thus $\Phi(x) = \phi(x) + \omega(x)$ will have all the desired properties.

We shall now prove the following very important theorem, which in the case of asymptotic stability of a closed invariant set M, characterizes the flow in the set $A(M) \setminus M$.

2.12.19 *THEOREM*

If a closed invariant set $M \subset X$ *is asymptotically stable, then for each* $x \in A(M), J^+(x) \subset M$, *and for each* $x \in A(M) \setminus M, J^-(x) \cap A(M) = \emptyset$.

Proof: Let, if possible, $x^* \in A(M)$ and $y \in J^+(x^*)$, $y \notin M$. Set $\rho(y,M) = \alpha\ (> 0)$. Since M is uniformly stable, there is a $\delta > 0$ such that $\gamma^+(S(M,\delta)) \subset S(M, \frac{\alpha}{2})$. Since $x^* \in A(M)$, there is a $T > 0$ such that $x^*T \in S(M,\delta)$. Since $S(M,\delta)$ is open, there is an $\eta > 0$ such that $S(x^*T,\eta) \subset S(M,\delta)$. Now $N \equiv S(x^*T,\eta)\ (-T)$ is a neighborhood of x^* such that for each $x \in N$, $xT \in S(x^*T,\eta)$ and consequently $y[T, +\infty) \subset S(M, \frac{\alpha}{2})$. Now since $y \in J^+(x)$, there exist sequences $\{x_n\}$ in X and $\{t_n\}$ in R, $t_n \to +\infty$, such that $x_n \to x^*$, $x_nt_n \to y$. We may assume without loss of generality, that $x_n \in N$, and $t_n \geq T$.

But then $x_n t_n \in S(M, \frac{\alpha}{2})$. Thus if $x_n t_n \to y$, we must have $\rho(y,M) \leq \frac{\alpha}{2}$. A contradiction as $\rho(y,M) = \alpha$. Thus $J^+(x) \subset M$. The second statement follows from the fact that if $y \in J^-(x)$, then $x \in J^+(y)$. Now let $x \in A(M) \setminus M$, and assume that $y \in J^-(x) \cap A(M)$. Then we have $y \in A(M)$, $x \in J^+(y)$, $x \notin M$, which has already been ruled out.

2.12.20 *COROLLARY*

If a closed invariant set M *is asymptotically stable and* A(M) \ M *(or in particular the space* X) *is locally compact and contains a countable dense subset in it, then the invariant set* A(M) \ M *is parallelizable.*

The proof follows from the above theorem, and Theorem 2.11.22.

2.12.21 *Remark*

The considerations in Section 5.8. show that if X is locally compact and $M \subset X$ is a compact, invariant, asymptotically stable set, then if $\phi(x)$ is a function satisfying conditions of Lemma 2.8.1 in a neighborhood N of M, the sets $\{x \in S[M,\varepsilon]: \phi(x) = \alpha\}$ for fixed $\varepsilon > 0$ such that $S[M,\varepsilon] \subset N$, and α sufficiently small, represent sections of the parallelizable flow in A(M) \ M (See proof of Theorem 2.7.9. How far the same method of construction can be extended to non-compact closed sets, depends naturally on whether the flow in A(M) \ M is parallelizable.

We shall now prove that uniform asymptotic stability of a closed set $M \subset X$ implies that the flow in A(M) \ M is parallelizable, even when the subspace A(M) \ M of X is assumed to be neither locally compact nor separable.

2.12.22 *PROPOSITION*

Let $M \subset X$ *be a closed, invariant, uniformly asymptotically stable set with* A(M) *as its region of attraction. Then* A(M) \ M *is parallelizable.*

Proof: Since M is asymptotically stable, we can find a function $\phi(x)$ defined on $A(M)$ and having the properties given in Theorem 2.12.18. Since M is uniformly asymptotically stable, there is an $\alpha > 0$ such that $S[M,\alpha] \subset A(M)$, and such that for any $\varepsilon > 0$, there is a $T > 0$ with the property that $\gamma^+(xT) \subset S(M,\varepsilon)$ for every $x \in S[M,\alpha]$. Now let

$$m_o = \inf\{\phi(x) : \rho(x,M) = \alpha\} \quad .$$

Indeed $m_o > 0$. Consider now any set

$$S_\eta = \{x \in S[M,\alpha] : \phi(x) = \eta\}.$$

We claim that if $\eta < m_o$, then S_η is a section of the flow in $A(M) \setminus M$ with the property that there is a continuous function $\tau(x), \tau : A(M) \setminus M \to R$, such that for each $x \in A(M) \setminus M$, $\tau(x)$ is unique, and $x\tau(x) \in S_\eta$. The existence of such a section indeed shows that the flow on $A(M) \setminus M$ is parallelizable (Theorem 2.11.20). To see that S_η has the properties enunciated above, we consider the set $P_\eta = \{x \in S[M,\alpha] : \phi(x) \leq \eta\}$. Indeed $P_\eta \subset A(M)$, and $P_\eta \supset M$. We note now that any trajectory in $A(M) \setminus M$ can intersect S_η at most at one point. This is so because if any trajectory in $A(M) \setminus M$ has two points x_1, x_2 on S_η, then we may assume that $x_2 = x_1 t$ where $t > 0$. But then since $S_\eta \cap M = \emptyset$, $\phi(x_2) = \phi(x_1 t) < \phi(x_1)$, which contradicts the definition of S_η. To see that every trajectory γ in $A(M) \setminus M$ intersects S_η we note first that if $x \notin P_\eta$, there is a $t > 0$ such that $xt \in P_\eta$. But then $x[0,t] \cap \partial P_\eta \neq \emptyset$. However, $S_\eta \equiv \partial P_\eta$. If $x \in P_\eta$, then we claim that there is a $t \leq 0$ such that $xt \in S_\eta$. For otherwise $\gamma^-(x) \subset IP_\eta$. In this case we can set $\delta = \inf\{\rho(xt,M) : t \leq 0\}$, and $\delta > 0$ (otherwise M will be unstable). If now $T > 0$ be such that $y \in S[M,\alpha]$ implies $\rho(yT,M) < \frac{\delta}{2}$ (such $T > 0$ exists by uniform asymptotic stability), then $x(-T) = y \in P_\eta \subset S[M,\alpha]$, but $yT = x \notin S(M, \frac{\delta}{2})$. This contradiction shows that every trajectory in $A(M)\setminus M$ intersects S_η exactly once. We now define the function $\tau: A(M)\setminus M \to S_\eta$ by the requirement that $x\tau(x)\in S_\eta$ for $x\in A(M)\setminus M$. Then $\tau(x)$ is uniquely defined and is continuous. The continuity follows in the same way as in the proof of Theorem 2.7.9. We have thus proved our proposition.

We shall now prove the following theorem.

2.12.23 *THEOREM*

A closed set M *is uniformly asymptotically stable and equi-attracting with an open set* N *containing* $S(M,\delta)$ *for some* $\delta > 0$, *as its region of attraction, if and only if there exists a continuous function* $\phi(x)$ *defined on* N *with the following properties:*

(i) $\phi(x) = 0$ *for* $x\in M, \phi(x) > 0$ *for* $x\notin M$;

(ii) there exists strictly increasing continuous functions $\alpha(r)$, $\beta(r)$, $\alpha(0) = \beta(0) = 0$ *such that*

$$\alpha(\rho(x,M)) \leq \phi(x) \leq \beta(\rho(x,M));$$

(iii) there is a sequence of closed sets $\{E_n\}$ *such that* $E_{n+1}\supset I(E_n)\supset S(M,\delta_n)$ *for some* $\delta_n > 0$; $\bigcup\{E_n : n = 1,2,3,\ldots\} = N$, *and this sequence has the property that for any* $\alpha > 0$ *there is an integer* n_0 *such that* $\phi(x) > \alpha$ *for* $x\notin E_{n_0}$;

(iv) $\phi(xt) = e^{-t}\phi(x)$ *for* $x\in N$, $t\in R$.

Proof: We shall not give complete details as the arguments are similar to those used in Section 2.8. For the proof of sufficiency we remark that (iii) and (iv) imply that N is invariant. (i), (ii) and (iv) ensure uniform stability, as well as attraction and show that $N\subset A(M)$. Since N is invariant neighborhood, it follows that $N = A(M)$. Uniform attraction and equi-attraction follow from (ii) and (iv). To prove necessity, we consider the region of attraction $A(M)$ and define a $\phi(x)$

on A(M) as in Theorem 2.12.18. We then consider a section S_η of the flow in $A(M) \setminus M$ defined by this $\phi(x)$ (Proposition 2.12.22), with the corresponding continuous map $\tau: A(M) \setminus M \to R$. Lastly, we define

$$\phi(x) = e^{\tau(x)} \qquad \text{for } x \in A(M) \setminus M,$$

and

$$\phi(x) = 0 \qquad \text{for } x \in M \quad .$$

This $\phi(x)$ is easily shown to have all the properties (i) - (iv). Note that to get the sequence $\{E_n\}$, we set $E_0 = \{x : \phi(x) \leq 1\}$. Then define $E_n = E_0[-n,0]$. These sets are closed and have the required properties.

Setting $\Phi(x) = -\dfrac{\phi(x)}{1 + \phi(x)}$, we obtain the following very important corollary.

2.12.24 *COROLLARY*

A closed invariant set M *is asymptotically stable and equi-attracting with an open set* N *containing* S(M,δ) *for some* $\delta > 0$, *as its region of attraction, if and only if there exists a continuous function* $\Phi(x)$ *defined on* N *with the following properties*

(*i*) $-1 < \Phi(x) < 0$ *for* $x \in N \setminus M$,

(*ii*) $\Phi(x) \to 0$ *as* $\rho(x,M) \to 0$,

(*iii*) *for any* $\varepsilon > 0$ *there is a* $\delta > 0$ *such that* $\Phi(x) \leq -\varepsilon$ *for* $\rho(x,M) \geq \delta$,

(*iv*) $\Phi(x) \to -1$, *as* $x \to y \in \partial N$,

(*v*) $\left.\dfrac{d\Phi(xt)}{dt}\right|_{t=0} = -\Phi(x)\,(1 + \Phi(x))$.

We shall now give a theorem on the lines of the Theorem 2.12.18 for the case of uniform asymptotic stability.

2.12.25 *THEOREM*

Let the space X *be locally compact and separable. Then a closed set* $M \subset X$ *is uniformly asymptotically stable with an open set* N *containing* $S(M,\delta)$ *for some* $\delta > 0$, *if and only if there exists a continuous function* $\phi(x)$ *defined on* N *and having the following properties:*

(i) $\phi(x) = 0$ *for* $x \in M, \phi(x) > 0$ *for* $x \notin M$,

(ii) there exist continuous strictly increasing functions $\alpha(r)$, $\beta(r)$, $\alpha(0) = \beta(0)$, *such that*

$$\alpha(\rho(x,M)) \leq \phi(x) \leq \beta(\rho(x,M)),$$

(iii) there exists a sequence of closed sets $\{E_n\}$, $E_n \subset I\ E_{n+1}$, $\bigcup_{n=1}^{\infty} E_n = N$, *such that given any* $\alpha > 0$ *there is an* n_0 *such that* $\phi(x) > \alpha$ *if* $x \notin E_{n_0}$, *and on every* $E_n, \phi(x)$ *is bounded,*

(iv) $\phi(xt) \leq e^{-t}\phi(x)$.

The conditions can easily be shown to be sufficient. To prove the necessity we need the following lemma.

2.12.26 *LEMMA*

Let $f(r,x)$ *be a function from* $(0,1] \times X \to [0,+\infty)$, *where* X *is locally compact separable metric space. Let* $f(r,x)$ *be bounded on every compact subset of* $(0,1] \times X$. *Then there exist two functions* $H(r)$ *and* $G(x)$ *defined on* $(0,1]$ *and* X *respectively, which are bounded on compact subsets of* $(0,1]$ *and* X *respectively (and may even be chosen continuous), such that*

$$f(r,x) \leq H(r) \cdot G(x) .$$

roof: Since X is locally compact and separable we can find a sequence of ompact sets U_n such that $U_n \subset U_{n+1}$ and $X = \bigcup_{n=1}^{\infty} U_n$. We now define

$$H(r) = \sup\{f(r,x) + 1 : x \in U_{n_o}, \text{ where } 1 + \frac{1}{r} \geq n_o \geq \frac{1}{r}\} \qquad ,$$

nd $G(x) = \sup\{\frac{f(r,x)}{H(r)} : 1 \geq r > 0\}$. The above defined $H(r)$ and $G(x)$ have the equired properties. Indeed $H(r)$ is defined here as a step function.

.12.27 *Proof of necessity of Theorem 2.12.25.* Let $A(M)$ be the region of ttraction of the set M. Since M is an attractor, there is an $\alpha > 0$ such that $(M,\alpha) \subset A(M)$. We might choose $\alpha \leq 1$. For each $r \in (0,\alpha)$ define $(r,x) = \inf\{\tau > 0 : xt \in S(M,r) \text{ for } t \geq \tau\}$. We assert that $T(r,x)$ is bounded on ny compact set $K \subset A(M)$ and for fixed r. To prove this we need to show that for ach compact $K \subset A(M)$, and $r > 0$, there exists a $T > 0$ such that $:\subset S(M,r)$ for $t \geq T$. We note first that by stability of M, there is a > 0 such that $y \in S(M,\delta)$ implies $\gamma^+(y) \subset S(M, r)$. For $x \in A(M)$, choose $T(x)$ ch that $xT(x) \in S(M,\delta)$. Since $S(M,\delta)$ is open, we can choose a $\sigma > 0$ such that $\overline{(xT(x),\sigma)}$ is compact and contained in $S(M,\delta)$. Then its inverse image

$$N_x = \overline{S(xT(x),\sigma)} \cdot (-T(x))$$

a compact neighborhood of x. N_x has, moreover, the property, that $T(x) \subset S(M,\delta)$. Thus we have in fact shown that for each $x \in A(M)$, there exists a $x)$ and a $\rho(x) > 0$, such that $y \in S(x, \rho(x))$ implies $yt \in S(M, r)$ for $\geq T(x)$. Consider now the open cover $\{S(x, \rho(x)) : x \in K\}$ of the compact set K. the Borel Theorem, there exist a finite number of sets, say, $x_1, \rho(x_1)), \ldots, S(x_n, \rho(x_n))$ which cover K. We can now choose $T = \max(T(x_1), \ldots, T(x_n))$ en $x \in K$ implies $xt \in S(M,r)$ for $t \geq T$.

For any given integer $n > \frac{1}{\alpha}$, define

$$\phi_n(x) = \sup\{\rho(x\tau, S(M, \tfrac{1}{n})) \cdot \exp(\tau) : \tau \geq 0\}.$$

We assert that $\phi_n(x)$ is continuous on $A(M)$. To see this, note that for $\rho > 0$ such that $\overline{S(x,\rho)}$ is a compact subset of $A(M)$, there exists a $T > 0$ such that

$$\rho(y\tau, S(M, \tfrac{1}{n})) = 0$$

for $y \in S(x,\rho)$ and $\tau \geq T$. Therefore, if $y \in S(x,\rho)$ we have

$$|\phi_n(x) - \phi_n(y)| = |\sup\{\rho(x\tau, S(M, \tfrac{1}{n})) \cdot \exp(\tau) : 0 \leq \tau \leq T\}$$

$$- \sup\{\rho(y\tau, S(M, \tfrac{1}{n})) \cdot \exp(\tau) : 0 \leq \tau \leq T\}|$$

$$\leq \exp(T) \cdot \sup\{|\rho(x\tau, S(M, \tfrac{1}{n})) - \rho(y\tau, S(M, \tfrac{1}{n}))| : 0 \leq \tau \leq T\} \quad .$$

This implies that

$$|\phi_n(x) - \phi_n(y)| \leq \exp(T) \cdot \sup\{\rho(x\tau, y\tau) : 0 \leq \tau \leq T\} \quad .$$

Using the continuity axiom we conclude that the right hand side tends to zero as $\rho(x,y) \to 0$. Thus $\phi_n(x)$ is continuous on $A(M)$. This $\phi_n(x)$ has further the followi[ng] important property

$$\phi_n(xt) \leq \exp(-t)\, \phi_n(x), \qquad \text{for } t > 0 \quad .$$

To see this, note that for $t \geq 0$

$$\phi_n(xt) = \sup\{\rho(x(t+\tau), S(M, \tfrac{1}{n}))\ \exp\ (\tau): \tau \geq 0\}$$

$$= \sup\{\rho(x\tau, S(M, \tfrac{1}{n}))\ \exp\ (\tau - t): \tau \geq t\}$$

$$= \exp\ (-t) \cdot \sup\{\rho(x\tau, S(M, \tfrac{1}{n}))\ \exp\ (\tau): \tau \geq t\}$$

$$\leq \exp\ (-t) \cdot \phi_n(x) \quad \text{as} \quad t \geq 0 \quad .$$

We now note that

$$\phi_n(x) = \sup\{\rho\ (x\tau, S(M, \tfrac{1}{n}))\ \exp\ (\tau): 0 \leq \tau \leq T(\tfrac{1}{n}, x)\}$$

as $\rho(x\tau, S(M, \frac{1}{n})) = 0$ for $\tau \geq T(\frac{1}{n}, x)$. Thus

$$\phi_n(x) \leq \exp(T(\tfrac{1}{n},x))\ \sup\{\rho(x\tau, S(M, \tfrac{1}{n})): \ \tau \geq 0\} \quad .$$

Since the function $\exp(T(\frac{1}{n},x))$ has the properties of the function $f(r,x)$ of Lemma 2.12.26, we can choose a function $H(r)$, such that $\phi_n(x) / H(\frac{1}{n})$ is uniformly bounded on each compact subset $K \subset A(M)$.

We now define

$$\Phi(x) = \sum_{n=n_0}^{\infty} \phi_n(x) / H(\tfrac{1}{n})n!, \quad \text{where} \quad n_0 \geq \frac{1}{\alpha} \quad .$$

Then $\Phi(x)$ is continuous on $A(M)$ and has

$$\Phi(xt) \leq \exp\ (-t)\ \Phi(x) \quad .$$

Note that $\Phi(x) = 0$ for $x \in M$, and $\Phi(x) > 0$ for $x \notin M$. By uniform asymptotic stability, $\Phi(x) \to 0$ if $\rho(x,M) \to 0$, there is thus a strictly increasing function $\beta(r)$, $\beta(0) = 0$ such that

$$\Phi(x) \leq \beta(\rho(x,M)) \quad .$$

Further if $\rho(x,M) \geq \varepsilon > 0$, then for sufficiently large n, $\phi_n(x) \geq \delta_n > 0$ for some δ_n. And hence $\Phi(x) > \delta > 0$. Thus there is a strictly increasing function $\alpha(r)$, $\alpha(0) = 0$, such that $\Phi(x) \geq \alpha(\rho(x,M))$. We now choose $k > 0$ such that

$$k < \inf\{\Phi(x): \rho(x,M) = \alpha\} \quad .$$

Consider the sets

$$P_k = \{x \in A(M): \Phi(x) < k\} \cap S[M,\alpha],$$

and

$$S_k = \{x \in A(M): \Phi(x) = k\} \cap S[M,\alpha] \quad .$$

Then as shown in Theorem 2.12.23, S_k is the section of the flow in $A(M)$ consisting of all those trajectories in $A(M)$ which are not in M. For each $x \in A(M)$, $x \notin P_k$ we can define $\tau(x)$ by the requirement that $x\tau(x) \in S_k$. Then $\tau(x)$ is continuous. Now define

$$\phi(x) = \Phi(x) \qquad \text{for } x \in P_k \quad ,$$

and

$$\phi(x) = ke^{\tau(x)} \qquad \text{for } x \notin P_k \quad .$$

This $\phi(x)$ has all the properties required in the theorem as may easily be verified.

2.12.27 *Remark*

Note that if in the above theorem we assume M to be invariant and would construct $\phi(x)$ as in Theorem 2.12.23, then this $\phi(x)$ need not satisfy the property (ii) in the above theorem. Indeed if that were the case, then construction of the $\phi_n(x)$ would be superfluous and then uniform asymptotic stability will imply equi-attraction which is indeed not the case.

2.12.28 *Notes and references*

The notion of equi-stability seems to be new. Theorem 2.12.19 seems to pave the way for the use of the theory of parallelizable flows in studying various problems on asymptotic stability, especially its connection with the existence of the so-called Liapunov Functions. The exposition in this section is not complete, but is more general than that of Antosiewicz and Dugundji.

Theorem 2.12.23 similar but better than a well known theorem of Zubov ([6], Translation page 52), and is in line with results in Section 2.7. See also end of Section 2.7 for further notes.

2.13 *Higher prolongations and stability*

The first positive prolongation, and the first positive prolongational limit set have been shown to be useful in characterizing various concepts in dynamical systems. Notable applications being the characterization of stability of a compact set in a locally compact metric space, and the characterization of a dispersive flow. The first positive prolongation may be thought of as an extension of the positive semi-trajectory. For example consider a dynamical system (E^2,R,π) which is geometrically described by the following figure

2.13.1 *Figure*

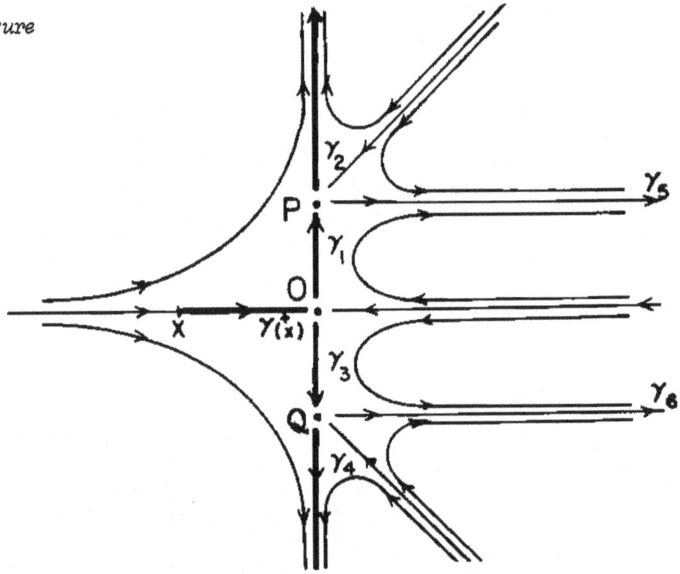

The first positive prolongation of the point x in the figure consists of the semi-positive trajectory $\gamma^+(x)$, the equilibrium points $0,P,$ and $Q,$ and the trajectories $\gamma_1,\gamma_2,\gamma_3,$ and γ_4. In a way, to get the prolongation of a point, we migh find ourselves arguing that we move along the positive semi-trajectory and approach the equilibrium point 0. So we transfer to the point 0. From 0 we transfer to a trajectory which leaves 0, e.g., we can transfer to γ_1 or γ_3. If we transfer to γ_1, then we approach the equilibrium point P. So we transfer to P, and thence to a trajectory leaving P, and so on. If indeed this procedure were laid down to define the prolongation of a point, then notice that we would have to include the trajectories γ_5 and γ_6 in the prolongation of x. The definition of a prolongation, however, excludes the trajectories γ_5 and γ_6 from the prolongation of x. If, however, we wished to include these in a prolongation, then either we must change the definition of prolongation, or in a sense introduce other prolongations which will do precisely what we did with the intuitive reasoning above. Just as the first positive prolongation is in fact a meaningful extension of the positive trajectory, the 2nd and higher prolongations which will be presently introduced, will be shown to be meaningful extensions of the first prolongation.

The description of higher prolongations is facilitated by the introduction of two operations $\mathcal{S}$ and $\mathcal{D}$ on the class of maps from X into 2^X.

2.13.2 *The operators* $\mathcal{S}$ *and* $\mathcal{D}$.

If $\Gamma : X \to 2^X$, we define $\mathcal{D}\Gamma$ by

2.13.3 $$\mathcal{D}\Gamma(x) = \bigcap \{\overline{\Gamma(U)} : U \in \mathcal{N}(x)\} ,$$

where $\mathcal{N}(x)$ denotes the neighborhood filter of x. Further, $\mathcal{S}\Gamma$ is defined by

2.13.4 $$\mathcal{S}\Gamma(x) = \bigcup \{\Gamma^n(x) : n = 1,2,\ldots\} ,$$

where $\Gamma^1(x) = \Gamma(x)$, and $\Gamma^n(x) = \Gamma(\Gamma^{(n-1)}(x))$, $n = 2,3,\ldots$.

In the sequel the following lemma will prove useful.

2.13.5 *LEMMA*

For any $x \in X$, *and* $\Gamma : X \to 2^X$, $\mathcal{D}\Gamma(x)$ *is the set of all points* $y \in X$ *such that there are sequences* $\{x_n\}, \{y_n\}$ *in* X, $y_n \in \Gamma(x_n)$, *and* $x_n \to x$, $y_n \to y$. *Further,* $\mathcal{S}\Gamma(x)$ *is the set of all points* $y \in X$ *such that there are points* $x_1, x_2, \ldots, x_k$, *with the property that* $x_{i+1} \in \Gamma(x_i), i = 1,2,\ldots,k-1$, *and* $x_1 = x$, $x_k = y$.

The proof is immediate and is left to the reader.

The following lemma gives some elementary properties of the operators $\mathcal{D}$ and S.

2.13.6 *LEMMA*

(a) $\mathcal{D}^2 = \mathcal{D}$, *and* $S^2 = S$. *Thus* $\mathcal{D}$ *and* S *are idempotent operators,*

(b) If $M \subset X$ *is compact, then*

$$\mathcal{D}\Gamma(M) = \bigcup\{\mathcal{D}\Gamma(x) : x \in M\}$$

is closed,

(c) If $v = \phi(x)$ *is a continuous real-valued function on* X, *such that* $\phi(y) \leq \phi(x)$, *whenever* $y \in \Gamma(x)$, *then* $\phi(y) \leq \phi(x)$ *whenever* $y \in \mathcal{D}\Gamma(x) \cup S\Gamma(x)$.

Proof: (a) Let $\Gamma: X \to 2^X$. If $y \in \mathcal{D}\Gamma(x)$, then indeed $y \in \mathcal{D}\mathcal{D}\Gamma(x)$. Thus $\mathcal{D}\Gamma(x) \subset \mathcal{D}\mathcal{D}\Gamma(x)$. If $y \in \mathcal{D}\mathcal{D}\Gamma(x)$, then there are sequences $\{x_n\}$, $\{y_n\}$, $y_n \in \mathcal{D}\Gamma(x)$, such that $x_n \to x$ and $y_n \to y$. For each x_k, y_k, $y_k \in \mathcal{D}\Gamma(x_k)$, there are sequences $\{x_k^n\}$, $\{y_k^n\}$, $y_k^n \in \Gamma(x_k^n)$, such that $x_k^n \to x_k$, $y_k^n \to y_k$. Now consider the sequences $\{x_n^n\}$, $\{y_n^n\}$. Clearly $y_n^n \in \Gamma(x_n^n)$. Further $x_n^n \to x$, and $y_n^n \to y$. Thus $y \in \mathcal{D}\Gamma(x)$. We have thus proved that $\mathcal{D}\mathcal{D}\Gamma(x) \subset \mathcal{D}\Gamma(x)$. This together with the previous observation shows that $\mathcal{D}\mathcal{D}\Gamma(x) = \mathcal{D}\Gamma(x)$. Hence $\mathcal{D}^2 = \mathcal{D}$. Proof of $S^2 = S$ is even simpler and is left as an exercise.

(b) Let $\{y_n\}$ be a sequence in $\mathcal{D}\Gamma(M)$ such that $y_n \to y$. Then there is a sequence $\{x_n\}$ in M such that $y_n \in \mathcal{D}\Gamma(x_n)$. Since M is compact, we may assume that $x_n \to x \in M$. Thus $y \in \mathcal{D}\mathcal{D}\Gamma(x) = \mathcal{D}\Gamma(x) \subset \mathcal{D}\Gamma(M)$. This shows that $\mathcal{D}\Gamma(M)$ is closed.

(c) If $y \in \mathcal{D}\Gamma(x)$, then there are sequences $\{x_n\}$, $\{y_n\}$, $y_n \in \Gamma(x_n)$, such that $x_n \to x$, and $y_n \to y$. It is given that $\phi(y_n) \leq \phi(x_n)$. Since ϕ is continuous we get by proceeding to the limit $\phi(y) \leq \phi(x)$. If $y \in S\Gamma(x)$, then there are points $x = x_1, x_2, \ldots, x_n = y$ such that $x_{i+1} \in \Gamma(x_i), i = 1,2,\ldots,n-1$. Hence $\phi(y) = \phi(x_n) \leq \phi(x_{n-1}) \leq \ldots \leq \phi(x_2) \leq \phi(x_1) = \phi(x)$. This completes the proof of the lemma.

2.13.7 *DEFINITION*

A map $\Gamma: X \to 2^X$ *will be called* *<u>transitive</u>, if* $S\Gamma = \Gamma$.

2.13.8 *EXERCISE*

(a) A map $\Gamma: X \to 2^X$ is transitive if and only if $\Gamma^2 = \Gamma$.

(b) Given $\Gamma: X \to 2^X$, $S\Gamma$ is transitive.

2.13.9 *DEFINITION*

A map $\Gamma: X \to 2^X$ *will be called a <u>cluster map</u> if* $\mathcal{D}\Gamma = \Gamma$.

2.13.10 *DEFINITION*

A map $\Gamma: X \to 2^X$ *will be called a* <u>c-c map</u> *if it has the following property: For any compact set* $K \subset X$ *such that* $x \in K$, *one has either* $\Gamma(x) \subset K$, *or* $\Gamma(x) \cap \partial K \neq \emptyset$.

2.13.11 *THEOREM*

Let the space X *be locally compact. Let* Γ *be a* c-c *map. Then if* $\Gamma(x)$ *is compact, then it is connected.*

Proof: If $\Gamma(x)$ is compact, but not connected, then we can write $\Gamma(x) = M_1 \cup M_2$, where M_1, M_2 are non-empty compact disjoint sets. Since X is locally compact, we can choose compact neighborhoods U_1, U_2 of M_1, M_2 respectively such that $U_1 \cap U_2 = \emptyset$. Note that $\Gamma(x) \cap U_1 \neq \emptyset$, but $U_1 \not\supset \Gamma(x)$, as $M_2 \cap U_1 \neq \emptyset$. However, $\partial U_1 \cap \Gamma(x) = \emptyset$, contradicting the fact that $\Gamma(x)$ is a c-c map. Thus $\Gamma(x)$ is connected.

2.13.12 *THEOREM*

Let the space X *be locally compact. Let* Γ *be a* c-c *map. If* $M \subset X$ *is compact, then* $\mathcal{D}\Gamma(M) = M$ *if and only if for each neighborhood* U *of* M, *there is a neighborhood* W *of* M *such that* $\Gamma(W) \subset U$.

Proof: Sufficiency. Note that for any c-c map Γ, $x \in \Gamma(x)$. Hence $\mathcal{D}\Gamma(M) \supset M$ always. Consider now a sequence of closed neighborhoods $\{U_n\}$ of M, such that $\bigcap U_n = M$. Then there is a sequence of neighborhoods $\{W_n\}$ such that $\overline{\Gamma(W_n)} \subset U_n$. Then $\mathcal{D}\Gamma(M) \subset \bigcap \overline{\Gamma(W_n)} \subset \bigcap U_n = M$. Thus we have proved that $\mathcal{D}\Gamma(M) = M$, and sufficiency is proved.

Necessity. Indeed assume, if possible, that there is a neighborhood U of M, such that for every neighborhood W of M, $\Gamma(W) \not\subset U$. We may assume without loss of generality that U is compact (because X is locally compact). Then there is a sequence $\{x_n\}$, $x_n \to x \in M$, and a sequence $\{y_n\}$, $y_n \in \Gamma(x_n)$ such that $y_n \notin U, n = 1,2,\ldots$. Indeed we may assume that $x_n \in U$, $n = 1,2,\ldots$. But then since Γ is a c-c map and $\Gamma(x_n) \not\subset U$, we must have $\Gamma(x_n) \cap \partial U \neq \emptyset$. Consequently, there is a sequence $\{z_n\}$, $z_n \in \Gamma(x_n) \cap \partial U$, $n = 1,2,\ldots$. Since ∂U is compact, we may assume that $z_n \to z \in \partial U$. But then $z \in \mathcal{D}\Gamma(x) \subset \mathcal{D}\Gamma(M)$, but $z \notin M$. A contradiction, as $\mathcal{D}\Gamma(M) = M$. This proves necessity and the theorem is proved.

We remark now that c-c maps have also the following properties, which help build families of c-c maps.

2.13.13 *LEMMA*

(i) Let $\{\Gamma_\alpha\}$, $\alpha \in A$, *be a family of* c-c *maps. Then* $\Gamma = \bigcup \Gamma_\alpha$ *is a* c-c *map.*

(ii) If Γ_1, Γ_2 *are* c-c *maps, then so is the map* $\Gamma = \Gamma_1 \circ \Gamma_2$.

(iii) If Γ *is a* c-c *map, then so are* $\mathcal{S}\Gamma$ *and* $\mathcal{D}\Gamma$.

Proof. (i) Let K be a compact set, and $x \notin K$. If $\Gamma(x) \not\subset K$, we must prove that $\Gamma(x) \cap \partial K \neq \emptyset$. Indeed if $\Gamma(x) \not\subset K$, then there is at least one map, say Γ_α, such that $\Gamma_\alpha(x) \not\subset K$. Since $x \in K$, and Γ_α is a c-c map, we must have $\Gamma_\alpha(x) \cap \partial K \neq \emptyset$. Thus $\Gamma(x) \cap \partial K \neq \emptyset$, and so Γ is a c-c map. (ii) We need consider the case that $\Gamma_2(x) \subset I(K)$, where $x \notin K$, and K compact. This is so, because $\Gamma_1 \circ \Gamma_2(x) = \Gamma_1(\Gamma_2(x)) \supset \Gamma_2(x)$. If $\Gamma_1 \circ \Gamma_2(x) \not\subset K$, and $\Gamma_2(x) \subset I(K)$, then there is a $y \in \Gamma_2(x)$ such that $\Gamma_1(y) \not\subset K$. But then $\Gamma_1(y) \cap \partial K \neq \emptyset$, and since $\Gamma_1(y) \subset \Gamma_1(\Gamma_2(x)) = \Gamma_1 \circ \Gamma_2(x)$, we have $\Gamma_1 \circ \Gamma_2(x) \cap \partial K \neq \emptyset$. Thus $\Gamma = \Gamma_1 \circ \Gamma_2$ is a c-c map. (iii) If Γ is a c-c map, then $\mathcal{S}\Gamma$ is indeed a c-c map by the assertions (i) and (ii). To show that $\mathcal{D}\Gamma$ is a c-c map. Let $x \in K$, where K is compact. If $\mathcal{D}\Gamma(x) \not\subset K$, we may assume without loss of generality that $x \in I(K)$. If $\mathcal{D}\Gamma(x) \not\subset K$, then there is a sequence $\{x_n\}$ in K, $x_n \to x$, and a sequence $\{y_n\}$, $y_n \notin K$, $y_n \to y \notin K$, $y_n \in \Gamma(x_n)$. Since $\Gamma(x_n) \not\subset K$, and $x_n \in K$, there is a sequence $\{z_n\}$, $z_n \in \Gamma(x_n)$, and $z_n \in \partial K$. Since ∂K is compact, we may assume that $z_n \to z \in \partial K$. But then $z \in \mathcal{D}\Gamma(x)$, so that $\mathcal{D}\Gamma(x) \cap \partial K \neq \emptyset$. This proves that $\mathcal{D}\Gamma$ is a c-c map.

We now prove the following interesting theorem.

2.13.14 *THEOREM*

Let X *be locally compact, and* M *a compact subset of* X. *Let* Γ *be a* c-c *map which is moreover a transitive map as well as a cluster map. Then* $\Gamma(M) = M$ *if and only if there exists a fundamental system of compact neighborhoods* $\{U_n\}$ *of* M *such that* $\Gamma(U_n) = U_n$.

Proof: Let $\{W_n\}$ be a fundamental system of neighborhoods of M. Since $\mathcal{D}\Gamma(M) = \Gamma(M) = M$, by Theorem 2.13.12, we have compact neighborhoods K_n of M such that $\Gamma(K_n) \subset W_n$. Now notice that $\Gamma(K_n)$ are closed and may be considered as compact. Setting $\Gamma(K_n) = U_n$, we get a fundamental system of compact neighborhoods of M such that $\Gamma(U_n) = U_n$. This is so because $\Gamma(U_n) = \Gamma(\Gamma(K_n)) = \Gamma(K_n) = U_n$, as Γ is transitive.

We shall now apply the theory constructed above to a dynamical system (X,R,π).

2.13.15 *The higher prolongations (Definition).*

Consider the map $\gamma^+ : X \to 2^X$ which defines the positive semi-trajectory through each point $x \in X$. Then since $\gamma^+(x)$ is connected, γ^+ is a c-c map. Further, notice that $\gamma^+(\gamma^+(x)) = \gamma^+(x)$, so that γ^+ is a transitive map, i.e., $S\gamma^+ = \gamma^+$. We now set $\mathcal{D}S\gamma^+ \equiv \mathcal{D}\gamma^+ = D_1^+$, and call $D_1^+(x)$ as the first positive prolongation of x. This is clearly the same as defined in Section 2.3, where it is denoted by D^+. Indeed D_1^+ is a cluster map as $\mathcal{D}$ is idempotent, but D_1^+ is not transitive as simple examples will show. We, therefore, consider the map $\mathcal{D}SD_1^+$ and denote it by D_2^+ and call $D_2^+(x)$ as the 2nd prolongation of x. Having defined D_n^+, we define $D_{n+1}^+ = \mathcal{D}SD_n^+$, and call $D_n^+(x)$ to be the nth-prolongation of x. This defines a prolongation of x for any positive integer n. Using transfinite induction, we define a prolongation $D_\alpha^+(x)$ of x for any ordinal number α as follows: If α is a successor ordinal, then having defined $D_{\alpha-1}^+$, we set $D_\alpha^+ = \mathcal{D}SD_{\alpha-1}^+$. If α is not a successor ordinal, then having defined D_β^+ for every $\beta < \alpha$, we set $D_\alpha^+ = \mathcal{D} \bigcup_{\beta < \alpha} SD^+$. This defines for each $x \in X$, a prolongation $D_\alpha^+(x)$ for every ordinal α.

Notice that each of the map considered above is a c-c map. Moreover, the prolongations $D_\alpha^+(x)$ for any ordinal α are closed.

We give below some properties of these prolongations.

2.13.16 *THEOREM*

Let Ω *be the first uncountable ordinal number. Then*

(i) $D_\Omega^+ = \{D_\alpha^+ : \alpha < \Omega\}$

(ii) D_Ω^+ *is a transitive map.*

Proof: Recall that $D^+_\Omega = \mathcal{D} \bigcup \{SD^+_\alpha : \alpha < \Omega\}$, as Ω is not a successor ordinal. Let for any $x \in X$, $y \in D^+_\Omega(x)$. Then there are sequences $\{x_n\}$, $\{y_n\}$, $x_n \to y$, $y_n \to y$, and $y_n \in \Gamma_{\alpha_n}(x_n)$, where $\Gamma_{\alpha_n} = SD^+_{\alpha_n}$ and α_n is some ordinal number, $\alpha_n < \Omega$. Let β be an ordinal number such that $\alpha_n < \beta < \beta + 1 < \Omega$ (such ordinals exist). Then indeed $y_n \in SD^+_\beta(x_n)$ for each n, so that $y \in \mathcal{D}SD^+_\beta(x) = D^+_{\beta+1}(x)$. Consequently $y \in \Gamma(x)$ where Γ stands for $\bigcup\{D^+_\alpha : \alpha < \Omega\}$. Indeed for any $\alpha < \Omega$, $D^+_\alpha(x) \subset D^+_\Omega(x)$, so that $\Gamma(x) \subset D^+_\Omega(x)$. This proves (i). To prove that D^+_Ω is transitive, we need show only that $D^+_\Omega \circ D^+_\Omega = D^+_\Omega$. Suppose that $y \in D^+_\Omega(x)$, and $z \in D^+_\Omega(y)$, so that $z \in D^+_\Omega \circ D^+_\Omega(x)$. Then there is a $\beta < \Omega$ such that $y \in D^+_\beta(x)$, and $z \in D^+_\beta(y)$. But then $z \in SD^+_\beta(x) \subset \mathcal{D}SD^+_\beta(x) = D^+_{\beta+1}(x) \subset D^+_\Omega(x)$. Hence $SD^+_\Omega = D^+_\Omega$. The theorem is proved.

2.13.17 *COROLLARY*

If $\alpha > \Omega$, *then* $D^+_\alpha = D^+_\Omega$. *This follows by induction as* D^+_Ω *is transitive, and indeed also a cluster map.*

We shall now define a host of stability concepts with the help of the higher prolongations introduced above.

2.13.18 *DEFINITION*

Let X *be locally compact. Let* M *be a compact subset of* X. *The set* M *will be called stable of order* α, *or* α*-stable, if* $D^+_\alpha(M) = M$. *If* M *is stable of order* α, *for every ordinal number* α, *then* M *is said to be* *absolutely stable.*

2.13.19 *REMARK*

Stability of order 1, is the same thing as stability defined in Section 2.6.1, as is evident from Theorems 2.6.5 and 2.6.6. Note also that absolute stability is the same as stability of order Ω, where Ω is the first uncountable ordinal, as is clear from Theorem 2.13.16 and the corollary 2.13.17.

2.13.20 *THEOREM*

A compact set $M \subset X$, *where* X *is locally compact, is* α-*stable, if and only if for every neighborhood* U *of* M, *there exists a neighborhood* W *of* M, *such that* $D^+_\alpha(W) \subset U$.

We now give a few simple examples to illustrate that the various higher prolongations introduced above are indeed different concepts.

2.13.21 *EXAMPLE*

Consider a dynamical system defined on the real line. The points of the form $\pm\frac{n}{1+n}$, $n = 0,1,2,\ldots,$ are equilibrium points, and so are the points -1, and $+1$. Between any two successive (isolated) equilibrium points p,q, such that $p < q$, there is a single trajectory which has q as its positive limit point, and p as its negative limit point. There is a single trajectory with -1 as its sole positive limit point, and it has no negative limit points, and there is a single trajectory with $+1$ as its negative limit point, and it has no positive limit points. See figure 2.13.22. For any point P such that $P < -1$. $D^+_1(P) = \{x \in R: P \leq x \leq -1\}$. $D^+_2(P) = D^+_1(P)$. But $D^+_3(P) = \{x \in R: P \leq x \leq +1\}$, and $D^+_4(P) = \{x \in R: P \leq x\}$. Note also that if P is the point -1, then $D^+_1(P) = P$, $D^+_2 = \{x \in R: -1 \leq x \leq 1\}$, $D^+_3(P) = \{x: x \geq -1\}$.

2.13.22 *Figure*

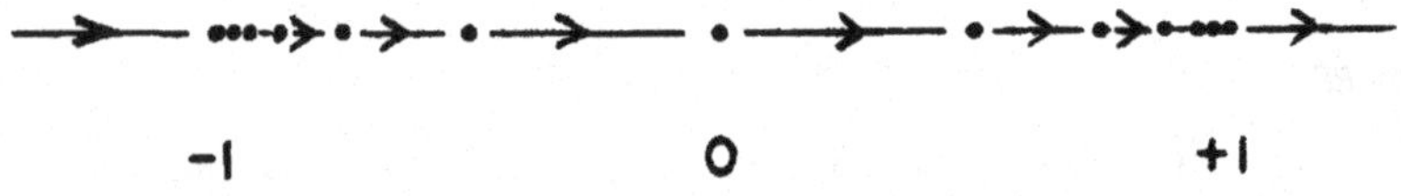

2.13.23 *EXAMPLE*

Consider again a flow on the real line, such that we have the equilibrium points as in the above example, and, moreover, between any two such successive equilibrium points, say q and p, there are two sequences of equilibrium points, say $\{p_n\}$, and $\{q_n\}$, $\ldots q_n \leq q_{n-1} \ldots \leq q_1 \leq p_1 \leq p_2 \leq \ldots$, $p_n \to p$, and $q_n \to q$. Then direction of motion on a trajectory between any two equilibrium points is again from left to right, as in the previous example. In this case, if we consider the point $P = -1$, then indeed $D_1^+(P) = P$, $D_2^+(P) = P$, but $D_3^+(P) = \{x \in R: -1 \leq x \leq +1\}$, and $D_4^+(P) = \{x \in R: -1 \leq x\}$.

Proceeding in this fashion it is easy to see that we can construct examples on the real line in which a point is stable of order n (n integer), but is not stable of order $n + 1$.

2.13.24 *EXAMPLE*

We now give an example of a dynamical system defined on the real line, in which an equilibrium point is stable of every integral order n, but not stable of order ω, where ω is the first countable ordinal. To obtain such an example we consider a sequence of points $\{P_n\}$, $P_n \to 0$, $P_1 > P_2 > P_3 > \ldots > 0$. To the right of P_1, we introduce a sequence $\{P_1^n\}$, such that $P_1^1 > P_1^2 > \ldots > P_1$, and $P_1^n \to P_1$. Between P_1 and P_2, we first introduce a sequence $\{P_{2n}\}$, $P_1 > P_{21} > P_{22} > P_{23} > \ldots > P_2$, $P_{2n} \to P_2$. Then for each P_{2k}, we introduce a sequence $\{P_{2k}^n\}$ between P_{2k} and $P_{2(k-1)}$, such that, $P_{2k}^1 > P_{2k}^2 > \ldots > P_{2k}$, and $P_{2k}^n \to P_{2k}$. Between P_2 and P_3 we first introduce a sequence of points as between P_1 and P_2, and then between any two successive points we introduce a monotone decreasing sequence converging to the point on the left. We now proceed inductively. Having introduced a suitable sequence between say P_{n-1} and P_n, we introduce a sequence between P_n and P_{n+1} similar to the one introduced between P_{n-1} and P_n, then between each pair of successive points of this sequence, we introduce a monotonic decreasing sequence converging to the point on the left. Now we are ready to introduce the dynamical system on the real line. Each point of the countable set of points introduced on the line is an equilibrium point. There are no other equilibrium point, and the motion between any two successive equilibrium points is from left to right. It is easy to see, that each point $\{P_n\}$ of the first sequence introduced above has the following property. P_1 is not stable, P_2 is stable of order 1, but not stable of order 2, P_3 is stable of order 2 (and hence also of order 1), but is not stable of order 3, P_{n+1} is stable of order n but not stable of order $n+1$. The point 0 is stable of every integral order n, but is not stable of order ω.

If we consider example 2.13.21, then it is an easy matter to show that no <u>continuous</u> scalar function satisfying conditions of Theorem 2.12.10 exists for the uniformly stable equilibrium point -1. An example in the plane, e.g., example 1.5.32(v) and figure 1.5.35 can be used to establish the same thing. In fact, even the point 0 in example 2.13.24 which is stable of every integral order n is such that no <u>continuous</u> function satisfying conditions of Theorem 2.12.10 can exist for this point. The question obviously arises, as to what are the implications of the existence of a continuous function satisfying the conditions of Theorem 2.12.10 for a given closed set M. The answer for a compact set M in locally compact spaces X is given by the following theorem.

2.13.25 *THEOREM*

Let X *be locally compact, and let* $M \subset X$ *be compact. Then the following are equivalent:*

(i) There is a real-valued function satisfying conditions of Theorem 2.12.10 *which is continuous in some neighborhood of* M,

(ii) M *possesses a fundamental system of absolutely stable compact neighborhoods,*

(iii) M *is absolutely stable.*

We shall need the following lemma, whose proof is immediate from the definitions.

2.13.26 *LEMMA*

Let $v = \phi(x)$ *be a real valued function satisfying conditions of Theorem* 2.12.10. *If* M *is compact, and the space* X *is locally compact, then the set* $\{U_\alpha : \alpha > 0\}$ *is a fundamental system of neighborhoods of* M, *where* $U_\alpha = \{x \in X : \phi(x) \leq \alpha\}$.

Proof of Theorem 2.13.25: (i) implies (ii). Let U be a compact neighborhood of M. Let $m_0 = \min\{\phi(x) : x \in \partial U\}$. Then $m_0 > 0$, and $\{U_\alpha : 0 < \alpha < m_0\}$, where $U_\alpha = \{x \in X : \phi(x) \leq \alpha\}$, is a fundamental system of compact, positively invariant neighborhoods of M. We will now show that each U_α is absolutely stable. We shall show this by using lemma 2.13.6(c). To do this, we consider the function $\Phi(x)$ defined on X, by means of $\Phi(x) = \phi(x)$ for $x \in U_{m_0}$, and $\Phi(x) = m_0$ for $x \notin U_{m_0}$. This is a continuous function which is decreasing along the trajectories. For $0 < \alpha < \beta < m_0$, U_β is indeed a compact neighborhood of U_α. Since $\Phi(x)$ is decreasing along the trajectories, we get $y \in D^+_\Omega(x)$, then $\Phi(y) \leq \Phi(x)$. If $D^+_\Omega(U_\alpha) \neq U_\alpha$, then there is a $\beta > 0$ such that $D^+_\Omega U_\alpha \not\subset U_\beta$. Since D^+_Ω is a c-c map, there is an $x \in U_\alpha$, and a $y \in D^+_\Omega(x) \cap \partial U_\beta$. On one hand, therefore, $\Phi(y) \leq \Phi(x) \leq \alpha$, and, on the other hand, $\Phi(y) = \beta > \alpha$. This contradiction shows that $D^+_\Omega(U_\alpha) = U_\alpha$, i.e., each U_α is absolutely stable.

(ii) implies (iii). This is immediate.

(iii) implies (i). Using Theorem 2.13.14 (since D^+_Ω is a c-c map which is moreover a transitive as well as a cluster map) we first construct a fundamental system of absolutely stable neighborhoods $U_{\frac{1}{2^n}}$, n = 0,1,2,..., such that $U_{\frac{1}{2^n}} \subset I(U_{\frac{1}{2^{n-1}}})$. We now extend this system of absolutely stable compact neighborhoods to one defined over the diadic rationals, i.e., numbers of the type $\alpha = j/2^n$, $n = 0,1,2,\ldots;\ j = 1,2,\ldots,2^n$, in such a way that (a) the compact neighborhood corresponding to any diadic rational is absolutely stable, (b) if $\alpha < \beta$ are diadic rationals, then $U_\alpha \subset I(U_\beta)$, (c) $M = \bigcap\{U_\alpha : \alpha$ diadic rational$\}$. Indeed this is possible by using Theorem 2.13.14. Now if $x \in U_1$, define $v = \phi(x) = \inf\{\alpha : x \in U_\alpha$, α diadic rational$\}$. Clearly $\phi(x) = 0$ if and only if $x \in M$. If $t > 0$, then $\phi(xt) \leq \phi(x)$. This is so, because if $x \in U_\alpha$, then since U_α is positively invariant, we have $xt \in U_\alpha$, hence $\phi(xt) \leq \phi(x)$. Finally, to see that $\phi(x)$ is continuous on U_1, we assume that this is not true. Then there is an $x \in U_\alpha$, and a sequence $\{x_n\}$ in U_1 such that $\phi(x_n) \to \alpha \neq \phi(x) = \alpha_x$. If $\alpha < \alpha_x$, then we can choose diadic rationals α_1, α_2, such that $\alpha < \alpha_1 < \alpha_2 < \alpha_x$. Then for large x_n, $x_n \in U_{\alpha_1}$, whereas $x \notin U_{\alpha_2}$. Since U_{α_1} is closed, and $x_n \to x$, $x \in U_{\alpha_1}$. This is a contradiction as $U_{\alpha_1} \subset I(U_{\alpha_2})$. If again $\alpha > \alpha_x$, then choose diadic rationals α_1, α_2, such that $\alpha > \alpha_1 > \alpha_2 > \alpha_x$. Then $x_n \notin U_{\alpha_1}$ for large n, whereas $x \in U_{\alpha_2}$. But $U_{\alpha_2} \subset I(U_{\alpha_1})$, which contradicts $x_n \to x$. This completes the proof of the theorem.

2.13.27 *Prolongations and stability of closed sets.*

Although Theorem 2.6.6 gives an excellent characterization of Liapunov stability of compact sets in locally compact spaces, a similar characterization is not available for closed (noncompact) sets, or in general metric spaces. Indeed we defined several concepts of stability of closed sets in Section 2.12, and it appears that if we are to reach at a characterization we must first change the definition of prolongation for noncompact sets.

The following lemma gives an insight into what may be done.

2.13.28 *LEMMA*

If the set $M \subset X$ *is compact, then*

$$D_1^+(M) = \bigcap\{\overline{\gamma^+(S(M,\delta))} : \quad \delta > 0\} \quad .$$

The proof is elementary and is left as an exercise. We only recall that $D_1^+(M)$ is by definition the set $\bigcup\{D_1^+(x) : x \in M\}$.

It is now to be noted that $D_1^+(M)$ need not even be closed for closed sets M, which are not compact. And further, in general, if for any closed set M, we have $D_1^+(M) = M$, then the set M need neither be stable or equi-stable. We now introduce the following definition.

2.13.29 *DEFINITION*

Given any non-empty set M *in* X, *we shall call the set* $\bigcap\{\overline{\gamma^+(S(M,\delta))} : \delta > 0\}$ *as the uniform (first) (positive) prolongation of* M *and denote it by* $D_u^+(M)$.

Lemma 2.13.28 says that if M is compact, then $D_u^+(M) = D_1^+(M)$.

The uniform prolongation has further the following properties

2.13.30 *LEMMA*

(i) For any non-empty set $M \subset X$, $D_u^+(M)$ *is closed and positively invariant,*

(ii) $D_u^+(M) = \{y \in X$: *there are sequences* $\{x_n\}$ *in* X *and* $\{t_n\}$ *in* R^+ *such that* $\rho(x_n, M) \to 0$, *and* $x_n t_n \to y\}$,

(iii) $D_u^+(M) \supset \overline{MR^+}$.

The proofs are immediate consequences of the definition.

The uniform prolongation is useful in characterizing the equi-stability of a closed set.

2.13.31 *THEOREM*

A closed set $M \subset X$ *is equi-stable if and only if* $D_u^+(M) = M$.

This is an immediate consequence of the definitions and we leave the details to the reader. We note that Theorem 2.6.6 of Ura falls as a corollary of this theorem, when we note Proposition 2.12.2.

2.13.32 *Notes and References*

The notion of higher prolongations is due to Ura [4] who also showed their close connection with stability and introduced the notion of stability of order α. The exposition here is based on Auslander and Seibert [2]. We have followed the enumeration of Auslander and Seibert for prolongations. Ura's enumeration is different.

For example the 2nd prolongation of Ura is $D_1^+ \circ D_1^+$. The prolongation D_2^+ here is what Ura labels as D_ω where ω is the first countable ordinal. Ura [4] (page 195) also showed that the prolongations introduced here are the only ones which lead to different concepts of stability. The notion of a c-c map is one of the axioms of Auslander and Seibert for an abstract prolongation. We show that this is the concept which leads to various properties which are needed for results on stability. Thus sections 2.13.2 to 2.13.14 are independent of the notion of a dynamical system. For example Theorem 2.13.12 contains as a particular case Ura's characterization of stability: Theorem 2.6.6. Theorem 2.13.25 is due to Auslander and Seibert.

2.14 *Higher prolongational limit sets and generalized recurrence.*

In Section 2.3 we introduced the first positive prolongation, and the first positive prolongational limit set, and we studied some of their properties. We introduced the higher prolongations in Section 2.13. We shall now introduce also the higher prolongational limit sets and study some of the properties. We shall then use these to characterize the notion of generalized recurrence introduced by Joseph Auslander.

2.14.1 *DEFINITION*

The first positive prolongational limit set $J_1^+(x)$ *of any point* $x \in X$ *is defined by* $J_1^+(x) = \{y \in X$: *there are sequences* $\{x_n\}$ *in* X, *and* $\{t_n\}$ *in* R *such that* $x_n \to x$, $t_n \to +\infty$, *and* $x_n t_n \to y\}$. *In Section 2.3 this set was simply denoted by* $J^+(x)$. *Using now the operators* S *and* $\mathcal{D}$ *introduced in Section 2.13.2, we define for any* $x \in X$

$$J_2^+(x) = \mathcal{D}SJ_1^+(x) \qquad ,$$

and if α *is any ordinal number, and* J_β^+ *has been defined for all* $\beta < \alpha$, *we set*

$$J_\alpha^+(x) = \mathcal{D}(\cup\{SJ_\beta^+ : \beta < \alpha\})\ (x) \qquad .$$

We have immediately the following lemma as a consequence of the definition. In the sequel we denote J_α^+ simply by J_α.

2.14.2 *LEMMA*

If $\alpha > 1$, *then* $y \in J_\alpha(x)$ *if and only if there are sequences* $\{x_n\}$, $\{y_n\}$, $y_n \in J_{\beta_n}^{k_n}(x_n)$, $x_n \to x$, $y_n \to y$, *where* β_n *are ordinal numbers less than* α, *and* k_n *are positive integers. Recall that for any map* $\Gamma : X \to 2^X$, $\Gamma^n = \Gamma \circ \Gamma^{n-1}$, *where* $\Gamma^1 = \Gamma$.

We leave the proof to the reader. It is also to be noted that

2.14.3 *LEMMA*

For any ordinal $\alpha > 1$, $y \in D^+_\alpha(x)$ *if and only if there are sequences* $\{x_n\}, \{y_n\}$ *in* X *such that* $x_n \to x$, $y_n \to y$, *and* $y_n \in D^{k_n}_{\beta_n}(x_n)$, *where for each* n, β_n *is an ordinal less than* α *and* k_n *is a positive integer. (In this lemma and hereafter* D^+_α *is simply written as* D_α *to facilitate the use of upper indices.)*

The following lemma now expresses some elementary properties of prolongations and prolongational limit sets.

2.14.4 *LEMMA*

For any $x \in X$, *and any ordinal* α

(i) $J_\alpha(x)$ *is closed and invariant,*

(ii) $J_\alpha(xt) = J_\alpha(x)t = J_\alpha(x)$, *for all* $t \in R$,

(iii) $D_\alpha(x) = \gamma^+(x) \cup J_\alpha(x)$,

(iv) $D_\alpha(x)$ *is closed and positively invariant,*

(v) If the space X *is locally compact, then* $D_\alpha(x)$, $J_\alpha(x)$ *are connected, whenever they are compact (if one is compact, then so is the other), and if* $D_\alpha(x)$ $(J_\alpha(x))$ *is not compact it does not possess any compact components.*

Proof: (i) $J_1(x)$ has been proved to be closed and invariant (Section 2.3). $J_\alpha(x)$ is closed by construction. To prove invariance, let $J_\beta(x)$ be invariant for all $\beta < \alpha$. Let $y \in J_\alpha(x)$, and $t \in R$. Let $x_n \to x$, $y_n \to y$, $y_n \in J^{k_n}_{\beta_n}(x_n)$, where $\beta_n < \alpha$ and k_n is a positive integer. Then by the induction hypothesis $y_n t \in J^{k_n}_{\beta_n}(x_n)$. Since $y_n t \to yt$, we have $yt \in J_\alpha(x)$, and the result follows. (ii) $J_\alpha(x)t = J_\alpha(x)$ is a trivial consequence of invariance of $J_\alpha(x)$. To see $J_\alpha(xt) = J_\alpha(x)t$, note that $J_1(xt) = J_1(x)t$ (this is an easy consequence of the definition). Now assume that $J_\beta(xt) = J_\beta(x)t$ for all $\beta < \alpha$. Let $y \in J_\alpha(xt)$. Let $y_n \in J^{k_n}_{\beta_n}(x_n t)$ (where

$\beta_n < \alpha$, and t_n positive integers) such that $x_n t \to xt$ and $y_n \to y$. Now $x_n \to x$, and $y_n(-t) \in J_{\beta_n}^{k_n}(x_n t)(-t) = J_{\beta_n}^{k_n}(x_n)$, by the induction hypothesis. Since $y_n(-t) \to y(-t)$, so $y(-t) \in J_\alpha(x)$, and $y \in J_\alpha(x)t$. Hence $J_\alpha(xt) \subset J_\alpha(x)t$. Now $J_\alpha(x)t = J_\alpha(xt(-t))t \subset J_\alpha(xt)(-t) = J_\alpha(xt)$. This proves (ii). (iii) In Section 2.3 we proved that $D_1(x) = \gamma^+(x) \cup J_1(x)$. Now assume that the result is true for all $\beta < \alpha$. Notice that if $y' \in D_\beta^k(x')$, then by (ii) $y' \in \gamma^+(x')$, or $y' \in J_\beta^m(x')$, where $m \leq k$. Now if $y \in D_\alpha(x)$, let $x_n \to x$, $y_n \to y$, $y_n \in D_{\beta_n}^{k_n}(x_n)$ (where $\beta_n < \alpha$, k_n positive integers). If $y_n \in J_{\beta_n}^{\ell_n}(x_n)$ ($\ell_n \leq k_n$) for infinitely many n, then $y \in J_\alpha(x)$. If $y_n \in \gamma^+(x_n)$ for infinitely many n, then $y \in D_1(x) = \gamma^+(x) \cup J_1(x)$. In either case $y \in \gamma^+(x) \cup J_\alpha(x)$. Thus $D_\alpha(x) \subset \gamma^+(x) \cup J_\alpha(x)$. Since $\gamma^+(x) \cup J_\alpha(x) \subset D_\alpha(x)$ is obvious, we have $D_\alpha(x) = \gamma^+(x) \cup J_\alpha(x)$. This completes the proof of (iii). (iv) Positive invariance is an immediate consequence of (iii), and $D_\alpha(x)$ is closed by definition. (v) The proof of this statement may easily be constructed by the method adopted for the proof of a similar statement about $\Lambda^+(x)$ in Section 2.2, and about $D_1(x)$ and $J_1(x)$ in Section 2.3. This we leave to the reader.

2.14.5 *Exercise*

Show that for any ordinal

$$J_\alpha(x) = \bigcap\{D_\alpha(xt) : t \in R\} \quad .$$

We now recall some of the notions of recurrence that have occurred earlier, namely, a rest point, a periodic trajectory (or periodic point), a positively or negatively Poisson stable motion (or point), a non-wandering point. We recall that these concepts are respectively equivalent to $x = xt$ for all $t \in R$, $x \in \Lambda^+(x)$ or $x \in \Lambda^-(x)$, and $x \in J_1^+(x)$ which is equivalent to $x \in J_1^-(x)$.

Now let V denote the class of real-valued continuous functions f on X such that $f(xt) \leq f(x)$, for all $x \in X$ and all $t > 0$.

2.14.6 *DEFINITION*

Let R *denote the set of all points* $x \in X$ *such that* $f(xt) = f(x)$, *for all* $f \in V$, *and all* $t \geq 0$. R *will be called the* ***generalized recurrent set.***

We have immediately

2.14.7 *LEMMA*

R *includes the non-wandering points in* X.

Proof: Let $x \in J_1(x)$. Let $t > 0$, and $f \in V$. Then indeed $x \in J_1(xt)$, and there are sequences $x_n \to xt$, $t_n \to +\infty$, and $x_n t_n \to x$. Then indeed $f(x_n t_n) \leq f(x_n)$, and since f is continuous, we have

$$f(x) \leq f(xt) \qquad .$$

As $f(xt) \leq f(x)$ holds by hypothesis, we get $f(xt) = f(x)$. Thus $x \in R$.

Now we have

2.14.8 *THEOREM*

R *is closed and invariant.*

Proof: That R is closed is clear. To see invariance, let first $x \in R$, and $\tau > 0$. Then for any $f \in V$ $f((x\tau)t) = f(x(\tau + t)) = f(x) = f(x\tau)$. Thus $x\tau \in R$. Secondly, let $\tau < 0$, and $x\tau \notin R$. Then there is an $f \in V$ and a $t_o > 0$ such that $f((x\tau)t_o) < f(x\tau)$. Define now $g \in V$ by $g(x) = f(x\tau)$ for any $x \in X$. Then $g(xt_o) = f((xt_o)\tau) = f((x\tau)t_o) < f(x\tau) = g(x)$. This contradicts $x \in R$, and the theorem is proved.

It is clear that if $f \in V$, then so are $\tan f$ and $cf + d$, where c and d are real numbers with $c \geq 0$. This remark and the above theorem yield

2.14.9 *LEMMA*

Let a, b, $a < b$ *be real numbers. Set* $V_{a,b} = \{f \in V : a \leq f(x) \leq b,$ *for all* $x \in X\}$. *Then* $x \in R$ *if and only if* $f(xt) = f(x)$ *for all* $f \in V_{a,b}$ *and all real* t.

From now on we shall assume that the space X is locally compact and separable.

The following theorem shows that in the class V of functions there is a function which is constant along any trajectory in the recurrent set, but is strictly decreasing along any trajectory which is not in the recurrent set.

2.14.10 *THEOREM*

There is an $f \in V$ *such that*

(i) If $x \in R$, *then* $f(x) = f(xt)$ *for all real* t, *and*

(ii) If $x \notin R$, *and* $t > 0$, *then* $f(xt) < f(x)$.

Proof: Let $C(X)$ denote the continuous real-valued functions on X, provided with the topology of uniform convergence on compact sets. Then $C(X)$ contains a countable dense subset and so does $V' = V_{-1,1}$. Let $\{f_k\}$, $k = 1,2,\ldots$, be a countable dense set in V'. Then $x \in R$ if and only if $f_k(xt) = f_k(x)$ for $k = 1,2,\ldots$, and real t. Set $g = \sum_{k=1}^{\infty} \frac{1}{2^k} f_k$. Since $|f_k(x)| \leq 1$, it follows that g is continuous and $|g(x)| \leq 1$. Thus $g \in V'$. If $g(xt) = g(x)$, for all $t > 0$, then $f_k(xt) = f_k(x)$ for $k = 1,2,\ldots$, and so $x \in R$. If $x \notin R$, there is a sequence $\{t_n\}$ in R^+ with $t_n \to +\infty$ such that $g(x) > g(xt_1) > g(xt_2)$ .... Define $f(x) = \int_0^{\infty} e^{-t} g(xt)dt$. Then indeed $f \in V'$, and f has the properties required in the theorem.

We shall now obtain a characterization of R by means of the prolongational limit sets. First, the following lemma.

2.14.11 *LEMMA*

If $f \in V$, *and* $y \in D_\alpha(x)$, *then*

$$f(y) \leq f(x).$$

This is an immediate consequence of Lemma 2.13.6 and the definition of $D_\alpha(x)$.

2.14.12 *DEFINITION*

The set of all points $x \in X$ *such that* $x \in J_\alpha(x)$ *will be denoted by* R_α. *And we set* $R' = \bigcup\{R_\alpha : \alpha$ *an ordinal number*$\}$.

The following theorem characterizes R.

2.14.13 *THEOREM*

$R = R'$. *That is,* $x \in R$ *if and only if* $x \in J_\alpha(x)$ *for some ordinal* α.

For the proof we need the following topological theorem.

2.14.14 *THEOREM*

Let X *be a locally compact, separable metric space and let* $\prec$ *be a closed quasi order on* X. *Let* x *and* y *in* X *such that* $x \prec y$ *does not hold. Then there is an* f *in* $C(X)$ *such that* (*i*) *if* $z \prec z'$, *then* $f(z) \leq f(z')$, (*ii*) $f(y) < f(x)$.

Proof of Theorem 2.14.13.

We first show that $R' \subset R$. Indeed let $x \in J_\alpha(x)$ for some α. Then for any real t, $x t \in J_\alpha(xt) = J_\alpha(x)$, and in particular this holds for $t > 0$. Then if $f \in V$, we have $f(x) \leq f(xt)$. However, we have $f(xt) \leq f(x)$ by definition of f. Thus for each $f \in V$, $f(xt) = f(x)$ for all $t > 0$, and, therefore, $x \in R$. This proves $R' \subset R$. To prove $R \subset R'$, we define a relation $\prec$ on X by $y \prec x$ if and only if $y \in D'(x) = \bigcup D_\alpha(x)$. Then $\prec$ is a closed quasi order on X. Observe hat $xt \prec x$, whenever $x \in X$, and $t > 0$. If $x \prec y$ but not $y \prec x$, we write $\prec\prec y$. Note now that if $x \notin R'$, and $t > 0$, then $xt \prec\prec x$. To see this note that $t \prec x$. If $x \prec xt$, then $x \in D'(xt)$. Thus $x \in D_\alpha(xt)$ for $t > 0$ and some ordinal . Then $x \in \gamma^+(xt) \cup J_\alpha(xt)$. Thus either $x \in J_\alpha(xt)$ or $x \in \gamma^+(xt)$. In the second ase $\gamma(x)$ is periodic and so $x \in J_1(xt) \subset J_\alpha(xt)$. In any case then $x \in J_\alpha(xt) = J_\alpha(x)$. hus $x \in R'$ and this is a contradiction. The rest follows from Theorem 2.14.14.

.14.15 *Remark*

By a quasi order on X one means a reflexive, transitive, but not necessarily ntisymmetric relation.

.14.16 *Notes and References*

This section is almost exclusively a reproduction of results of Joseph Auslander 3]. The only exception is the statement (v) in Lemma 2.14.4. Notice that first part f the statement about $D_\alpha(x)$ follows from Theorem 2.13.11 as D_α is a c-c map. wever, the remaining parts do need a separate proof. Indeed it is not too difficult construct examples of c-c maps Γ such that $\Gamma(x)$ is closed but not compact, and (x) has a compact component.

2.15 *Relative Stability and Relative Prolongations.*

We shall assume in this section that the phase space X is locally compact.

2.15.1 *DEFINITION*

Given a point $x \in X$, *and a set* $U \subset X$, *the (first) (positive)* <u>*relative prolongation*</u> *of* x *with respect to the set* U *is the set* $D^+(x,U)$ *given by* $D^+(x,U) = \{y \in X$: *there is a sequence* $\{x_n\}$, *and a sequence* $\{t_n\}$, $x_n \in U$, *and* $t_n \geq 0$ *for each* n, *such that* $x_n \to x$, *and* $x_n t_n \to y\}$.

2.15.2 *DEFINITION*

Given a compact set $M \subset X$, *and a set* $U \subset X$, *the set* M *is said to be (positively)* <u>*relatively stable*</u> *with respect to the set* U, *if given an* $\varepsilon > 0$, *there exists a* $\delta > 0$, *such that* $\gamma^+(S(M,\delta) \cap U) \subset S(M,\varepsilon)$.

It is clear that if in the definition 2.15.1, U is a neighborhood of x, then $D^+(x,U) = D_1^+(x)$. Further, in definition 2.15.2, if U is a neighborhood of M, then relative stability of M with respect to U is the stability of M as defined in Section 2.6.

We have now the following theorem.

2.15.3 *THEOREM*

A compact set $M \subset X$ *is relatively stable with respect to the set* $U \subset X$ *if and only if* $M \supset D^+(M,U)$. *Here* $D^+(M,U) = \bigcup\{D^+(x,U) : x \in M\}$.

2.15.4 *Remark*

In the above theorem or definitions, the set U need not contain the point x or the set M. If, however, $U \supset M$, then in the above theorem one may replace the condition $M \supset D^+(M,U)$ by $M = D^+(M,U)$. Further, in case U is a neighborhood of M one obtains Theorem 2.6.6.

oof of Theorem 2.15.3: Sufficiency: Let $M \supset D^+(M,U)$, and let, if possible, M not relatively stable with respect to U. Then there is an $\varepsilon > 0$, a sequence $\{x_n\}$ in U, $x_n \to x \in M$, and a sequence $\{t_n\}$, $t_n \geq 0$, such that $\rho(x_n t_n, M) = \varepsilon$. We y assume that $H(M,\varepsilon)$ is compact. Thus the sequence $\{x_n t_n\}$ may be assumed to converg a point $y \in H(M,\varepsilon)$. Then $y \in D^+(x,U) \subset D^+(M,U)$, but $y \notin M$. This contradiction oves sufficiency.

cessity: Let M be relatively stable with respect to U. Then clearly $(M,U) \subset S[M,\varepsilon]$ for arbitrary $\varepsilon > 0$. Hence $D^+(M,U) \subset \cap\{S[M,\varepsilon]:\varepsilon > 0\} = M$. This oves the theorem.

The concept of relative stability may be motivated by considering the example a limit cycle C in the plane, with the property that all trajectories outside e disc bounded by the limit cycle C, have C as their sole positive limit set, and l trajectories in the interior of the disc bounded by C tend to an equilibrium point. tice that if U is the complement of the disc bounded by C, then C is relatively able with respect to U. Notice also that if C is an asymptotically stable limit cle, then C is stable with respect to every component of $R^2 \setminus C$. These considera- ons lead to the following definition and theorem.

15.5 *DEFINITION*

Let $M \subset X$ *be compact. We say that* M *is* <u>*component-wise stable*</u> *if* M *is latively stable with respect to every component of* $X \setminus M$.

We have then

15.6 *THEOREM*

Let a compact set $M \subset X$ *be positively stable. Then* M *is component-wise ble.*

The proof is obvious and is ommitted.

The converse of Theorem 2.15.6 is in general not true. To see this, we consider a simple example.

2.15.7 *Example*

Let $X \subset E^2$ (the euclidean plane) be given by $X = \{(x,y) \in E^2 : y = \frac{1}{n},$ n any integer, or $y = 0\}$. The space X is a metric space with the distance between any two points being the euclidean distance between the points in E^2. We define a dynamical system on X by the differential equations

$$\dot{x} = 0, \quad \dot{y} = 0 \quad \text{if} \quad y = 0$$

and

$$\dot{x} = 1, \quad \dot{y} = 0 \quad \text{if} \quad y \neq 0 \quad .$$

Then the set $\{(0,0)\} \subset X$ is component-wise stable, but is not stable.

The question now arises, as to when the converse of Theorem 2.15.6 is true. For this purpose the following definition is convenient.

2.15.8 *DEFINITION*

Let $M \subset X$ *be compact. We shall say that the pair* (M, X) *is* <u>*stability-additive*</u> *if the converse of Theorem 2.15.6 holds for every dynamical system defined on* X *which admits* M *as an invariant set.*

In this connection the following theorems are important.

2.15.9 *THEOREM*

The pair (M, X) *is stability-additive if* $X \setminus M$ *has a finite number of components.*

2.15.10 *THEOREM*

The pair (M, X) *is stability-additive if* $X \setminus M$ *is locally connected.*

The proof of Theorem 2.15.9 is immediate and is left as an exercise. We prove Theorem 2.15.10.

Proof of Theorem 2.15.10: Let M be a compact invariant set for a given dynamical system on X and let M be component-wise stable. Since M is locally compact, there is an $\varepsilon > 0$ such that $S[M,\varepsilon]$ and hence also $H(M,\varepsilon)$ is compact. We claim that only a finite number of components of $X \setminus M$ can intersect $H(M,\varepsilon)$. For otherwise, if an infinite number of components of $X \setminus M$ intersect $H(M,\varepsilon)$, then we may choose a sequence of points $\{x_n\}$ in $H(M,\varepsilon)$ such that no two points of the sequence are in the same component. Since $H(M,\varepsilon)$ is compact we may assume that $x_n \to x \in H(M,\varepsilon)$. Since X is locally connected, there is a neighborhood of x, say N, such that N is a subset of a component of $X \setminus M$. Now there is an integer n_o such that $x_n \in N$ for $n \geq n_o$ and hence all x_n for $n \geq n_o$ belong to the same component of $X \setminus M$, which is a contradiction. Now notice that every component of $X \setminus M$ is an invariant set. If $C_1, C_2, \ldots, C_p$ are the components of $X \setminus M$ which intersect $H(M,\varepsilon)$, then by component stability we have positive numbers $\delta_1, \delta_2, \ldots, \delta_p$ such that $\gamma^+(S(M,\delta_i) \cap C_i) \subset S(M,\varepsilon)$, $i = 1,2,\ldots,p$. If now $\delta = \min(\delta_1, \delta_2, \ldots, \delta_p)$ we get $\gamma^+(S(M,\delta)) \subset S(M,\varepsilon)$, i.e., M is stable. To see this last assertion, note that if $x \in S(M,\delta)$, then either $x \in C_i$ for some $i = 1,2,\ldots,p$ and hence $\gamma^+(x) \subset S(M,\varepsilon)$, or x is an element of a component of $X \setminus M$ which does not intersect $H(M,\varepsilon)$ and hence is contained in $S(M,\varepsilon)$, or $x \in M$. In either of the last two cases $\gamma^+(x) \subset S(M,\varepsilon)$. The theorem is proved.

2.15.11 *Notes and References*

The concepts introduced here are from Ura [7]. We refer the reader to this paper for a detailed discussion of these concepts and their relation to saddle sets. We remark that one can in a similar fashion define the concept of relative asymptotic stability and discuss many similar problems.

CHAPTER 3

THE SECOND METHOD OF LIAPUNOV FOR ORDINARY DIFFERENTIAL EQUATIONS.

3.1 *Dynamical systems defined by ordinary differential equations.*

In this section we shall prove theorems for existence, uniqueness, and extendability of solutions of ordinary differential equations.

Consider the autonomous differential equation

3.1.1 $$\dot{x} = f(x)$$

where x and f are n-vectors. Under certain conditions, given any point x^o in the n-dimensional Euclidean space E, spanned by the components of the vector x, the differential equation 3.1.1 defines a differentiable function (solution)

3.1.2 $$x = x(t, x^o)$$

such that

3.1.3 $$\dot{x}(t, x^o) = f(x(t, x^o))$$

on all points t of a certain interval (a, b) which is such that $t_o \in (a,b)$ and

3.1.4 $$x^o = x(t_o, x^o) \quad .$$

If for any point $x^o \in E$ there exists a unique solution 3.1.2 of 3.1.1 which satisfies 3.1.4, which is a continuous function of x^o and t and which is defined for all $t \in R$, then, clearly such solutions 3.1.2 induce on E a flow satisfying the axioms 1.1.2 and thus define a dynamical system.

In this section various sufficient conditions for an ordinary differential equation to define a dynamical system will be given. Some theorems are standard and may be found in any modern work on differential equations, others have a more specialized purpose.

For the sake of convenience we shall derive these conditions for existence, uniqueness, and continuity of solutions in the formally more general case of the differential equation

3.1.5 $$\dot{x} = f(x, t)$$

with initial condition

3.1.6 $$x(t_0, x^0, t_0) = x^0$$

From now on we shall denote with $x = x(t) = x(t, x^0, t_0)$ a solution of the equation 3.1.5 which satisfies the initial condition 3.1.6.

We shall proceed next with the proofs of the basic existence theorems.

The first existence theorem that we shall present is the classical result due to Peano and its proof is based upon the following basic lemma on uniformly bounded and equicontinuous families of functions.

Note that a family $F = \{f(t)\}$ of functions $f(t)$ defined on a bounded interval $[a,b]$ is called equicontinuous if for each $\varepsilon > 0$, there is a $\delta = \delta(\varepsilon) > 0$ such that $t_1, t_2 \in [a,b]$, $|t_1 - t_2| < \delta$ implies $|f(t_1) - f(t_2)| < \varepsilon$ for all $f(t) \in F$. In particular if a sequence of continuous functions on a compact set Ω is uniformly convergent on Ω, then it is uniformly bounded and equicontinuous.

3.1.7 *LEMMA (Arzela', Ascoli).*

Let $F = \{f(t)\}$ *be an equicontinuous, uniformly bounded family of functions* $f(t)$, *defined on a bounded interval* $[a, b]$. *Then there exists a uniformly convergent sequence* $\{f_n(t)\}$ *of functions* $f_n(t) \in F$.

We are now in the position of proving the basic existence theorem.

3.1.8 *THEOREM (Peano's existence theorem).*

Let $f(x,t)$ *be continuous on a parallelepiped* $\Omega \subset E \times R$ *defined by the relations:*

$$t_0 \leq t \leq t_0 + a , \qquad |x - x^0| \leq b .$$

Let

$$M = \max_{(t,x) \in \Omega} |f(x,t)|$$

and

$$\alpha = \min (a, b/M) .$$

Then the ordinary differential equation 3.1.5 *has at least one solution on* $[t_0, t_0 + \alpha]$.

Proof. Let T_{ab} denote the closed interval $[t_0 - a, t_0 + b]$. Consider an n-dimensional continuously differentiable vector $x^0(t)$, defined on the interval $T_{\varepsilon o}$, ($\varepsilon > 0$, sufficiently small) which satisfies the conditions $\dot{x}(t_0) = f(x^0, t_0)$, $x^0(t_0) = x^0$,

$|x^0(t) - x^0| \leq b$, and $|f(x^0(t), t)| \leq M$ for all $t \in T_{\varepsilon o}$. Next we shall construct a vector $x^\delta(t)$ on the interval $T_{\varepsilon\alpha}$ as follows:

$$x^\delta(t) = x^0(t) \qquad t \in T_{\varepsilon o}$$

while $x^\delta(t)$ is a solution of the integral equation

3.1.9 $$x^{\delta}(t) = x^{o} + \int_{t_o}^{t} f(x^{\delta}(\tau-\delta),\tau)\, d\tau$$

with $0 < \delta \leqslant \varepsilon$ on the interval $T_{o\alpha}$. It must now be shown that such a solution $x^{\delta}(t)$ of the integral equation above indeed exists on the whole interval $T_{o\alpha}$. Clearly, such $x^{\delta}(t)$ exist, on $T_{o\alpha_1}$ where $\alpha_1 = \min(\alpha,\delta)$. Then on the interval $T_{\varepsilon\alpha_1}$, $x^{\delta}(t) \in C^1$ and satisfies the conditions $|x^{\delta}(t) - x^{o}| \leqslant b$. Then clearly this solution can be extended in the same fashion on the interval $T_{\varepsilon\alpha_2}$ where $\alpha_2 = \min(\alpha, 2\delta)$ etc. By repeated application of this procedure it is possible to construct on $T_{\varepsilon\alpha}$ a continuously differentiable function $x^{\delta}(t)$, with $x^{\delta}(t_o) = x^{o}$ and $|x^{\delta}(t)-x^{o}| \leqslant b$, which satisfies the integral equation 3.1.9.

Now since $|f(x^{\delta}(t), t)| = |\dot{x}^{\delta}(t)| \leqslant M$ it follows that the family of continuously differentiable functions $x^{\delta}(t)$, $0 < \delta \leqslant \varepsilon$ is equicontinuous. Then from Lemma 3.1.7 there exists a sequence $\{\delta(n)\}$ such that $\lim_{n\to\infty} x^{\delta(n)}(t) = x(t)$ exists uniformly on $T_{\varepsilon\alpha}$.

From the uniform continuity of $f(x,t)$ it follows that $f(x^{\delta(n)}(t - \delta(n)),t)$ tends uniformly to $f(x(t),t)$ as $n \to \infty$, thus equation 3.1.9 with solution $x^{\delta(n)}$ tends to integral equation

3.1.10 $$x(t) = x^{o} + \int_{t_o}^{t} f(x(\tau), \tau)\, d\tau$$

which in the domain Ω is equivalent to the differential equation 3.1.5 together with the initial condition 3.1.6. Thus $x(t)$ so constructed is a solution of 3.1.5, which proves the theorem.

The integral at the right hand side of expression 3.1.10 is defined for a much larger class than the one of continuously differentiable functions $x(t)$. This fact allows us to define solutions "in the Carathèodory sense" of the differential equation 3.1.5 which exist under less restrictive conditions than the one required by the previous theorem. In the next theorem we shall state the classical conditions for the existence of such solutions in the "Carathéodory sense".

3.1.11 *THEOREM*

Let $f(x,t)$ *be defined on* Ω *, with* Ω *defined as in* 3.1.8 *, continuous in* x *for each fixed* t *and measurable in* t *for each fixed* x*. If on the interval* $[t_0, t_0 + a]$ *there exists a function* $m(t) \in L^1[t_0, t_0 + a]$, *such that*

$$|f(t,x)| \leq m(t) \quad \textit{for} \quad (x,t) \in \Omega$$

then there exists on some interval $[t_0, t_0 + \beta]$ $(\beta > 0)$ *an absolutely continuous function* $x(t)$ *such that* $x(t_0) = x^0$ *and which satisfies the differential equation* 3.1.5 *for all* $t \in [t_0, t_0 + \beta]$, *but a set of Lebesgue-measure zero.*

3.1.12 *COROLLARY*

Let $f(x,t)$ *be continuous in* x *(for fixed* t*) and piecewise continuous in* x,t *in* $E \times R$ *. Then for all* (x^0, t_0) *in* $E \times R$ *, there exists a solution* $x(t)$ *which satisfies equation* 3.1.5 *on an open interval and is such that* $x(t_0) = x^0$ *.*

In the case of corollary 3.1.12 the solution of the differential equation 3.1.5 relative to the initial condition 3.1.6 is equivalent to the integral equation 3.1.10 at all points of continuity of $f(x,t)$.

We shall now investigate the relations between the rectangle Ω in which the system is defined and the number α defined in Theorem 3.1.8 which defines the interval of definition of the solution. We are, in particular, concerned about the properties of $x(t)$ when $f(x,t)$ is defined in the whole space $E \times R$. This problem is called in the literature: "Problem of the extension of solutions of an ordinary differential equation." Suppose that $f(x,t)$ is defined on $E \times R$ and let $x^o(t)$ be a solution of the ordinary differential equation 3.1.5. Assume that $x^o(t)$ is defined on an interval $[\alpha,\beta]$. Then the point $(x^o(\beta), \beta)$ is in $E \times R$ and it is possible to find a solution $x^1(t)$ such that $x^1(\beta) = x^o(\beta)$ and which is defined on some interval $[\beta,\delta]$ with $\beta < \delta$. Clearly the function $x^2(t)$, defined on the interval $[\alpha,\beta] \cup [\beta,\delta]$ by the relations

$$x^2(t) = x^o(t) \qquad \text{for} \quad t \in [\alpha,\beta]$$
$$x^2(t) = x^1(t) \qquad \text{for} \quad t \in [\ ,\delta]$$

is a solution of the differential equation 3.1.5 which is defined on an interval which is larger than either $[\alpha,\beta]$ or $[\beta,\delta]$. Such a solution $x^2(t)$ is called an extension of either one of the solutions $x^o(t)$ and $x^1(t)$. This process of extension may be applied at either end of a closed interval and a given solution extended to a larger interval. By repeated application of the above process a *maximal interval of existence* of any given solution can be constructed. Obviously such a maximal interval of existence is open.

For the case of solutions defined on the maximal interval of existence the following theorem holds.

3.1.13 *THEOREM*

Let $\Omega \subseteq E \times R$. *Let* $\Gamma \subset \Omega$ *be compact. Fix* $(x^o,t_o) \in \Gamma$. *Let*

$I_x = (t^-, t^+)$ *be the maximal interval of existence of a solution* $x(t)$ *of the differential equation* 3.1.5 *where* (x,t) *is defined and continous in* Ω, *for* (x^0, t_0). *Then there exists* $\tau \in I_x$, $\tau > t_0$, *such that* $x(t, x^0, t_0) \in C(\Gamma)$ *for* $\tau \leq t < t^+$ $(t^- < t \leq \tau)$.

Proof. We need show only, that if t^+ is finite and if $(t, x(t))$ lies in a compact subset N of Ω for $t \in [t_0, t^+)$ where $t^- < t_0 < t^+$, then the interval (t^-, t^+) cannot be a maximal interval of existence. We will show first that in such a case $\lim_{t \to t^+ - 0} x(t)$ exists.

For this purpose, let $\{t_n\}$, $\{\tau_n\}$ be any two sequences, such that $x(t_n) \to z_1$ and $x(\tau_n) \to z_2$ and $z_1 \neq z_2$. Clearly $z_1, z_2 \in N \subset \Omega$. We have of course

$$x(t_n) = x_0 + \int_{t_0}^{t_n} f(\tau, x(\tau))\, d\tau$$

$$x(\tau_n) = x_0 + \int_{t_0}^{\tau_n} f(\tau, x(\tau))\, d\tau \quad ,$$

so that

$$||x(t_n) - x(\tau_n)|| \leq \left|\int_{\tau_n}^{t_n} ||f(\tau, x(\tau))||\, d\tau\right| \leq M|t_n - \tau_n| \ ,$$

where $M = \max_{(t,x) \in N} ||f(t,x)||$.

Proceeding to the limit as $n \to \infty$ we find that $||z_1 - z_2|| \leq 0$, which contradicts $z_1 \neq z_2$. Hence $\lim_{t \to t^+ - 0} x(t)$ exists. Set $z = \lim_{t \to t^+ - 0} x(t)$, then the point $(t^+, z) \in N \subset \Omega$. Hence there exists a solution $\theta(t)$, $\theta(t^+) = z$, defined on some interval $[t^+, d]$, $t^+ < d$.

Consider the function $y(t)$ defined on $(t^-, d]$ such that

$$y(t) = x(t)\ ,\ t \in (t^-, t^+);\ y(t) = \theta(t),\ t \in [t^+, d] \quad .$$

We claim that $y(t)$ is a solution. Since $\lim_{t \to t^+ - 0} x(t) = z = \theta(t^+)$,

all that we need verify is that the derivative $\dot{y}(t)$ exists and $\dot{y}(t^+) = f(t^+, \tau)$. But this is so, for

$$\lim_{t \to t^+ - 0} \dot{y}(t) = \lim_{t \to t^+ - 0} \dot{x}(t) = \lim_{t \to t^+ - 0} f(x(t), t) = f(\tau, t^+)$$

and

$$\lim_{t \to t^+ + 0} \dot{y}(t) = \lim_{t \to t^+ + 0} \dot{\theta}(t) = \lim_{t \to t^+ + 0} f(\theta(t), t) = f(\tau, t^+)$$

as $f(t,x)$ is continuous. Hence $y(t)$ is indeed an extension of the solution $x(t)$, which contradicts that (t^-, t^+) was the maximal interval of existence of $x(t)$. This shows that there exists a sequence $\{t_n\}$, $t_n \to t^+ - 0$, such that $x(t_n) \to y$, where $(y, t^+) \in \partial\Omega$. In order to complete the proof of the theorem we have to prove now that no limit point of a sequence $\{x(t_n), t_n\}$ where $t_n \to t^+$ can be an interior point of Ω . This statement follows from:

3.1.14 *LEMMA*

Let $f(x,t)$ *be continuous on* $\Omega \subseteq E \times R$. *Let* $x(t)$ *be a solution of* 3.1.5 *on* $[a, b)$, $b < \infty$ *such that there exists a sequence* $\{t_n\}$ *with* $a \leq t_n \to b$ *as* $n \to \infty$ *and such that* $\lim_{t_n \to b} x(t_n)$ *exists and is* x^o.

If $f(x,t)$ *is bounded on* $\Omega \cap N(x^o, b)$ *where* $N(x^o, b)$ *is a neighborhood of* (x^o, b), *then* $\lim_{t_n \to b} x(t_n) = \lim_{t \to b} x(t)$.

Proof. Let $\varepsilon > 0$ and $m_\varepsilon > 1$ be such that $|f(x,t)| \leq m_\varepsilon$ for $(x,t) \in \Omega \cap D$ where $D = \{(x,t): |x - x^o| \leq \varepsilon, \ 0 \leq b - t \leq \varepsilon\}$. Let n be such that $0 < b - t_n \leq \varepsilon/\tau m_\varepsilon$ and $|x(t_n) - x^o| \leq \varepsilon/\tau$. Then

$$|x(t) - x(t_n)| < m_\varepsilon(b - t_n) \leq \varepsilon/2 \quad \text{for} \quad t_n \leq t < b \quad .$$

Assume that this is false let then $\tau: t_n < \tau < b$ be the smallest τ for which $|x(\tau) - x(t_n)| = m(b - t_n) \leq \varepsilon/2$. Thus

$$|x(t) - x^o| \leq \varepsilon/2 + |x(t_n) - x^o| \leq \varepsilon \quad \text{for} \quad t_n \leq t < \tau$$

and then $|\dot{x}(t)| \leq m_\varepsilon$ for $t_n \leq t \leq \tau$, which implies that $m_\varepsilon(\tau - t_n) \leq m_\varepsilon(b - t_n)$ which contradicts the above assumption and proves the theorem.

3.1.15 *Remark*

Notice that if $t^+ = +\infty$, then the result of Theorem 3.1.13 is true. Consider the special case $\Omega = \Gamma \times R$ when $\Gamma \subset E$ is open, then the above results say that if $t^+ \neq +\infty$, then for any compact set $\Theta \subset \Gamma$, $x(\tau;\, t_o, x^o) \not\in \Theta$ for τ sufficiently close to t^+. If also $\bar{\Gamma}$ is compact, then $x(\tau, t_o, x^o) \to \partial\Gamma$ as $\tau \to t^+$. While if $t^+ \neq +\infty$ and if $\Gamma = E$, then $\|x(\tau, t_o, x^o)\| \to \infty$ as $\tau \to t^+$.

Results corresponding to Theorem 3.1.13 can also be proved for the case of "Carathéodory systems".

We shall now proceed with the study of conditions on the differential equation so that the flow defined by its solutions satisfies the second requirement for it to induce a dynamical system, that of uniqueness. A condition for uniqueness of solutions of the equation 3.1.5 is based upon the notion of a Lipschitz-continuous function which is given below.

3.1.16 *DEFINITION*

A vector-valued function $f(x,t)$ *is said to be Lipschitz-continuous with respect to* x *if there exists a piecewise continuous function* $\chi(t)$, *such that for all* $x, y \in E$,

3.1.17 $$\|f(x,t) - f(y,t)\| \leq \chi(t)\,\|x - y\|$$

We shall now prove the classical theorem of existence and uniqueness. This theorem goes back to Lipschitz. The simple proof that we present, based upon the idea of the norm 3.1.19 , is taken from Jones [7] . Since in the theory of dynamical systems we are essentially interested in global properties, we shall prove this theorem only in the global case, i.e., the vector $f(x,t)$ is defined for all $(x,t) \in E \times R$. A similar local theorem for the case in which $f(x,t)$ is defined in the set $\Omega \subset E \times R$ defined in Theorem 3.1.8 can be easily proved in the same way.

3.1.18 *THEOREM (Existence and Uniqueness)*

Let $f(x,t)$ *be piecewise continuous in* t *and suppose that condition* 3.1.17 *holds. Then for any* (x^o, t_o) *in* $E \times R$, *there exists a unique solution of* 3.1.5.

$$x = x(t, x^o, t_o)$$

such that

$$x^o = x(t_o, x^o, t_o)$$

Proof. For arbitrary $a > 0$ and $\lambda > 0$ let $C_a = C[t_o - a, t_o + a]$ be the space of continuous functions defined on the interval

$[t_0 - a, t_0 + a]$ and for all elements x in C_a let

3.1.19 $||x|| = \max\{e^{-\lambda|t-t_0|}\,|x(t)| : t \text{ in } [t_0 - a, t_0 + a]\}$

serve as a norm. For any fixed choice of a and λ, C_a and $||\cdot||$ form a complete normed linear space. Let us define a function $C_a \to C_a$ by the formula

$$f(x)(t) = x_0 + \int_{t_0}^{t} f(\tau, x(\tau))d\tau, \quad t \text{ in } [t_0 - a, t_0 + a] .$$

Clearly $f(x)(t)$ is continuous on $[t_0 - a, t_0 + a]$ in the sense of $\|\cdot\|$. For arbitrary x and y in C_a we have that

$$(f(x) - f(y))(t) = \int_{t_0}^{t} (f(\tau, x(\tau)) - f(\tau, y(\tau)))\, d\tau .$$

Using 3.1.17 we have that

$$\|(f(x) - f(y))(t)\| \leq \left|\int_{t_0}^{t} \chi(\tau)\, \|x(\tau) - y(\tau)\|\, d\tau\right| ,$$

$$\|(f(x) - f(y))(t)\| \leq \left|\int_{t_0}^{t} \chi(\tau)e^{\lambda|\tau-t_0|}\, e^{-\lambda|\tau-t_0|}\, \|x(\tau) - y(\tau)\| d\tau\right| ,$$

and

$$\|f(x) - f(y))(t)\| \leq \left|\int_{t_0}^{t} \chi(\tau)e^{\lambda|\tau-t_0|}d\tau\right|\, \|x - y\| .$$

Multiplying through by $e^{-\lambda|t-t_0|}$ we have

$$e^{-\lambda|t-t_0|}\, \|(f(x) - f(y))(t)\| \leq \left|e^{-\lambda|t-t_0|} \int_{t_0}^{t} \chi(\tau)e^{\lambda|\tau-t_0|}\, d\tau\right|\, ||x-y|| .$$

Now $\chi(t)$ is piecewise continuous on $[t_0 - a, t_0 + a]$ and is bounded by some constant K on this interval. Hence

$$e^{-\lambda|t-t_o|}\,||(f(x) - f(y))(t)|| \leqslant K|e^{-\lambda|t-t_o|}\int_{t_o}^{t} e^{\lambda|\tau-t_o|}d\tau|\,||x-y|| \ .$$

Furthermore, one can easily verify that on the interval $[t_o - a, t_o + a]$

$$|e^{-\lambda|t-t_o|}\int_{t_o}^{t} e^{\lambda|\tau-t_o|}\,d\tau| \leqslant \frac{1}{\lambda}\ (1-e^{-\lambda a}) \quad ,$$

so for all t in $[t_o - a, t_o + a]$

$$e^{-\lambda|t-t_o|}\,\|(f(x) - f(y))(t)|| \leqslant \frac{K}{\lambda}\ (1-e^{-\lambda a})\,||x - y|| \ .$$

Thus it follows that

$$|f(x) - f(y)| \leqslant \frac{K}{\lambda}\ (1 - e^{-\lambda a})\,||x - y|| \ .$$

Since we are free to choose λ as large as we please, we choose it so that

$$\frac{K}{\lambda}\ (1 - e^{-\lambda a}) < 1 \quad .$$

Thus the function f is a contraction, and it follows from the contraction principle that there exists a unique function x such that $f(x) = x$. That is, there exists a unique function x such that

$$x(t) = x^o + \int_{t_o}^{t} f(\tau, x(\tau))d\tau \quad ,$$

for t in $[t_o - a, t_o + a]$. Since a was chosen arbitrarily, it follows that x is defined and unique on the whole real line R . Therefore, our theorem is proved.

3.1.20 *Remark*

Clearly the Lipschitz condition could have given on any norm on E , not necessarily the Euclidean norm or the norm $|x|$. In those cases the proof of the previous theorem should have been suitably changed.

3.1.21 *LEMMA*

Let x , y , *and* z *be real-valued piecewise continuous functions defined on a real interval* [a, b] , *and let* z *be nonnegative on this interval. If for all* t *in* [a, b]

3.1.22
$$x(t) \leq y(t) + \int_a^t z(\tau)\, x(\tau)d\tau \quad ,$$

then

3.1.23
$$x(t) \leq y(t) + \int_a^t z(\tau)\, y(\tau)\, \exp\, (\int_\tau^t z(\nu)d\nu)d\tau \quad ,$$

for all t *in* [a, b] .

Proof. Let

$$h(t) = \int_a^t z(\tau)\, x(\tau)d\tau \quad .$$

Then 3.1.22 implies that

$$h'(t) - z(t)\, h(t) \leq z(t)\, y(t)$$

at all points of continuity of $z(t)\, x(t)$. Multiplying through by $\exp\, (-\int_a^t z(\tau)d\tau)$, we observe that the left-hand side of the resulting expression is an exact differential (is $\frac{d}{dt}\, (\exp(- \int_a^t z(\tau)d\tau)\, h(t))$) . Integrating we have that

$$\exp(-\int_a^t z(\tau)d\tau)\, h(t) \leq \int_a^t z(\tau)\, y(\tau)\, \exp\, (-\int_a^\tau z(\nu)d\nu)d\tau)$$

or

$$\int_a^t z(\tau)\, x(\tau)d\tau \leq \int_a^t z(\tau)\, y(\tau)\, \exp(\int_\tau^t z(\nu)d\nu)d\tau) \, ,$$

and this relation obviously implies 3.1.23. Hence our lemma is proved.

3.1.24 *COROLLARY*

In the special case that in the inequality $y(t) = K > 0$, K *constant, and* $x(t)$ *and* $z(t)$ *are both non-negative, real-valued piecewise continuous functions defined on* $[a, b]$ *, the inequality* 3.1.23 *takes the simpler form:*

3.1.25 $$x(t) \leq K \exp \int_a^t z(\tau)d\tau.$$

Let us now use the Lemma 3.1.21 to show continuous dependence on initial data for solutions at 3.1.5.

3.1.26 *THEOREM*

Under the hypotheses of Theorem 3.1.18 *, it is also true that the solutions* 3.1.18 *are continuous functions of* x^o *and* t_o *.*

Proof. Let x and y be solutions of 3.1.5 corresponding to x^o at t_o and y^o at τ_o respectively. That is,

$$x(t) = x^o + \int_{t_o}^t f(\tau, x(\tau))d\tau \quad ,$$

and, because of the group property:

$$y(t) = y^o + \int_{\tau_o}^t f(\tau, y(\tau))d\tau = y(t_o) + \int_{t_o}^t f(\tau, y(\tau))d\tau \quad .$$

Then

$$\|x(t) - y(t)\| \leq \|x^o - y(t_o)\| + \int_{t_o}^t \|f(\tau, x(\tau)) - f(\tau, y(\tau))\|d\tau$$

$$\leq \|x^o - y(t_o)\| + \int_{t_o}^t \chi(\tau) \ \|x(\tau) - y(\tau)\| \, d\tau \quad .$$

Applying the Corollary 3.1.24 we have that

$$\|x(t) - y(t)\| \leqslant \|x^o - y(t_o)\| \quad \exp\left(\int_{t_o}^{t} \chi(\tau)d\tau\right)$$

$$\leqslant \left(\|x^o - y^o\| + \|y^o - y(t_o)\| \right) \quad \exp\left(\int_{t_o}^{t} \chi(\tau)d\tau\right).$$

Since y is continuous, clearly we can make $\|x^o - y^o\| + \|y^o - y(t_o)\|$ as small as we like by choosing $\|x^o - y^o\|$ and $|\tau_o - t_o|$ sufficiently small. Thus for t on an arbitrary finite interval we can make $\|x(t) - y(t)\|$ arbitrarily small by choosing $\|x^o - y^o\|$ and $|\tau_o - t_o|$ sufficiently small, and our theorem is proved.

Other existence and uniqueness theorems may be proved under various conditions formally different from the Lipschitz condition (3.1.17). For instance, we can prove the following result.

3.1.27 *THEOREM*

Let $f(t,x)$ *be continuous in* x *for fixed* t *and piecewise continuous in* x *and* t . *Let*

3.1.28 $$|\langle (x - y), [f(t,x) - f(t,y)] \rangle| \leqslant \chi(t)\,\|x - y\|^2$$

where $\chi(t)$ *is as in* 3.1.17 . *Then for each* $(t_o, x^o) \in R \times E$, *there exists a unique function* $x(t)$ *for* $t \geqslant t_o$ *which satisfies the differentia[l] equation* 3.1.5 *on an open interval and such that* $x(t_o) = x^o$. *Furthermore these solutions* $x(t, t_o, x^o)$ *are continuous functions of* t_o *and* x^o .

Proof. From the previous theorem we have the existence of at least one solution corresponding to each pair (t_o, x_o) . Hence we need only to concern ourselves with uniqueness and continuous dependence on initial data.

Suppose we have two solutions $x(t)$ and $y(t)$ of equation 3.1.5 having the values x_0 and y_0 respectively at t_0 and τ_0 respectively. We observe that at every point where

$\frac{d}{dt}\|x(t) - y(t)\|^2$ exists, that

$$\frac{d}{dt}\|x(t) - y(t)\|^2 = 2\langle x(t) - y(t), f(x(t), t) - f(y(t), t)\rangle .$$

Hence it follows that

$$\|x(t) - y(t)\|^2 = \|x_0 - y(t_0)\|^2 + 2\int_{t_0}^{t} \langle x(\tau) - y(\tau), f(x(\tau), \tau) - f(y(\tau), \tau)\rangle d\tau ,$$

and by 3.1.28

$$\|x(t) - y(t)\|^2 \leq \|x^0 - y(t_0)\|^2 + 2\int_{t_0}^{t} \chi(\tau)\, \|x(\tau) - y(\tau)\|^2 d\tau .$$

From the Corollary 3.1.24 it follows that

$$\|x(t) - y(t)\|^2 \leq (\|x^0 - y^0\| + \|y^0 - y(t_0)\|)^2 \exp (2\int_{t_0}^{t} \chi(\tau)d\tau).$$

Clearly it follows that, if $x^0 = y^0$ and $t_0 = \tau_0$, then $x(t) = y(t)$ for all t. Furthermore, y is continuous, so we can make $\|x^0 - y^0\| + \|y_0 - y(t_0)\|$ as small as we like if $\|x^0 - y^0\|$ and $|\tau_0 - t_0|$ are chosen sufficiently small. Thus for t on an arbitrary finite interval, we can make $\|x(t) - y(t)\|^2$ arbitrarily small by choosing $\|x^0 - y^0\|$ and $|\tau_0 - t_0|$ sufficiently small. Hence our theorem is proved.

3.1.29 *Remark*

Notice that the Lipschitz condition 3.1.17 implies condition 3.1.28. This can be seen by applying Schwartz's inequality to condition 3.1.28 as follows

3.1.30 $$|\langle (x-y, (f(x,t) - f(y,t)\rangle| \leq ||x-y|| \; ||f(x,t) - f(y,t)|| \; .$$

Thus 3.1.17 implies 3.1.28 .

The converse is not true, so 3.1.28 is a weaker condition than 3.1.17.

We shall now present some additional conditions for uniqueness of solutions of ordinary differential equations. These conditions are essentially based upon differential inequalities and the comparison with the properties of the solutions of suitably defined first order differential equations. We must then first investigate some particular properties of the solutions of first order differential equation $\dot{\chi} = \mu(\chi,t)$ in the plane (χ,t) , in the case in which such equations satisfy the Peano existence condition. We shall in particular be interested in the case in which there exists more than one solution $\chi = \chi(t, \chi_0, t_0)$ of the above equations through the point (χ_0, t_0) . In this case we are interested in studying the properties of the set of all solutions through (χ_0, t_0) and in particular those of two important solutions within this set: the minimal and the maximal solutions defined below.

3.1.31 *Definition*

Consider the scalar differential equation

3.1.32 $$\dot{\chi} = \mu(\chi,t) \; , \quad \chi(t_0) = \chi_0$$

where $\mu(\chi, t)$ *is continuous on* $\Omega \subseteq E^2$. *If* $\chi = \chi_M(t) = \chi_M(t, \chi_0, t_0)$ *is a solution of* 3.1.32 *on a maximal interval of existence such that for all other solutions* $\chi(t, \chi_0, t_0)$

3.1.33 $$\chi(t, \chi_0, t_0) \leq \chi_M(t, \chi_0, t_0)$$

at all points of the interval of existence common to $\chi_M(t)$ *and* $\chi(t)$, *then* $\chi_M(t)$ *is called* *__maximal solution.__* *If* $\chi = \chi_m(t) = \chi_m(t, \chi_0, t_0)$ *is a solution of* 3.1.32 *on a maximal interval of existence such that for all other solutions* $\chi(t, \chi_0, t_0)$

$$\chi(t, \chi_0, t_0) \geq \chi_m(t, \chi_0, t_0)$$

at all points of the common interval of existence, then $\chi_m(t)$ *is called* *__minimal solution.__*

We want now to prove that all equations 3.1.32 have one maximal and one minimal solution through each point (χ_0, t_0). Such solutions obviously coincide in the case of uniqueness. The proof of this will be based upon the following three lemmas.

3.1.34 *LEMMA*

The equation 3.1.32 *can have at most one maximal solution, and at most one minimal solution.*

The proof is obvious and is left as an exercise.

3.1.35 *LEMMA*

Let $\mu_1(\chi, t)$, $\mu_2(\chi, t)$ *be defined and continuous in a region* $\Omega \subset E^2$ *and let*

3.1.36 $$\mu_1(\chi, t) < \mu_2(\chi, t), \quad (\chi, t) \in \Omega .$$

Let $\chi_1(t)$ *be a solution of* $\dot{\chi} = \mu_1(\chi, t)$, *and* $\chi_2(t)$ *be a solution of* $\dot{\chi} = \mu_2(\chi, t)$ *with* $\chi_1(t_o) = \chi_2(t_o) = \chi_o$.

If (a,b), $t_o \in (a,b)$, *is the common interval of existence of* $\chi_1(t)$ *and* $\chi_2(t)$, *then*

$$\chi_1(t) < \chi_2(t) \quad \textit{for} \quad t_o < t < b$$

and

$$\chi_1(t) > \chi_2(t) \quad \textit{for} \quad a < t < t_o \ .$$

Proof. We have $\dot{\chi}_1(t_o) = \mu_1(\chi_o, t_o) < \mu_2(\chi_o, t_o) = \dot{\chi}_2(t_o)$. Thus there is a $\tau > t_o$ such that

$$\chi_1(t) < \chi_2(t) \quad \text{for} \quad t_o < t < \tau \ .$$

Let τ be the largest number for which this inequality holds. If $\tau = b$, the result is proved. In the other case, $\tau < b$, we must have

$$\chi_1(\tau) = \chi_2(\tau) \ .$$

But then for $t_o < t < \tau$ we have:

$$\frac{\chi_1(t) - \chi_1(\tau)}{t - \tau} > \frac{\chi_2(t) - \chi_2(\tau)}{t - \tau} \ .$$

Proceeding to the limit as $t \to \tau - 0$, we get

$$\dot{\chi}_1(\tau) \geq \dot{\chi}_2(\tau) \ .$$

Thus

$$\mu_1(\chi_1(\tau), \tau) \geq \mu_2(\chi_2(\tau), \tau) \ ,$$

which is a contradiction as $\chi_1(\tau) = \chi_2(\tau)$. This establishes the result for $t_o < t < b$. The other part can be proved similarly.

3.1.37 *LEMMA*

Let in the differential equation

3.1.38 $$\dot{\chi} = \mu(\chi,t) + \varepsilon \quad ,$$

$\varepsilon > 0$ *and* $\mu(\chi,t)$ *be continuous in a region* $\Omega \subset E^2$. *Let* $a > 0,\ b > 0$ *be chosen such that*

$$N_0 = \{(\chi,t) : |\chi-\chi_0| \leq b\ ,\ \ |t-t_0| \leq a\} \subset \Omega .$$

Lastly set $\tau = \min\,(a, \frac{b}{M})$, *where* $M = \max\limits_{(\chi,t)\in N_0} |\mu(\chi,t)|$.

Then for each τ' , $0 < \tau' < \tau$, *the equation* 3.1.38 *admits a solution* $\chi(t,\varepsilon)$, $\chi(t_0,\varepsilon) = \chi_0$, *for all sufficiently small* $\varepsilon > 0$. *Further,* $\lim\limits_{\varepsilon\to 0+} \chi(t,\varepsilon)$ *exists uniformly on* $[t_0, t_0 + \tau']$ *and is a solution of* 3.1.32.

Proof. By Theorem 3.1.8 , the equation 3.1.32 admits a solution on $[t_0 - \tau, t_0+\tau]$. Since $\max\limits_{(\chi,t)\in N_0} |\mu(\chi,t) + \varepsilon| \leqq M + \varepsilon$, the equation 3.1.38 admits a solution on $[t_0 - \nu, t_0 + \nu]$, where $\nu = \min\,(a, \frac{b}{M+\varepsilon})$. Thus if $0 < \tau' < \tau$, we can choose ε_0 sufficiently small such that $\tau' \leq \nu < \tau$, so that solutions of 3.1.38 are defined on $[t_0 - \tau' , t_0 + \tau']$ for all small ε , say $0 < \varepsilon \leqq \varepsilon_0$.

The family $F = \{\chi(t,\varepsilon)\}$, $0 < \varepsilon \leqq \varepsilon_0$, is uniformly bounded, and equicontinuous on $[t_0, t_0 + \tau']$. To see this note that

$$\chi(t,\varepsilon) = \chi_0 + \int_{t_0}^{t} \mu(\chi(\tau,\varepsilon), \tau)d\tau + \varepsilon(t-t_0) ,$$

which gives

$$|\chi(t,\varepsilon) - \chi_0| \leqq (M + \varepsilon)(t-t_0) \leqq (M + \varepsilon_0)\,\tau' \quad ,$$

showing that the family F is uniformly bounded. Further

$$|\chi(t,\varepsilon) - \chi(s,\varepsilon)| \leqq |\int_s^t \mu(\chi(\tau,\varepsilon),\tau)d\tau| \leqq M|t - s| \; ,$$

showing that the family F is equicontinuous on $[t_0, t_0 + \tau']$. Thus by Lemma 3.1.7 every sequence $\{\varepsilon_n\}$, $0 < \varepsilon_n \leqq \varepsilon_0$, $\varepsilon_n \to 0$ as $n \to +\infty$, contains a subsequence $\{\varepsilon_{nk}\}$, such that the sequence of functions $\{\chi(t, \varepsilon_{nk})\}$ converges uniformly on $[t_0, t_0 + \tau']$, to a function $\chi_0(t)$. Any function $\chi_0(t)$ obtained in this way is a solution of 3.1.32 . For

$$\chi(t,\varepsilon_{nk}) = \chi_0 + \int_{t_0}^t \mu(\tau, \chi(\tau,\varepsilon_{nk}))d\tau + \varepsilon_{nk}(t-t_0) \; ,$$

and since we can now proceed to the limit, we see that

$$\chi_0(t) = \chi_0 + \int_{t_0}^t \mu(\tau, \chi_0(\tau))d\tau \quad .$$

This shows that $\chi_0(t)$ is a solution of 3.1.32 . This solution $\chi_0(t)$ of 3.1.32 has the property, that if $\chi(t)$, $\chi(t_0) = \chi_0$, is any other solution defined on $[t_0, t_0 + \tau']$, then

$$\chi(t) \leqq \chi_0(t) \; .$$

This is so, for by Lemma 3.1.35 we have

$$\chi(t) < \chi\,(t,\varepsilon_{nk}) \; , \; t_0 < t \leqq \tau'+t_0 \; ,$$

and proceeding to the limit, we find the truth of our assertion.

$\chi_o(t)$ is therefore the maximal solution of 3.1.32 on the interval $[t_o, t_o + \tau']$. Thus $\chi_o(t)$ is unique. This shows that $\lim_{\varepsilon \to 0+} \chi(t,\varepsilon)$ exists uniformly on $[t_o, t_o + \tau]$ and is equal to $\chi_o(t)$.

In a similar way one can prove

3.1.39 *COROLLARY*

Let the functions $\mu^i(\chi,t)$ *be continuous in a region* $\Omega \subset E^2$. *Assume that the* $\{\mu^i\} \to \mu$ *uniformly. Let* $\chi = \chi^i(t, \chi_o, t_o)$ *be a solution of*

$$\dot{\chi} = \mu^i(\chi,t)$$

with $\chi_o = \chi^i(t_o, \chi_o, t_o)$, *then* $\{\chi^i(t, \chi_o, t_o)\} \to \chi(t, \chi_o, t_o)$ *where* $\chi(t, \chi_o, t_o)$ *is a solution of* 3.1.32 *with* $\chi_o = \chi(t_o, \chi_o, t_o)$.

Finally, we can prove the desired result:

3.1.40 *THEOREM*

Through each point (χ_o, t_o) *of* E^2, *there exist a maximum and a minimum solution of equation* 3.1.32.

Proof. Since in Lemma 3.1.37, τ' was arbitrary we can assume that $\chi_o(t)$ is defined on $[t_o, t_o + \tau]$. Using Lemma 3.1.37 $\chi_o(t)$ can be extended to the right of $t_o + \tau$. In a similar way one can construct the extensions of $\chi_o(t)$ to the left of t_o, say considering the solution of equation 3.1.38 with $\varepsilon < 0$. Same consideration for the minimal solutions.

In view of Lemma 3.1.35, we have assumed for $0 < |\varepsilon| \leqq \varepsilon_o$, solutions $\chi(t,\varepsilon)$ of 3.1.38 are defined on $[\alpha,\beta]$, $\alpha < t_o < \beta$,

$$\lim_{\varepsilon \to 0+} \chi(t,\varepsilon) = \chi_M(t), \quad t_o \leq t \leq \beta \; .$$

$$\lim_{\varepsilon \to 0+} \chi(t,\varepsilon) = \chi_m(t), \quad \alpha \leq t \leq t_o \; ,$$

$$\lim_{\varepsilon \to 0-} \chi(t,\varepsilon) = \chi_m(t), \quad t_o \leq t \leq \beta \; ,$$

and

$$\lim_{\varepsilon \to 0-} \chi(t,\varepsilon) = \chi_M(t), \quad \alpha \leq t \leq t_o \; .$$

As the last noteworthy property of maximal and minimal solutions, we shall state the following result due to Montel (2).

3.1.41 *THEOREM*

Let $\chi_M(t, \chi_o, t_o)$ *be the maximal solution with* $\chi_M(t_o, \chi_o, t_o) = \chi_o$. *Consider the sequence* $\{\chi_n\}$: $\chi_n > \chi_o$, *then* $\chi_M(t, \chi_n, t_o) \to \chi_M(t , \chi_o, t_o)$ *as* $\chi_n \to \chi_o$. *Similar property holds for the minimal solutions.*

We are now in the position of proving the following basic result.

3.1.42 *THEOREM*

Let $\mu(\chi,t)$ *be defined and continuous in a region* $\Omega \subset E^2$. *Let* $\omega(t)$ *be a continuous function such that* $(t, \omega(t)) \in \Omega$ *for* $t_o \leq t < b \leq +\infty$ *and*

3.1.43 $$D_r\, \omega(t) = \lim_{h \to 0+} \sup \frac{\omega(t+h)-\omega(t)}{h} \leq \mu(\omega(t), t)$$

Then if $\omega(t_o) \leq \chi_o$, *we have* $\omega(t) \leq \chi_M(t)$ *for all* $t \geq t_o$ *for which* $\omega(t)$ *and the maximal solution* $\chi_M(t)$ *of* 3.1.32 *are both defined, while if* $\omega(t_o) \geq \chi_o$, *then* $\chi_m(t) \leq \omega(t)$ *for all* $t \leq t_o$ *for which* $\omega(t)$ *and the minimal solution* $\chi_m(t)$ *of* 3.1.32 *are both defined.*

Proof. Let $\chi_M(t)$ be defined on (a,b) , $a < t_o < b$. Then the solutions $\chi(t,\varepsilon)$, $\chi(t_o, \varepsilon) = \chi_o$ of

3.1.38 $$\dot{\chi} = \mu(\chi,t) + \varepsilon$$

are defined on $[t_o, t_o + \tau]$ for some $\tau > 0$ and all sufficiently small $\varepsilon > 0$ (Lemma 3.1.37) . Let further $\omega(t)$ be defined on $[t_o, t_o + \tau]$. Then

$$\omega(t) \leqslant \chi(t,\varepsilon) \quad \text{for} \quad t_o \leqslant t \leqslant t_o + \tau \quad .$$

For otherwise, there will exist an interval (α,β) , $t_o < \alpha < \beta \leqslant t_o + \tau$ such that

$$\omega(t) > \chi(t,\varepsilon) \quad \text{for} \quad \alpha < t < \beta \quad ,$$

and $\omega(\alpha) = \chi(\alpha,\varepsilon)$. In this event, however, we must have

$$\frac{\omega(\alpha+h) - \omega(\alpha)}{h} > \frac{\chi(\alpha+h,\varepsilon) - \chi(\alpha)}{h}$$

and proceeding to the limit

$$\lim_{h \to 0+} \sup \frac{\omega(\alpha+h) - \omega(\alpha)}{h} \geqslant \dot{\chi}(\alpha,\varepsilon) = \mu(\chi(\alpha,\varepsilon),\alpha) + \varepsilon = \mu(\omega(\alpha),\alpha) + \varepsilon \quad .$$

This contradicts the assumption

$$\lim_{h \to 0+} \sup \frac{\omega(t+h) - \omega(t)}{h} \leqslant \mu(\omega(t),t)$$

for $t_o \leqq t \leqslant t_o + \tau$. Hence we must have

$$\omega(t) \leqslant \chi(t,\varepsilon) \quad \text{for} \quad t_o \leqslant t \leqslant t_o + \tau \ .$$

Since this holds for all sufficiently small $\varepsilon > 0$, we conclude

$$\omega(t) \leq \lim_{\varepsilon \to 0+} \chi(t,\varepsilon) = \chi_M(t)$$

on $t_o \leq t \leq t_o + \tau$. The assertion that the inequality holds on the common interval of existence on the right of t_o follows now from the extension process previously explained.

In a similar way one can prove the following theorem

3.1.44 *THEOREM*

Let $\mu(\chi,t)$ *be defined and continuous in a region* $\Omega \subset E^2$. *Let* $\omega(t)$ *be a continuous function such that* $(\omega(t), t) \in \Omega$ *for* $t_o \leq t < b \leq +\infty$, *and let*

3.1.45 $$D_\ell\, \omega(t) = \lim_{h \to 0+} \inf \frac{\omega(t+h) - \omega(t)}{h} \geq \mu(\omega(t), t)$$

Then if $\omega(t_o) \geq \chi_o$, *we have* $\omega(t) \geq \chi_m(t)$ *for all* $t \geq t_o$ *for which* $\omega(t)$ *and the minimal solution* $\chi_m(t)$ *of* 3.1.32 *are both defined, while if* $\omega(t_o) \leq \chi_o$, *then* $\omega(t) \leq \chi_M(t)$ *for all* $t \leq t_o$ *for which* $\omega(t)$ *and the maximal solution* $\chi_M(t)$ *of* 3.1.32 *are both defined.*

We are now in the position of proving the additional uniqueness theorems. We shall start the presentation of these results from the statement of the so-called Kamke general uniqueness theorem.

3.1.46 *THEOREM*

Assume that:

i) $f(\chi, t)$ *is continuous in the parallelepiped* Ω *as in Theorem* 3.1.8,

ii) $\mu(\chi,t)$ *is continuous on the rectangle* D: $t_o < t \leq t_o + a$;

$0 \leq \chi \leq 2b$,

iii) $\mu(0,t) = 0$ *for all* $t \in [t_o, t_o + a]$,

iv) *The only solution* $\chi = \chi(t)$ *of the differential equation* $\dot{\chi} = \mu(\chi,t)$ *on any interval* $(t_o, t_o + \zeta)$ *, such that* $\lim_{t\to t_o+} \chi(t) = 0$ *and*

$$\lim_{t\to t_o+} \frac{\chi(t)}{t-t_o} = 0 \quad \textit{is} \quad \chi(t) \equiv 0,$$

v) $|f(x^1,t) - f(x^2,t)| \leq \mu(|x^1 - x^2|,t)$ *for* $t > t_o$ *and* $(x^1,t), (x^2,t) \in \Omega$.

Then the differential equation 3.1.5 *has at most one solution on any interval* $[t_o, t_o + \zeta]$ *satisfying the initial condition* 3.1.6.

This theorem is a particular case of a more general result due to F. Brauer and S. Sternberg which follows next.

3.1.47 *THEOREM*

Assume that

i) $\theta(x,t)$ *is a weakly positive definite**, *real-valued continuous function with one-sided derivatives with respect to* t *and the components* x_i *of the vector* x ,

ii) $f(x,t)$ *is continuous in a bounded region* $t_o \leq t \leq b$, $\theta(x,t) < d$,

iii) $\mu(\chi,t)$ *is a continuous non-negative function defined on* $\chi \geq 0$, $t_o < t \leq b$,

iv) *the only solution* $\chi(t)$ *of the differential equation*

3.1.48 $$\dot{\chi} = \mu(\chi,t)$$

on any interval $t_o \leq t \leq \beta$ *with* $\beta \in (t_o,b)$, *such that*

3.1.49 $$\chi(0) = \dot{\chi}(0) = 0$$

---

*Def. If $\theta(x,t)$ is a real-valued function it is called weakly positive definite if $\theta(x,t) \geq 0$ for all x,t and $\theta(x,t) = 0$ if and only if $x = 0$.

is $\chi(t) \equiv 0$,

v) $$\frac{\partial\theta}{\partial t}(x-y,\ t) + \sum_i^n \frac{\partial\theta}{\partial x_i}(f_i(x,t) - f_i(y,t)) \leq \mu[\theta(x-y,\ t),\ t]$$

in the region $t_o < t < b$, $\theta(x,t) < d$. *Then there exists at the most one solution of the differential equation*

3.1.5 $$\dot{x} = f(x,t)$$

in the interval $t_o \leq t \leq b$.

Proof. Let $x^1(t)$ and $x^2(t)$ be two solutions of the differential equation 3.1.5 with $x^1(t_o) = x^2(t_o)$ and let

$$y(t) = x^2(t) - x^1(t) .$$

Let $\omega(t) = \theta[y(t),\ t]$ and $D_r\ \omega(t) = \lim_{h\to 0+} \sup[\omega(t+h) - \omega(t)]/h$. Clearly from the hypothesis iii) and v) it follows that:

$$D_r\ \omega(t) \leq \mu[\omega(t),\ t] .$$

Assume that there exists τ, $t_o < \tau \leq b$, such that $\omega(\tau) > 0$. If such a τ should not exist, then from the property i) it would follow that $x^1 \equiv x^2$ and the theorem would be proved.

Because of Theorem 3.1.42 there exists then a minimal solution $\chi_m(t)$ of 3.1.48 , an interval $(\tau',\tau]$, $\tau' \geq 0$ with $\omega(\tau) = \chi_m(\tau)$ and $\chi_m(t)$ satisfies

3.1.50 $$0 < \chi_m(t) \leq \omega(t) .$$

The strict positivity of $\chi_m(t)$ follows from the fact that if for all $\theta \in (\tau',\tau]$ we had $\chi_m(\theta) = 0$, then $\chi_m(t)$ could be extended to (t_o,τ)

as a solution of 3.1.48 with the property 3.1.49, and $\chi_m(t)$ would identically vanish and $\chi_m(\tau) = \omega(\tau) = 0$ contradicting the previous assumption. Clearly $\chi_m(t)$ can be continued on the whole interval $(t_0, \tau]$. In addition, since $y(t_0) = 0$ because of the assumption i) and the inequality 3.1.50, we may define $\chi_m(t_o) = 0$.

From the hypothesis ii) it follows that, given any $\varepsilon > 0$ there exists $\delta > 0$ such that

$$||y(t)|| = \left|\left|\int_{t_0}^{t} [f(x^2(\sigma), \sigma) - f(x^1(\sigma), \sigma)]d\sigma\right|\right| < \varepsilon t$$

for $t_0 \leq t < t_0 + \delta$.

From the continuity of $\theta(x,t)$ it follows that this last inequality implies that, given any $\eta > 0$ there exists $\delta > 0$ such that $\theta(y(t), t) < \eta t$ for $t_0 \leq t \leq t_0 + \delta$. Thus $0 < \omega(t)/t < \eta$ for $t_0 < t < t_0 + \delta$ and $\lim_{t \to 0+} \omega(t)/t = 0$. Hence $\dot{\chi}_m(0) = 0$. Then from hypothesis iv) it follows that $\chi_m(t) \equiv 0$ on $[t_0, \tau]$ which again contradicts the hypothesis that $\omega(\tau) = \chi_m(\tau) > 0$. Thus $y(t) \equiv 0$ on $[t_0, a]$ which proves uniqueness.

3.1.51 *Remark.* If we let $\theta(x,t) = |x|$, Theorem 3.1.46 follows immediately from 3.1.47.

If we let $\theta(x,t) = \langle x,x \rangle$ and $\mu(\chi,t) = \chi(t) \langle x,x \rangle$ then we obtain the result presented in Theorem 3.1.27. If we let $\mu(\chi,t) = \theta(\chi)\ \lambda(t)$ with $\int_{0+}^{\infty} \frac{d\chi}{\theta(\chi)} = \infty$ and $\int_{0+}^{b} \lambda(t)dt < \infty$, then as a particular case, the uniqueness conditions due to Osgood follows from Theorem 3.1.47.

In the literature one can find mentioned other uniqueness theorems of different kinds, for instance, the one due to Von Kampen [2] and Perron [1]. Several works (see 3.1.70) are available in which the various uniqueness theorems are compared.

Consider now the case in which the differential equation 3.1.5 is defined and continuous on $E \times R$. In this case, in order for the flow defined by the differential equation to define a dynamical system in E it is necessary that all solutions of 3.1.5 can be extended backward and forward, i.e., $t^+ = +\infty$ and $t^- = -\infty$. We shall now present some criteria for a differential equation to have such a property. It will be seen that this property called *extendability* is closely related to that of uniqueness. The first theorem relates this property to a global Lipschitz condition.

3.1.52 *THEOREM.*

Consider the differential system

3.1.5 $$\dot{x} = f(x,t)$$

where x,t *are* n-*vector,* $f(x,t)$ *is defined and continuous on* $E \times R$ *and (it is uniform-Lipschitz-continuous) satisfies condition* 3.1.17 *with* $\chi(t) = \chi$ *at all points of* $E \times R$. *Then the solutions of such equations have global existence properties for all* $(x^0, t_0) \in E \times R$.

Proof.

From 3.1.17 we have as usual

$$\|x(t, x^0, t_0)\| \leq \|x^0\| + \left\|\int_{t_0}^{t} f(x(\tau; x^0, t_0), \tau)d\tau\right\|$$

$$\leq \|x^0\| + \chi\int_{t_0}^{t} \|x(\tau; x^0, t_0)\| d\tau .$$

By the Lemma 3.1.21 it follows that

$$\|x(t;\, x^o,\, t_o)\| \leqslant [\exp \chi(t - t_o)]\ \|x^o\|$$

which is finite for all finite $t - t_o$. Q.E.D.

We shall now present another theorem on extendability due to Conti [3], which is essentially related to the conditions for uniqueness based upon differential inequalities. The proof of this theorem is based upon the following theorems, which are essentially corollaries of Theorems 3.1.42 and 3.1.44 respectively, but are important enough to be labeled as theorems.

3.1.53 *THEOREM*

Consider the differential system

3.1.5 $$\dot{x} = f(x,t) ,$$

where x, f *are* n-*vectors, and* $f(x,t)$ *is defined and continuous on* $E \times R$. *Let* $\mu(\chi, t)$ *be defined and continuous on* E^2 *and let there exist a functional* $\theta(t,x)$, *defined, continuous, and locally Lipschitzian on* $E \times R$, *with the property that*

3.1.54 $$\limsup_{h \to 0+} \frac{\theta(x+hf(x,t),\, t+h) - \theta(x,t)}{h} \leqq \mu(\theta(x,t),\, t) .$$

Then if $\theta(x^o,\, t_o) \leqq \chi_o$, *and* $x(t)$, $x(t_o) = x^o$, *is any solution of* 3.1.5 *and* $\chi_M(t)$, $\chi_M(t_o) = \chi_o$, *is the maximal solution of the equation* $\dot{\chi} = \mu(\chi, t)$, *we have*

3.1.55 $$\theta(x(t),\, t) \leqq \chi_M(t) , \quad t_o \leqq t < t^+ \leqq +\infty$$

where $\chi_M(t)$ *and* $x(t)$ *are defined on* $[t_o,\, t^+)$.

Proof. This follows from Theorem 3.1.42 when we note that 3.1.54 implies

$$\limsup_{h \to 0+} \frac{\theta(x(t+h), t+h) - \theta(x(t), t)}{h} \leq \mu(\theta(x(t), t), t) .$$

The details are left to the reader.

Exactly in the same way Theorem 3.1.44 implies

3.1.56 *THEOREM*

Let in Theorem 3.1.53 *the inequality*

3.1.57 $$\liminf_{h \to 0+} \frac{\theta(x + hf(x,t), t+h) - \theta(x,t)}{h} \geq \mu(\theta(x,t), t)$$

hold instead of 3.1.54 . *Then if* $\theta(x^o, t_o) \geq \chi_o$, *and* $\chi_m(t)$, $\chi_m(t_o) = \chi_o$ *is the minimal solution of* $\dot{\chi} = \mu(\chi, t)$, *we have*

3.1.58 $$\theta(x(t), t) \geq \chi_m(t), \quad t_o \leq t < t^+ \leq +\infty ,$$

where $\chi_m(t)$ *and* $x(t)$ *are defined on* $[t_o, t^+)$.

From Theorem 3.1.53 the following criterion for extendability due to Conti can be immediately proved.

3.1.59 *THEOREM.*

Under the same hypotheses of Theorem 3.1.53 , *if in addition it is assumed that*

i) $\mu(\chi, t)$ *has the property that any solution of* 3.1.32 *is such that*

ii) $\lim_{||x|| \to \infty} \theta(x,t) = +\infty$ *uniformly in* t *on any finite interval* $[a,b] \subset R$.

Then all solutions of 3.1.5 *are such that* $t^+ = +\infty$.

Proof. If one such solution $x(t)$ were not extendable to $+\infty$, then $\lim_{t \to t^+} \|x(t)\| = +\infty$, as proved in Theorem 3.1.13, for the particular case $\Omega = E \times R$. Then from the hypothesis ii), θ would take arbitrarily large values of any left neighborhood of t^+. This would contradict the inequality 3.1.55 of Theorem 3.1.53.

3.1.60 *Remark.* In a similar way, using Theorem 3.1.56, it is possible to prove the analogous theorem of Theorem 3.1.59 on "backward extendability". Clearly then, if we replace the inequality signs with an equality sign in the relation 3.1.54 and 3.1.57 respectively one could easily derive a theorem on global extendability of the solutions on equation 3.1.5 on the whole interval $(-\infty, +\infty)$.

3.1.61 *Remark.* If we let in Theorem 3.1.59 $\theta(x,t) = |x|$, we obtain a well known criterion due to Wintner [1]. If in addition we choose $\mu(\chi, t) = \theta(\chi)\,\lambda(t)$ again with $\int_{0^+}^{\infty} d\chi/\theta(\chi) = \infty$ then another result of Wintner [4] can be obtained. This result is closely related to the previously mentioned uniqueness criterion of Osgood.

Another important criteria for extendability is the following:

3.1.62 *THEOREM*

If the differential system $\dot{x} = f(x,t)$ *with* $f(x,t)$ *defined in* $E \times R$ *satisfies conditions for existence and in addition for all* $t \in R$

3.1.63 $$|f_i(x,t)| < m \qquad i = 1, \ldots, n$$

where m *is a constant, then any solution of* 3.1.5 *is defined in the whole interval* $(-\infty, +\infty)$.

Proof. Because of Theorem 3.1.13, if some solution is not defined for all t, then we have $t^+ < +\infty (t^- > -\infty)$ and

3.1.64 $$\lim_{t\to t^+} \|x(t)\| = +\infty \qquad \left(\lim_{t\to t^-} \|x(t)\| = +\infty\right).$$

On the other hand existence (even in the Caratheodory sense) implies that the differential equation 3.1.5 together with the initial condition 3.1.6 is equivalent to the integral equations

3.1.65 $$x_i(t) = x_i^o + \int_{t_o}^{t} f_i(x(\tau), \tau)d\tau \qquad (i = i, \ldots, n)$$

and from the inequality 3.1.63 we obtain the estimates

$$|x_i(t) - x_i^o| < mt^+ \qquad (i = 1, \ldots, n)$$

which contradict the limits 3.1.64. Q.E.D.

Conditions 3.1.63, which seem extremely restrictive are,on the contrary,extremely useful since one can always find a suitable reparametrization of the variable t, which transforms an autonomous differential system having existence and uniqueness

3.1.1 $$\dot{x} = f(x)$$

into an equivalent one whose solutions exist in the large. Before stating and proving this very important theorem which clarifies the meaning of the property of existence in the large and its relationship with uniqueness, the following definition must be given.

3.1.66 *DEFINITION*

Two autonomous differential equations 3.1.1 *are called* ***equivalent*** *if the trajectories defined by the flow defined in the space* E *by their solutions coincide.*

3.1.67 *THEOREM*

Given an autonomous differential equation $\dot{x} = f(x)$ *whose solutions have property of existing forall points* $x^0 \in E$, *it is possible to define an autonomous differential equation* $\dot{x} = g(x)$, *equivalent to the given equation and such that the flow induced by the solutions of this latter equation in the space* E *are defined for all* $t \in R$.

Proof . This can be done by a suitable reparametrization of the variable t . For instance let

3.1.68 $$d\tau = [1 + \sum_{1}^{n} f_i^2(x)]dt$$

and instead of the system $\frac{dx}{dt} = f(x)$ consider the system

3.1.69 $$\frac{dx}{d\tau} = \frac{f(x)}{1 + \sum_{1}^{n} f_i^2(x)} = g(x)$$

Clearly the system 3.1.69 satisfies the condition 3.1.63 and therefore its solutions exist in the large. In addition, since the only difference between the given systen and 3.1.69 is a different "time-scale," the trajectories of both systems clearly coincide. Q.E.D.

Loosely speaking, this theorem allows us to conclude that when one has to investigate the qualitative (geometrical) properties of flows of autonomous differential equations with uniqueness, one does not have to worry about the properties of extendability, but one can apply immediately the results presented in the previous chapter.

3.1.70 *Notes and References*

In addition to the problem of finding conditions for the flow defined by the solutions of a differential equation to define a dynamical system, the inverse problem of finding conditions for a dynamical system to be equivalent to the flow defined by the solutions of a differential equation has also been posed. This problem is not elementary and has close relationships with the problem of structural stability (see, for instance, M. Peixoto) some preliminary results due to Grabar [1,2] are available. Applying some results from Bebutov [2,4], Grabar proves that every dynamical system in a separable, locally compact metric space can be topologically imbedded in an Hilbert space H so that the flow is defined by a differential equation $\dot{x} = f(x)$ in H. By this is meant that along each trajectory $x(t)$ the quotient $[x(t+h) - x(h)] / h$ converges strongly as $h \to 0$ to an element $\dot{x}(t)$ in H, and this derivative has a value $f(x)$, where f is a continuous operator in H. Recently the whole problem has been reformulated by O. Hajek [7]. He proves that continuous flows (with unicity and global existence, but not necessarily stationary) on differentiable manifolds are homeomorphic, in the sense described below, to flows defined by differential equations. More precisely, let $t(\alpha,x,\beta)$ denote the value at time $\alpha \in R^1$ of the solution of the given flow with initial condition $(x,\beta) \in P \times R^1$; it is assumed that P is a differentiable n-manifold, that the set $D = \{(x,\theta): (\theta,x,\theta) \in \text{domain } t\}$ of permissible initial data is open in $P \times R$, and that $t: D \times R \to P$ is continuous. Then there exists a differential n-manifold P^*, a homeomorphism $h: D \to P^* \times R$ into and such that always $h(x,\theta) = (h^*(x,\theta),\theta)$ for some $h^*: D \to P^*$, and finally a flow t^* on P^* defined by a differential equation (with domain $h[D]$ necessarily open in $P^* \times R$), with with the property that h takes solutions into solutions, i.e.,

$$h^*(t(\alpha,x,\beta),\alpha) = t^*(\alpha,h^*(x,\beta),\beta) \quad .$$

Theorems for the "Carathèodory systems" can be found in the book by Coddington and Levinson, Chapter 2.

Lemma 3.1.14 can be found in the book by Hartman [4].

The proof of Theorem 3.1.18 is taken from G. S. Jones [7] and is due to A. Bielecki [1].

Lemma 3.1.21 can be found, for instance, in the book by E. Beckenback and R. Bellman. Corollary 3.1.24 was originally proved by Gronwall.

Theorem 3.1.27 and Remark 3.1.29 are due to J. Yorke, (private communication).

Most of the results on scalar equation 3.1.34-3.1.41 are due to P. Montel [1,2]. They are also reported in the book by Kamke [1].

The proof of Kamke's general uniqueness theorem can be found in the book by Kamke [1]. It is reported also in the book by Hartman [4].

A comparison of various uniqueness theorems can be found in the papers by C. Olech [2], W. Walter, and F. Brauer and S. Sternberg.

Theorem 3.1.59 is due to Conti [3].

A converse of this theorem is due to J. Kato and A. Strauss in the paper "On the global existence of solutions and Liapunov Functions" to appear in the Annali di Matematica Pura ed Applicata.

Theorems 3.1.62 and 3.1.67 are due to Vinograd.

3.2 *Further properties of the solutions of ordinary differential equations without uniqueness.*

We shall now present some important qualitative properties of the flows defined by the solutions of ordinary differential equations of the type

3.2.1
$$\dot{x} = f(x,t)$$

satisfying the Peano's existence condition ($f(x,t) \in C^0$). We shall begin this section with the discussion of some theorems which are essentially the generalization to the order n of the results proved for the case of first order differential equations in Theorems 3.1.35 , 3.1.39 , 3.1.40 and 3.1.41 . The first theorem that we shall present is a convergence theorem, which we state without proof.

3.2.2 *THEOREM*

Assume that the vector-valued functions $f^k(x,t)(k = 0,1,2, \ldots)$ *are continuous in a region* $\Omega \subset E \times R$ *and that on each compact subset of* Ω

3.2.3
$$f^k(x,t) \to f^0(x,t)$$

holds uniformly as $k \to \infty$.

Let $x^k(t) = x^k(t,x^{ko},t_{ko})$ *be a solution of the ordinary differential equation*

3.2.4
$$\dot{x} = f^k(x,t)$$

with $x^k(t_{ko}) = x^{ko}$ *and* $(x^{ko}, t_{ko}) \subset \Omega$ *and maximum interval of existence* $I_{x^{ko}}$.

Assume that

$$(x^{ko},t_{ko}) \to (x^o,t_o) \subset \Omega \quad \text{as} \quad k \to \infty .$$

Then there exists a subsequence of solutions $x^{k(i)}(t) = x^{k(i)}(t,x^{k(i)o},t_{k(i)o})$ *of the equations* 3.2.4, *and a solution* $x(t,x^o,t_o)$ *of the ordinary differential equation* $\dot{x} = f^o(x,t)$ *with maximal interval of existence* I_{x^o}, *such that*

$$x^{k(i)}(t) \to x(t,x^o,t_o) \quad \text{as} \quad i \to \infty \qquad \text{for} \quad t \in I_{x^o} \quad ,$$

uniformly for each compact subset of I_{x^o}. *Furthermore for each compact subset* C *of* I_{x^o}, *there exists an* N *such that*

3.2.5 $$C \supseteq \bigcap_{i>N} I_{x^{k(i)0}} .$$

If in addition the solutions of $\dot{x} = f^o(x,t)$ *are unique, then* $x(t,x^o,t_o) = \lim_{k\to\infty} x^k(t,x^{ko},t_{ko})$ *and the convergence is uniform on compact sets in* I_{x^o} .

A version of this theorem on a *given* closed interval $[a,b] \subseteq I_{x^o}$ is due to Kamke [2] , while the proof of the complete theorem on I_{x^o} can be, for instance, found in Hartman [4, Ch. II, Theor. 3.2].

The local version of this theorem is on the other hand an immediate consequence of Theorem 3.1.8 [Hartman 4, Ch. I , Theor. 2.4].

We shall now illustrate some qualitative properties of the solutions of the differential equation 3.2.1 when the hypotheses of Peano's existence theorem are satisfied, but the flow does not have property of uniqueness. We shall begin our presentation with the following

3.2.6 *DEFINITION*

Consider the differential equation 3.2.1 *which satisfies the Peano existence condition. Through all points* $(x^o,t_o) \in E \times R$ *there exist at least one solution* $x(t,x^o,t_o)$ *of the differential equation* 3.2.1 . *Consider now the set of all such solutions* $x(t,x^o,t_o)$ *through a given point* (x^o,t_o) . *For a given* $(x^o,t_o) \in E \times R$ *we shall define:*

$$t_o^{+M} = \sup t^+ (\textit{for all } x(t,x^o,t_o) \textit{ with } x(t_o,x^o,t_o) = x^o)$$

$$t_o^{-M} = \inf t^- (\textit{for all } x(t,x^o,t_o) \textit{ with } x(t_o,x^o,t_o) = x^o)$$

3.2.7
$$t_o^{+m} = \inf t^+ (\textit{for all } x(t,x^o,t_o) \textit{ with } x(t_o,x^o,t_o) = x^o)$$

$$t_o^{-m} = \sup t^- (\textit{for all } x(t,x^o,t_o) \textit{ with } x(t_o,x^o,t_o) = x^o)$$

3.2.8
$$I_o^M = [t_o^{-M}, t_o^{+M}]$$

and

3.2.9
$$I_o^m = [t_o^{-m}, t_o^{+m}]$$

3.2.10 *DEFINITION**

The set $T(\tau,x^o,t_o) \subset E$ *which is the union of all points* $x(\tau,x^o,t_o)$ *reached by some solution* $x(t,x^o,t_o)$ *of the differential equation* 3.2.1 *at the time* $t = \tau$ *is called the* *__reachable (or attainable) set__* *of the differential equation* 3.2.1 *from the point* (x^o,t_o) *at the time* $t = \tau$. *The set* $T(x^o,t_o,I_o^M) = \bigcup_{t \in I_o^M} T(t,x^o,t_o)$ *is called* *__solution funnel__** *of the differential equation* 3.2.1 *through the point* (x^o,t_o) .

---

* Our definition of solution funnel is different from the original definition due to Kamke [2] in the sense that Kamke defines the solution funnel only on the interval I_o^m .

The set $T(\tau,x^o,t_o)$ *can also be called a* *__cross-section of the solution funnel__* *of Equation* 3.2.1 *through* (x^o,t_o) *at the instant* $t = \tau$.

The set $T(x^o,t_o,[a,b]) = \bigcup_{t\in[a,b]\subset I_o^M} T(t,x^o,t_o)$ *is called a* *__segment of the solution funnel.__*

Next we shall investigate some qualitative properties of solution funnels, their cross sections and their boundaries. The first theorems are due to Kamke [2].

3.2.11 *THEOREM*

Let the vector-valued function $f(x,t)$ *be continuous in the region* $\Omega \subset E \times R$. *Let* $(x^o,t_o) \in \Omega$. *Then if* $[a,b] \subset I_o^m$, *the segment of the solution funnel* $T(x^o,t_o,[a,b])$ *is a compact set.*

Proof. If $T(x^o,t_o,[a,b])$ is not bounded, then through the point x^o there exists a sequence of solution curves $x^1(x^o,t_o,t)$, $x^2(x^o,t_o,t)$, ... of Equation 3.2.1 such that for each of these curves there exists at least one instant $t_k \in [a,b] \subset I_o^m$ such that

$$|x^k(x^o,t_o,t_k)| > k .$$

From Theorem 3.2.2 , however, there must exist a subsequence of the sequence $\{x^n(x^o,t_o,t)\}$ above which converges uniformly to a solution curve $x = x(x^o,t_o,t)$ of the Equation 3.2.1 and such that for some $T \in [a,b]$

$$|x(x^o,t_o,t)| \to \infty \quad \text{as} \quad t \to T^-$$

Thus, because of Theorem 3.1.13, the solution $x(x^o,t_o,t)$ does not exist on the whole interval $[a,b]$ which contradicts the hypothesis. Hence $T(x_o,t_o,[a,b])$ is bounded.

Consider now a sequence $\{x^n\} \subset T(x^o,t_o,[a,b])$, $x_n \to x$. Through each point x^n of such sequence there exist at least one solution $x^n = x^n(x^o,t_o,t)$ of the equation 3.2.1. Because of Theorem 3.2.2 it follows that there exists a subsequence of $\{x^n\}$ which converges to a solution curve of 3.2.1 which joins the points x^o and x. Thus $x \in T(x^o,t_o,[a,b])$ and the theorem is proved.

3.2.12 *THEOREM*

Let μ *be a real number. Consider the vector-valued functions* $f(x,t,\mu)$ *continuous in the set* $\Omega \subset E \times R$ *for each* $\mu \in [\mu_o,\mu_1]$. *Assume that on each compact subset of* Ω

$$f(x,t,\mu) \to f(x,t,\mu_o)$$

holds uniformly in (x,t) *as* $\mu \to \mu_o$. *Denote with* $T_\mu(x,t,[a,b])$ *the segment of the solution funnel through the point* (x,t) *for the ordinary differential equation* $\dot{x} = f(x,t,\mu)$. *Consider the segment* $T_{\mu_o}(x^o,t_o,[c,d])$. *Then for all points* (x,t,μ) *sufficiently near to* (x^o,t_o,μ_o) *there exists a segment* $T_\mu(x,t,[c,d])$ *and* $T_\mu(x,t,[c,d]) \to T_{\mu_o}(x^o,t_o,[c,d])$ *as* $(x,t,\mu) \to (x^o,t_o,\mu_o)$ *in the sense that for each* $\varepsilon > 0$ *there exists a* $\tau > 0$ *such that* $(x,t,\mu) \in S((x,t,\mu),\tau)$ *implies* $T_\mu(x,t,[c,d]) \subseteq S(T_{\mu_o}(x^o,t_o,[c,d]),\varepsilon)$.

Proof. If the theorem were not true there would exist a real number $\varepsilon > 0$ and a sequence $(x^k,t_k,\mu_k) \to (x^o,t_o,\mu_o)$ so that for each k at least one of the integral curves $x = x(t,x^k,t_k,\mu_k)$ either would not exist in $[c,d]$ or would not belong to an ε-neighborhood of $T_{\mu_o}(x^o,t_o,[c,d])$. This contradicts Theorem 3.2.2.

From the Theorem 3.2.12 it immediately follows that:

3.2.13 *COROLLARY*

Consider points $x \in T_{\mu_o}(x^o,t_o,[c,d])$. *Consider the set* $M_\mu(x,t) \subset T_\mu(x,t,[c,d])$ *such that* $M_\mu(x,t) \cap \partial T_{\mu_o}(x^o,t_o,[c,d]) = \emptyset$, *then* $M_\mu(x,t) \to \partial T_{\mu_o}(x^o,t_o,[c,d])$ *as* $\mu \to \mu_o$.

We can now proceed with the proof of the renowned Kneser Theorem on the structure of the cross-sections of solution funnels. See Theorem 3.4.37 for a full proof using the concepts of weak invariance.

3.2.14 *THEOREM*

Let the vector-valued function $f(x,t)$ *be continuous in a region* $\Omega \subseteq E \times R$, *then each cross-section of the solution funnel through* (x^o,t_o) : $T(x^o,t_o,\tau), \tau \in I_o^m$, *is a compact, connected set.*

The proof of this theorem is usually based upon the following local theorem, which we state without proof since a more complete proof is in 3.4.

3.2.15 *LEMMA*

Let the vector-valued function $f(x,t)$ *be continuous in the set* $D: |t - t_o| \leq \alpha, |x - x^o| \leq \beta$. *Let* $|f(x,t)| \leq M$ *in* D, $\gamma = \min(\alpha, \frac{\beta}{M})$ *and* $|c - t_o| \leq \gamma$. *Then the cross-section of the solution funnel* $T(x^o,t_o,c)$ *is a compact, connected set.*

3.2.16 *Proof* of Theorem 3.2.14

From Lemma 3.2.15 it follows that for a certain $\mu \in (t_o,t_o^{+m})$ and all $\tau \in (t_o,\mu)$ the theorem is true. Because of continuity it follows that the theorem is also true for all $\tau \in (t_o,\mu+\varepsilon)$, $(\varepsilon > 0)$. In fact, since $T(x^o,t_o,\mu)$ is a compact subset of Ω there exist real numbers α_i and β_i such that each set

$$D_i : |t - \mu| \leq \alpha_i , |x - x'| \leq \beta_i$$

with $(\mu,x') \in T(x^o,t_o,\mu)$, is such that $D_i \subset T(x^o,t_o,\mu)$ and that $H = \cup D_i \subset T(x^o,t_o,\mu)$. By continuity $|f(x,t)| < M$, $(x,t) \in H$ and for all $(x,t) \in T(x^o,t_o,\mu)$ there exist $T(x,t,[\mu,\mu+\gamma))$ with γ as in Lemma 3.2.15.

Consider now the cross-section $T(x,t,\tau)$ with $\tau \in [\mu,\mu+\gamma)$. Clearly

$$T(x^o,t_o,\tau) = \cup \{T(x,t,\tau): (x,t) \in T(x^o,t_o,\mu)\} , \tau \in [\mu,\mu+\gamma) .$$

Suppose now that $T(x^o,t_o,\tau) = T_1 \cup T_2$, where $T_i \neq \emptyset$ $(i = 1,2)$ and $\bar{T}_1 \cap \bar{T}_2 = \emptyset$. If $T(x,t,\tau) \cap T_i \neq \emptyset$, then, because of Lemma 3.2.15, we have $T(x,t,\tau) \subseteq T_i$. Thus one can divide the points of $T(x^o,t_o,\mu)$ into two classes according to whether $T(x,t,\tau)$ belongs to T_1 or to T_2 . Since $T(x^o,t_o,\eta)$ is a continuuum then the two classes must have a common boundary, point b. Assume that b belongs to the first class. Clearly there exist then points c_i, arbitrarily close to b which belong to the second class. From Corollary 3.2.13, then the segments $T(c_i,[\mu,\tau])$ converge to $T(b,[\mu,\tau])$ as $c_i \to b$. There exist then points $T(b,\tau) \subset T_1$, which are arbitrarily close to $T(c_i,\tau) \subset T_2$, which contradicts the assumptions made on T_1 and T_2 and proves the theorem on the whole interval $[t_o,t_o^{+m})$.

3.2.17 *Remark.* Theorem 3.2.14 does not imply that, given (x^o,t_o) and $(x^1,t_1) \in \partial T(x^o, t_o, I_o^M)$, there exist a solution $x = x(t, x^o,t_o)$ of the differential equation 3.2.1 with $x^1 = x(t^1, x^o,t_o)$, such that $x(t,x^o,t_o) \subset \partial T(x^o,t_o,I_o^{+M})$ for all $t \in [t_o,t_1 + \varepsilon]$, $\varepsilon > 0$. This fact was pointed out by Nagumo and Fukuhara with the aid of an example.

An additional example was proposed by Digel.

The following example is to show how for $t \notin I_o$ the funnel section need not be closed nor connected.

3.2.19 *Example*

$$\dot{x}(t) = f(t,x,y)$$

$$\dot{y}(t) = g(t,x,y)$$

where $g \equiv 0$ for $t \in [0,5]$
and for $t \in [0,4]$

$$f(t,x,y) = \begin{cases} (2-x)^{1/2} & 1 \leq x \leq 2 \\ x^{1/2} & 0 \leq x \leq 1 \\ 0 & x \leq 0 \quad \text{or} \quad x \geq 2 \end{cases}$$

and for $t \in [4,5]$

$$f(t,x,y) = (5-t)\, f(0,x,y)$$

and for $t \in [5,\infty)$

$$f(t,x,y) = 0$$

$$g(t,x,y) = \frac{2\sin\left(\frac{\pi}{2}x\right)\sin(t+5)\cos(t+5)}{1-\sin\left(\frac{\pi}{2}x\right)\sin^2(t+5)}$$

The solution funnel of $(0,0,0)$ has cross-section which is not connected for $t \geq 4+\frac{\pi}{2}$ and is not closed for $t > 4+\frac{\pi}{2}$.

The (x,y)-cross-section is:

$$[0,t^2/4]\times\{0\} \quad \text{for} \quad t\in[0,2],$$

$$[0,-t^2/4+2t-2]\times\{0\} \qquad t\in[2,4],$$

$$[0,2]\times\{0\} \qquad t\in[4,5],$$

$$\{(x,(1-\sin\left(\tfrac{\pi}{2}x\right)\sin^2(t+5))^{-1}-1)\ |x\in[0,2]\} \qquad t\in[5,5+\tfrac{\pi}{2}),$$

$$\{(x,(1-\sin\left(\tfrac{\pi}{2}x\right)\sin^2(t+5))^{-1}-1)\ |x\in[0,1)\cup(1,2]\} \qquad t\geq 5+\tfrac{\pi}{2}.$$

Note in particular for $t = 5+\pi$, the cross-section is

$$[0,1)\times\{0\}\cup(1,2]\times\{0\}$$

which is a bounded but not closed set.

3.2.20 *Remark.* The following remarks on cross-section of solution funnels are due to Pugh [1] and Nagumo and Fukuhara. Let $\tau \in I_0^m$.

i) There exists a $T(\tau,x^0,t_0)$ which is not arcwise connected.

ii) There exists a non-simply-connected continuum which is a $T(\tau,x^0,t_0)$.

iii) There exists a continuum which is not a $T(\tau,x^0,t_0)$ for any differential equation.

iv) Any C^1-polyhedron is a funnel-section.

We shall now discuss a theorem due to Fukuhara on the qualitative properties of the boundary of the solution-funnel. See 3.4.33 for a simple proof using results on weak invariance.

3.2.21 *THEOREM*

Let the vector-valued function $f(x,t)$ *be continuous in the region* $\Omega \subset E \times R$. *Let* $(x^o,t_o) \in \Omega$. *Let* $(x^1,t_1) \in \partial T(x^o,t_o,I_o^m)$, $(x^1,t_1) \neq (x^o,t_o)$. *Then there exists a solution of the differential equation* 3.2.1, $x = x(t,x^o,t_o)$ *such that* $x^1 = x(t_1,x^o,t_o)$ *and such that for all* $t \in [t_o,t_1]$, $x(t,x^o,t_o) \in \partial T(x^o,t_o,I_o^m)$.

This theorem is true for even more general flows than the ones defined by the solutions of an ordinary differential equation without uniqueness. This will be proved in Section 3.4. The proof of Theorem 3.2.21 is therefore omitted.

Another property of differential equations which we must mention is that of the differentiability of the solutions with respect to their initial conditions. This property is illustrated in Theorems 3.2.24 and 3.2.29. Its proof is based upon the following lemma.

3.2.22 *Lemma*

Let the conditions of Theorem 3.1.8 *hold. Let* $x_1(t)$, $x_2(t)$ *be* ε_1*-approximate* and* ε_2*-approximate solutions of* 3.2.1 *defined on an interval* $[a,b]$ *containing* t_o . *And let* N *be a compact subset of* Ω *such that* $(t,x_1(t))$, $(t,x_2(t))$ *remain in* N *for* $t \in [a,b]$. *Then*

---

* Def. A function $\psi(t)$, defined and continuous on an interval I_τ is called an ε-approximate solution of the differential equation 3.2.1 if the following conditions hold: i) $(\psi(t),t) \in \Omega$ for $t \in I_\tau$. ii) $\psi(t)$ is continuously differentiable on $I_\tau \setminus S$, where S is a finite set of numbers, iii) $\|\dot\psi(t) - f(\psi(t),t)\| \leq \varepsilon$ for $t \in I_\tau \setminus S$.

3.2.23 $$\|x_1(t) - x_2(t)\| \leq \|x_1(t_o) - x_2(t_o)\| \exp k\,|t - t_o|$$

$$+ (\varepsilon_1 + \varepsilon_2)\,|t - t_o| \exp k\,|t - t_o,|$$

where k *is a Lipschitz constant on* N .

Proof. There exist continuous functions $\theta_1(t)$, $\theta_2(t)$ defined on $[a,b]$ such that $\|\theta_1(t)\| \leq \varepsilon_1$, $\|\theta_2(t)\| \leq \varepsilon_2$ and

$$x_1(t) = x_1(t_o) + \int_{t_o}^{t} f(\tau,x_1(\tau))d\tau + \int_{t_o}^{t} \theta_1(\tau)d\tau \quad ,$$

$$x_2(t) = x_2(t_o) + \int_{t_o}^{t} f(\tau,x_2(\tau))d\tau + \int_{t_o}^{t} \theta_2(\tau)d\tau \quad ,$$

for $t \in [a,b]$. Hence

$$x_1(t) - x_2(t) = x_1(t_o) - x_2(t_o) + \int_{t_o}^{t} [f(\tau,x_1(\tau)) - f(\tau,x_2(\tau))]d\tau$$

$$+ \int_{t_o}^{t} [\theta_1(\tau) - \theta_2(\tau)]d\tau \ .$$

This yields for $t > t_o$,

$$\|x_1(t) - x_2(t)\| \leq \| x_1(t_o) - x_2(t_o)\| + (\varepsilon_1 + \varepsilon_2)(t - t_o)$$

$$+ k\int_{t_o}^{t} \|x_1(\tau) - x_2(\tau)\|\, d\tau \ .$$

And in particular, if $t_o \leq t \leq T$, we get the inequality

$$\|x_1(t) - x_2(t)\| \leq \| x_1(t_o) - x_2(t_o)\| + (\varepsilon_1 + \varepsilon_2)(T - t_o)$$

$$+ k\int_{t_o}^{t} \|x_1(\tau) - x_2(\tau)\|\, d\tau \ .$$

And now by an application of Corollary 3.1.24 we get

$$\|x_1(t) - x_2(t)\| \leqq \|x_1(t_0) - x_2(t_0)\| \exp k(t - t_0)$$

$$+ (\varepsilon_1 + \varepsilon_2)(T - t_0) \exp k(t - t_0) ,$$

and since this holds for all $t_0 \leqq t \leqq T$, we get

$$\|x_1(T) - x_2(T)\| \leqq \|x_1(t_0) - x_2(t_0)\| \exp k(T - t_0)$$

$$+ (\varepsilon_1 + \varepsilon_2)(T - t_0) \exp k(T - t_0) .$$

This is inequality 3.2.23 with t replaced by T, $T \geqq t_0$. When $t \leqq t_0$, the same process yields 3.2.23 again. This proves the lemma.

We are now in the position to prove the result on differentiability of solutions.

3.2.24 *THEOREM*

Let $f(t,x)$ *be continuous and possess continuous partial derivatives with respect to all its arguments in a region* $\Omega \subset E \times R$. *Then any solution* $x(t,x^0,t_0)$ *of* 3.2.1 *considered as a function of* t,t_0,x^0 *possesses continuous partial derivatives with respect to all its arguments and* $\frac{\partial^2 x}{\partial t^2}$ *also exists and is continuous. In particular, if* x_k^0 *is the* kth *component of* x^0, *then* $\frac{\partial x(t,x^0,t_0)}{\partial x_k^0}$ *is the solution* $z(t)$, $z(t_0) = e_k$, *of the linear system*

3.2.25 $$\dot{z} = Jf(t, x(t,t_0,x_0)) \cdot z ,$$

where Jf *is the Jacobian matrix of* $f(t,x)$ *with respect to* x *and* e_k *is the* kth *column of the unit matrix* I .

Proof. Let $(x^o,t_o) \in \Omega$, and let $\Delta x^o = h\, e_k$. For sufficiently small h, $(x^o + \Delta x^o, t_o) \in \Omega$. Consider the two solutions $x(t) \equiv x(t, x_o, t_o)$ and $x(t,h) \equiv x(t, x^o + \Delta x^o, t_o)$ of 3.2.1 . We have

$$x(t,h) = x^o + h\, e_k + \int_{t_o}^{t} f(\tau, x(\tau,h))d\tau \quad ,$$

and

$$x(t) = x^o + \int_{t_o}^{t} f(\tau, x(\tau))d\tau \quad .$$

Thus

3.2.26 $$\frac{x(t,h)-x(t)}{h} = e_k + \int_{t_o}^{t} \frac{f(\tau,x(\tau,h)) - f(\tau,x(\tau))}{h}\, d\tau \ .$$

If the limit as $h \to 0$ of the above expression exists, then it is the partial derivative of $x(t,x^o,t_o)$ with respect to x_k^o . Consider now the linear system 3.2.25 . Solutions of this system exist and are unique as the coefficient matrix is a continuous function of t . If $z(t), z(t_o) = e_k$ is a solution of 3.2.25 , we have

3.2.27 $$z(t) = e_k + \int_{t_o}^{t} Jf(\tau, x(\tau)) \cdot z(\tau)d\tau \quad .$$

Using the mean-value theorem one can conclude that

$$f(\tau,x(\tau,h)) - f(\tau,x(\tau)) = [Jf(\tau,x(\tau)) + \Gamma](x(\tau,h) - x(\tau))$$

where $\|\Gamma\| \to 0$ as $h \to 0$ uniformly on any compact set $N \subset \Omega$.

Thus 3.2.26 and 3.2.27 yield

3.2.28
$$\frac{x(t,h)-x(t)}{h} - z(t) = \int_{t_o}^{t} Jf(\tau,x(\tau)) \left[\frac{x(\tau,h)-x(\tau)}{h} - z(\tau)\right] d\tau + \int_{t_o}^{t} \Gamma\left[\frac{x(\tau,h)-x(\tau)}{h}\right] d\tau \ .$$

Using now the inequality 3.2.23 we see that

$$\|x(t,h) - x(t)\| \leqq |h| \exp k|t - t_o| \ ,$$

where k is the Lipschitz constant in $N \subset \Omega$. Hence $\|\frac{x(t,h)-x(t)}{h}\|$ is bounded for all sufficiently small h .

Using this fact the relation 3.2.28 yields

$$\left\|\frac{x(t,h)-x(t)}{h} - z(t)\right\| \leqq \varepsilon(h) + \int_{t_o}^{t} \|Jf(\tau,x(\tau))\| \left\|\frac{x(\tau,h)-x(\tau)}{h} - z(\tau)\right\| d\tau$$

where $\varepsilon(h) \to 0$ as $h \to 0$. Using now Corollary 3.1.24 we get

$$\left\|\frac{x(t,h)-x(t)}{h} - z(t)\right\| \leqq \varepsilon(h) \exp \int_{t_o}^{t} \|Jf(\tau,x(\tau))\| d\tau \ .$$

Proceeding to the limit, we see that

$$\frac{\partial x(t,x^o,t_o)}{\partial x_k^o} = \lim_{h\to 0} \frac{x(t,h)-x(t)}{h} = z(t) \ .$$

This proves the result. Differentiability with respect to t_o follows by considering the system

$$\dot{x} = f(t,x)$$

$$\dot{t} = 1 \ .$$

The details are left to the reader.

The following more comprehensive theorem can be proved by repeated application of the processes explained in the above proof.

3.2.29 *THEOREM*

Let $f(t,x)$ *possess continuous partial derivatives of order* r *in all its arguments in a region* $\Omega \subset E \times R$. *Then the solution* $x(t,x^o,t_o)$ *of* 3.2.1 *possesses continuous partial derivatives of order* r *with respect to all its arguments and*

$$\frac{\partial^{r+1} x(t,x^o,t_o)}{\partial t^{r+1}}$$

also exists and is continuous.

The last property that we want to mention about the solutions of differential equations is the continuity properties of the maximal interval of existence I_o. This result is proved for the special case of differential equations with uniqueness.

3.2.30 *THEOREM*

Let $f(x,t,u)$ *be continuous in a region* $\Omega \subset E \times R \times E^m$ *such that for all* $(x^o,t_o,u) \in \Omega$ *(fixed* u*) the differential equation* 3.2.1 *has a unique solution* $x = x(t,x^o,t_o,u)$. *Let* I_o *be the maximal interval of existence of such a solution, then, in addition to the fact that* $x(t,x^o,t_o,u)$ *is continuous in* I_o, $t^+ = t^+(x^o,t_o,u)$ $[t^- = t^-(x^o,t_o,u)]$ *is a lower [upper] semicontinuous function of its arguments.*

3.2.31 *Notes and References*

Most of the results presented in this section will be discussed again in Section 3.4.

Theorem 3.2.2, 3.2.11, 3.2.12 and the proof 3.2.16 are due to Kamke [2].

Theorem 3.2.14 is due to Kneser [1].
Theorem 3.2.21 is due to Fukuhara [3].

3.3 *Continuous flows without uniqueness.*

In many situations, for instance in control problems, one has to cope with differential equations which do not have the property of uniqueness of solutions. In this section we shall extend some of the results presented in Chapter 1 to cover these more general cases. In the literature not much work has been done along these lines and the results that will be presented are not complete. In addition, some of the references have been impossible to consult. We present these preliminary notes since we believe that this will be a future fruitful research area.

In this presentation we will not define abstractly the properties of flows without uniqueness, but simply introduce a suitable notation, derive from the theorems presented in the previous section the suitable properties that the flow must have and discuss and extend those properties presented in Chapter 1 and which have particular interest for stability theory.

We shall present some systems of axioms defining dynamical systems which are more general then the one presented in Chapter 1.

We shall first define the concept of local dynamical systems.

This is essentially a generalization of the flow defined by an autonomous differential equation whose solutions have the uniqueness but not the necessarily global existence property. Local dynamical systems were introduced first by T. Ura [7].

We shall give next the following definition due to G. Sell [5,I].

3.3.1 *DEFINITION*

A transformation $\Pi: X \times I_x \to X$, *where* $I_x = (t^-, t^+)$ *is such that* $0 \subset I_x \subseteq R$, *is said to define a local dynamical system on* X *if it has the following properties:*

i) Π *is continuous,*

ii) $\Pi(x,0) = x$ *for all* $x \in X$,

iii) if $t \in I_x, s \in I_x$ *and* $t + s \in I_x$, *then* $\Pi(\Pi(x,t),s) = \Pi(x,t+s)$,

iv) either $t_x^+ = +\infty$ $(t_x^- = -\infty)$ *or for all compact sets* $\Gamma \subset X, x \in \Gamma$, *there exists* $\tau \in I_x$ *such that* $\Pi(x,t) \in C\Gamma$ *for* $\tau \leq t < t^+$ $(t^- < t \leq \tau)$,

v) the interval I_x is *lower semi-continuous, in* x, *i.e., if* $x^n \in X$ *and* $x_n \to x$. *Then* $I_x \subset \liminf I_{x_n}$.

The relationship between the flow defined by the solutions of an ordinary differential equation with uniqueness property, but not necessarily with global existence, can be clarified as follows: The property i) follows from Theorem 2.1.26, the property iii) from Theorem 3.1.18 or any theorem on uniqueness (3.1.46), and the property ii), for example, from Theorem 3.1.8 with the usual conversion $t_0 = 0$. The property iv) follows from Theorem 3.1.13, while v) is expressed by Theorem 3.2.30.

3.3.2 *Remark*

Notice that the local dynamical system (X, I_x, Π) defined in 3.3.1 and even its particular form (E, I_x, Π) is more general than the dynamical system (E, R, Π). In fact, Theorem 3.1.67 does not necessarily hold for a general dynamical system (E, I_x, Π).

We shall now discuss a few axiomatic systems for flows without uniqueness.

M. I. Minkevich, for instance, considers the flow without uniqueness (X, R, P), where X is a compact metric space and $P: X \times R \to C$ is a multivalued ma C is a set of nonempty closed subsets of X and it is metrized by the Hausdorff metric.

3.3.3 *DEFINITION*

A multivalued map $P: X \times R \to C$ *is said to define a flow without uniqueness if the following conditions are satisfied:*

i) $P(x,0) = \{x\}$ *for all* $x \in X$,

ii) $P(P(x,t),s) = P(x,t+s)$ *for all* $x \in X$, *and* $t,s \in R$ *with* $ts \geq 0$,

iii) $y \in P(x,t)$ *implies* $x \in P(y,-t)$ *for all* $x,y \in X$ *and* $t \in R$,

iv) the map $P(x,t),(x \in X, t \in R)$ *is continuous in* t *for each fixed* x.

These axioms are similar to the ones used by E. A. Barbashin [5,7,8]. Other axioms for a flow without uniqueness (dispersive flow) have been proposed by B. M. Budak. They are the following:

3.3.4 *DEFINITION*

Let X *be a metric space,* $A,B \subset X$ *and* $N(B,\varepsilon)$ *the* ε*-neighborhood of* B. *Let*

$$\tilde{\alpha}(A,B) = \inf\{\varepsilon : A \subset N(B,\varepsilon)\}$$

and

$$\alpha(A,B) = \max\{\tilde{\alpha}(A,B), \tilde{\alpha}(B,A)\}$$

A mapping $P: X \times R \to X$ *is said to define a* <u>*dispersive dynamical system*</u> *if the following conditions are satisfied:*

i) $P(x,0) = \{x\}$ *for all* $x \in X$,

ii) $P(x,t)$ *is a nonempty compactum for all* $x \in X$, $t \in R$,

iii) $y \in P(x,t)$ *implies* $x \in P(y,-t)$ *for all* $x,y \in X$, $t \in R$,

iv) $P(P(x,t),s) = P(x,t+s)$,

v) $x \to y$ *and* $t \to s$ *implies* $\tilde{\alpha}(P(x,t),P(y,s)) \to 0$ *for all* $x,y \in X$ *and* $t,s \in R$,

vi) $t \to s$ *implies* $\alpha(P(x,t),P(x,s)) \to 0$ *for all* $x \in X$ *and* $t,s \in R$.

vii) A <u>*motion*</u> *through* $x \in X$ *is a mapping* $P_x : R \to X$ *such that*

a) $P_x(0) = \{x\}$

b) $t < s$ *implies* $P_x(s) \subset P(P_x(t), s-t)$. *The set* $P_x(T)$ *is the* <u>*trajectory*</u> *through* x.

Quite recently I. V. Bronshtein [1,2,3,4,6], K. S. Sibirskii and and I. V. Bronshtein [1], K. S. Sibirskii, V. I. Krecu and I. V. Bronshtein [1] and K. S. Sibirskii and A. M. Stakhi have presented a series of works in which a class of generalized dynamical systems defined as semigroups of multivalued mappings is investigated.

We shall present next the definition given by Bronshtein of semigroups of multivalued mappings.

3.3.5 *DEFINITION*

Let T *be a topological space,* S *a topological semigroup** *with an identity element* e, P *a mapping such that for each point* $x \in T$, *and each element* $s \in S$, *the image set* $P(x,s) \subset T$ *is a nonempty compactum. The triplet* (T,S,P) *will be called* <u>*semigroup of multivalued mappings*</u> *if the following conditions are satisfied*

i) $P(x,e) = \{x\}$ *for all* $x \in T$,

ii) $P(P(x,s_1),s_2)) = P(x,s_1 + s_2)$ *for all* $s_1,s_2 \in S$ *and all* $x \in T$,

iii) for all $x \in T$ *and* $s \in S$ *and for any neighborhood* N(P(x,s)) *of the set* P(x,s) *in* T, *there exists a neighborhood* Q(x) *of the point* x *in* T *and a neighborhood* U(s) *of the element* s *in* S, *such that*

$$P(Q(x), U(s)) \subset N(P(x,s)).$$

From these axioms Bronshtein [2] derives various interesting theorems. In particular, he shows that the axiom (vi) of the Definition 3.3.4 is a consequence of the first five axioms.

---

*i.e., a topological space with binary associative multiplication operation which is continuous in the family of components.

Quite recently E. Roxin [5,7,8] has introduced a set of axioms defining a "general control system." These systems may have rather important application in the study of the qualitative properties of differential equations without uniqueness. In what follows, we shall briefly present some of the results obtained by Roxin.

3.3.6 *DEFINITION*

Let X *be a locally compact metric space. Let for* $x,y \in X$

$$\rho(x,y) = \frac{d(x,y)}{1+d(x,y)}$$

where $d(x,y)$ *is the given metric on* X.

Let for $x,y \in X$ *and* $A,B \subset X$

$$\rho(x,B) = \rho(B,x) = \inf\{\rho(x,y); y \in B\}$$

$$\rho^*(A,B) = \sup\{\rho(x,B); x \in A\}$$

$$\rho(A,B) = \rho(B,A) = \max\{\rho^*(A,B), \rho^*(B,A)\}.$$

$$S(A,\varepsilon) = \{x \in X; \rho(x,A) < \varepsilon\}$$

The triplet (X,R,F) *is called* *__general control system__* *if the following axioms are satisfied.*

i) $F(x,t_0,t)$ *is a closed nonempty subset of* X, *defined for all* $x \in X$ *and* $t_0,t \in R$,

ii) $F(x,t_0,t_0) = \{x\}$ *for all* $x \in X$ *and* $t_0 \in R$,

iii) for $t_0 \leq t_1 \leq t_2$ *and* $x \in X$

$$F(x,t_0,t_2) = \bigcup\{F(y,t_1,t_2) : y \in F(x,t_0,t_1)\},$$

iv) for each $y \in X$, $t_0 \leq t_1$ *there exists some* $x \in X$ *such that*

$$y \in F(x,t_0,t_1),$$

v) for each $x \in X$, $t_o \leq t_1$, $\varepsilon > 0$ *there exists a* $\delta > 0$ *such that* $|t-t_1| < \delta$ *implies*

$$\rho(F(x,t_o,t\ ),\ F(x,t_o,t_1)) < \varepsilon\ ,$$

vi) for each $x \in X$, $t \leq s$, $\varepsilon > 0$, *there exists a* $\delta > 0$ *such that*

$$\rho(x,y) < \delta,\ |t - t^1| < \delta,\ |s - s^1| < \delta,\ t^1 \leq s^1$$

implies

$$\rho^*(F(y,t^1,s^1),\ F(x,t,s)) < \varepsilon\ .$$

The principal results proved by Roxin [5] for the general control system (X,R,F) are the following:

3.3.7 *THEOREM*

If $A \subset X$ *is compact and* $t \geq t_o$, *then* $F(A,t_o,t)$ *is compact.*

3.3.8 *THEOREM*

If $A \subset X$ *is a continuum and* $t_o \leq t_1$, *then* $F(A,t_o,[t_o,t_1]) = \bigcup\{F(A,t_o,t) : t \in [t_o,t_1]\}$ *is a continuum.*

Notice that $F(x,t_o,t)$ has been so far defined only for $t \geq t_o$. Both for the theory of control systems as well as for the study of the qualitative properti of differential equations it is important to define the multivalued map F also for $t < t_o$.

3.3.9 *DEFINITION*

Let $G(x,t_o,t)$ *be defined by*

$$y \in G(x,t_o,t) \longleftrightarrow x \in F(y,t,t_o)\ .$$

It then follows that

3.3.10 *THEOREM*

i) If $x \in X$ *and* $t \leq t_0$, $G(x,t_0,t)$ *is a closed nonempty subset of* x.

ii) $G(x,t_0,t_0) = \{x\}$.

iii) If $x \in X$ *and* $t_0 \geq t_1 \geq t_2$, *then*

$$G(x,t_0,t_2) = \bigcup \{G(y,t_1,t_2);\ y \in G(x,t_0,t_1)\}.$$

iv) If $x \in X$ *and* $t_0 \geq t_1$, *there exists a* $y \in X$ *such that* $x \in G(y,t_0,t_1)$.

3.3.11 *Remark*

Notice that $G(x,t_0,t)$ does not satisfy a continuity condition as strong as axiom v) of Definition 3.3.6.

Notice that if $s \leq t_0$ and $A \subset X$ and $G(A,t_0,s)$ are compact sets, then $G(A,t_0,t)$ is compact for all $t \in [s,t_0]$, since

3.3.12 $$G(A,t_0,t) \subset F(G(A,t_0,s),s,t)$$

3.3.13 *Remark*

Notice then the flow defined by the solution of an ordinary (autonomous) differential equation satisfies the Definition 3.3.6. The set $F(x,t_0,t)$ is in this case the cross section of the solution funnel through (x,t_0). The trajectory defined in the usual way for the mapping $F(x,t_0,t)$ is then the solution funnel through (x,t_0).

We shall now proceed with the definition of the usual element for multivalued flows. On this subject there are some differences between the terminology used by Roxin [5] and that used in the Russian Literature. We shall adopt the terminology used by Roxin [5].

3.3.14 *DEFINITION*

A set $A \subset X$ *is called strongly invariant, if for all* $t \geq t_o$, $F(A,t_o,t) \subset A$ *and* $G(A,t,t_o) \subset A$, *positively strongly invariant, if for all* $t \geq t_o$, $F(A,t_o,t) \subset A$, *negatively strongly invariant if for all* $t \geq t_o$, $G(A,t,t_o) \subset A$, *weakly invariant if for all* $t \geq t_o$ *and all* $x \in A$ $F(x,t_o,t) \cap A \neq \emptyset$ *and* $G(x,t,t_o) \cap A \neq \emptyset$, *positively weakly invariant, if for all* $t \geq t_o$ *and all* $x \in A$, $F(x,t_o,t) \cap A \neq \emptyset$ *and negatively weakly invariant if for all* $t \geq t_o$ *and all* $x \in A$, $G(x,t,t_o) \cap A \neq \emptyset$.

Roxin [5] proves the following important property of invariant sets. Many similar results follow in Section 3.4.

3.3.15 *THEOREM*

If a set $A \subset X$ *is positively weakly invariant, so is its closure* $\bar{A}$.

For the case of weakly invariant set for a semigroup of multivalued mappings (3.3.5), Bronshtein [2] proves that

3.3.16 *THEOREM*

Every (weakly) invariant bicompact set contains a (weakly) minimal bicompact set.

Clearly one can define weak and strong stability properties of sets, as well as weak and strong limit sets, attractors, asymptotically stable sets etc.

As an example we give next the definition of weak and strong stability for compact sets.

3.3.17 *DEFINITION*

A set $A \subset X$ *is said to be strongly stable if for all* $t_o \in R$ *and* $\varepsilon > 0$, *there exists a* $\delta = \delta(\varepsilon,t_o) > 0$, *such that* $F(S(A,\delta),t_o,t) \subset S(A,\varepsilon)$ *for all* $t \geq t_o$

and <u>*weakly stable*</u> *if for all* $t_o \in R$ *and* $\varepsilon > 0$, *there exists a* $\delta = \delta(\varepsilon,t_o) > 0$, *such that* $\rho(y,A) < \delta$, *there exists a trajectory* $\phi(x)$ *with* $\phi(t_o) = x_o$, *such that* $\rho(\phi(t),A) < \varepsilon$ *for all* $t \geq t_o$.

Notice that the stability properties defined so far have always been strong properties, and the Liapunov stability theory that we have discussed in these notes characterizes strong properties. In Section 3.4 and 3.9 we shall present theorems for the characterization of weak stability properties, for ordinary differential equations. A few general results for the case of ordinary differential equations without uniqueness can also be found in the paper by G. Sell [5].

3.4 *Further results on nonuniqueness**

3.4.1 *Notations and terminology*

We will let E denote n-dimensional Euclidean space with some norm $|\cdot|$ and distance $d(\cdot,\cdot)$ given by $d(x,y) = |x - y|$. Let $B(b,x) = \{y: |x - y| \leq b\}$.

Sequences will always be subscripted by positive integers. When discussing the convergence of a sequence it will always be assumed we mean "convergence as the subscripts tend to $+\infty$." Hence "$x_n \to x$" means "$x_n \to x$ as $n \to \infty$." We say a real-valued or vector valued function $p(x)$ is $o(x)$ if $|p(x)| / |x| \to 0$ as $|x| \to 0^+$.

A set $X (\subset Y \subset E)$ is open (or closed) relative to Y if there is an open (or closed) set $X_0 \subset E$ such that $X = X_0 \cap Y$. For a function of several variables we often write "$\cdot$" in place of a variable to denote the variable. Hence $\phi(\cdot, x)$ will denote a solution with values $\phi(t,x)$ where x is the initial parameter held fixed. We discuss autonomous systems (A) and nonautonomous

(A) $$\dot{x} = g(x)$$

(B) $$\dot{y} = f(t,y)$$

(B) with solutions denoted by ϕ and ψ respectively, perhaps with subscripts or superscripts attached.

3.4.2 *DEFINITION*

A <u>*curve*</u> γ *is a continuous function mapping some interval* D_γ *into* E. *We always let* D_γ *represent the domain of the curve* γ. *If* F *is a family of curves, we say* $\gamma^* \in F$ *is an* <u>*extension*</u> *of* γ *if* $D_\gamma \subset D_\gamma^*$ *and* $\gamma = \gamma^*$ *on* D_γ. *We say* γ *is* <u>*maximal*</u> *in* F *if the only extension of* γ *is* γ *itself; that is, maximal refers to the domain and* <u>*not*</u> *to the function values. For any curve* γ, *the notation* $\gamma(\cdot;t,x)$ *means that* $t \in D_\gamma$ *and* $\gamma(t;t,x) = x$.

---

*This section is due to James A. Yorke.

We also write $\gamma(\cdot,x)$ *for* $\gamma(\cdot;0,x)$. *We say* F *is a* *<u>(right) Λ-family</u>* *if* F *is a family of curves such that for each* $x \in \Lambda$ *there exists a* $\delta(x) > 0$ *and a* $\gamma(\cdot,x) \in F$ *such that* $[0,\delta(x)) \subset D_{\gamma(\cdot,x)}$; *that is, there exists an element of* F *beginning at each point of* Λ. *We do <u>not</u> assume* $\gamma(t,x) \in \Lambda$ *for* $t \neq 0$. *We define* P(F) *to be the family of curves with domain* $[0,T]$ *or* $[0,T)$ *for some* $0 < T \leq \infty$, *which are <u>piecewise elements</u> of* F; *that is,* P(F) *is the smallest family of curves such that if* γ_1 *and* γ_2 *are in* F *and* $\tau_1 = \sup D_{\gamma_1}$ *and* $\gamma_1(\tau) = \gamma_2(0)$, *then the curve* α *is in* P(F) *where we define* $\alpha(t) = \gamma_1(t)$ *for* $t \in D_{\gamma_1}$ *and* $\alpha(t + \tau) = \gamma_2(t)$ *for* $t \in D_{\gamma_2}$.

3.4.3 *LEMMA*

For any set Λ *let* F *be a* Λ*-family of curves. Then for any* $\gamma \in F$ *there exists an extension* $\gamma^* \in P(F)$ *of* γ *which is maximal for* P(F).

The proof by Zorn's lemma is similar to the proof of Lemma 4.3 in Strauss and Yorke [2].

3.4.4 *DEFINITION*

We say γ *is an* *<u>ε-solution</u>* *or* *<u>(a right ε-approximate solution)</u>* *for* (A) *if* γ *is a curve with* $0 = \inf D_\gamma < \sup D_\gamma$, $0 \in D_\gamma$, γ *is absolutely continuous and, for almost all* $t \in D_\gamma$, $\frac{d\gamma(t)}{dt}$ *exists and*

$$\left|\frac{d\gamma(t)}{dt} - g(\gamma(t))\right| \leq \varepsilon:$$

Note that a ε-solution for $\varepsilon = 0$ *is actually a solution (on some* $[0,T)$ *or* $[0,T]$).

3.4.5 *LEMMA*

Let Λ *be any subset of* R^n *and* g *be continuous on* Λ. *Assume the set* F_ε *of all ε-solutions of* (A) *is a* Λ*-family. For any compact set* $C \subset \Lambda$ *and any maximal* $\gamma \in F_\varepsilon$, *there either exists a* $t \in D_\gamma$, $t > 0$, *such that*

Proof. Suppose $\gamma(t) \in C$ for $t \in D_\gamma$. Since g is bounded on C, $\frac{d\gamma}{dt}$ is also bounded and so γ is uniformly continuous. Therefore, if $D_\gamma = [0,T)$ where $T \neq \infty$, γ can be extended to $[0,T]$ continuously. If $D_\gamma = [0,T]$, then γ can be extended by "piecing" γ together with some $\alpha(\cdot,\gamma(T)) \in F_\varepsilon$. Therefore if $D_\gamma \neq [0,\infty)$, γ is not maximal.

3.4.6 *DEFINITION*

We say Λ *is* _*locally compact (locally closed)*_ *if for each* $x \in \Lambda$ *there exists* $b = b(x) > 0$ *such that* $B(b,x) \cap \Lambda$ *is compact. It is easy to see that if* C *is a compact subset of a locally compact set* Λ, *then there is a closed neighborhood* N *of* C *such that* $\Lambda \cap N$ *is compact. Let* $G(K) = \sup_{x \in K} |g(x)|$.

3.4.7 *LEMMA*

Let C *be any compact subset of the locally compact set* Λ. *Assume that the family* F_ε *of* ε*-solutions of* (A) *is a* Λ*-family of curves. Then there exists a compact neighborhood* N *of* C *and a* $\kappa = \kappa(C) > 0$ *such that for each* $x \in C$ *and each maximal* $\phi(\cdot,x)$, *we have* $[0,\kappa] \subset D_{\phi(\cdot,x)}$ *and* $\phi(t,x) \in N$ *for* $t \in [0,\kappa]$.

Proof. Choose $b > 0$ such that if N is $\{y:d(y,c) \leq b\}$ then $K_b = N \cap \Lambda$ is compact. By 3.4.5 if $\sup D_\phi < \infty$ for $x \in C$ and $\phi = \phi(\cdot,x) \in F_\varepsilon$, then $T = \inf\{t > 0:\phi(t) \notin K_b\} < \infty$. In fact $b \leq |\phi(T) - \phi(0)| \leq \int_0^T |\frac{d\phi}{dt}| \leq [G(K_b) + \varepsilon]T$. Hence if $\kappa = b[G(K_b) + \varepsilon]^{-1}$, then $T \geq \kappa$.

3.4.8 *LEMMA*

Let g *be continuous on the set* Λ *and let* $\{\phi_n\}$ *be a sequence of* ε*-solutions on* $[0,\kappa]$. *Let* ϕ *be a function,* $\phi = [0,\kappa] \to \Lambda$. *Suppose* $\phi_n(t) \to \phi(t)$ *for all* $t \in \Lambda$. *Then* $\phi_n \to \phi$ *uniformly on* $[0,\kappa]$.

Proof: The sequence $\{\phi_n\}$ is equibounded since $\phi_n(t)$ is in the compact set N for all n and all $t \in [0,\kappa]$. Also $|g|$ is bounded in N so , $\{\phi_n\}$ is equibounded. Hence every subsequence has a uniformly convergent sub-subsequence. Since $\phi_n(t) \to \phi(t)$ pointwise it follows that convergence must be uniform. (See for example Hartman [4, p. 4]).

The following Lemma is a special case of the Tietze's extension theorem. Though it is not essential for later work it does insure smooth progress. See Kelley [1].

3.4.9 *LEMMA. (Tietze).*

If g *is a continuous function from* Λ *relatively closed in the open set* $U \subset R^n$, *then there exists a continuous function* $g^*: U \to R^n$ *such that* $g \equiv g^*$ *on* $\Lambda \cap N$.

The usual existence theorem for the differential equation (A) where $g: \Lambda \to R^n$ says that if g is continuous and Λ is open, then for each $x \in \Lambda$ there exists an $\varepsilon = \varepsilon(x) > 0$ and a solution ϕ such that ϕ is defined on $[0,\varepsilon)$ and $\phi(0) = x$.

We shall prove next an existence theorem for the case in which Λ is locally compact (so in particular Λ may be open or closed).

3.4.10 *DEFINITION*

For $S \subset E$ *and* $x \in S$ *and* $v \in E$, *we say* v *is* <u>*subtangential*</u> *to* S *at* x *if*

$$d(S,x + tv) / t \to 0 \quad \text{as} \quad t \to 0^+$$

If g is continuous and for all $x \in \partial\Lambda$ (boundary of Λ) $g(x)$ is subtangential to Λ at x, then as above there exists a solution ϕ such through x on some $[0,\varepsilon)$. When Λ is open the following theorem is the Peano existence theorem since the boundary of Λ is empty.

3.4.11 *THEOREM*

Let Λ *be a locally compact subset of* E *and let* $g:\Lambda \to E$ *be continuous. Assume* $g(x)$ *is subtangential to* Λ *at* x *for all* $x \in \Lambda \cap \partial\Lambda$. *Then for each* x_0 *there exists a* $\delta = \delta(x_0) > 0$ *and a solution* $\phi(\cdot,x_0)$ *in* Λ *defined on* $[0,\delta)$.

3.4.12 *Examples and Comments*

(i) If $n = 1$ and Λ is the compact interval $[a,b]$, $a \leq b$, and $g:\Lambda \to R$ is continuous, then a necessary and sufficient condition for there to exist for each $x \in \Lambda$ a solution $\phi(\cdot,x)$ on some $[0,\varepsilon)$ is that $g(a) \geq 0$ and $g(b) \leq 0$; that is, that $g(a)$ and $g(b)$ are subtangential to Λ.

(ii) If $n = 2$ and $\Lambda = \{x: |x| = 1\}$, then the necessary and sufficient condition is that $g(x)$ is tangent to Λ. For any set Λ, if x is in the interior of $\bar{\Lambda}$, then every vector is subtangential to Λ at x. If we consider $\Lambda = \{\text{rationals}\}$, a set which is not locally compact, then for any g, $g(x)$ is subtangential to x for all $x \in \Lambda$ yet there exists no solution remaining in Λ through any x unless $g(x) = 0$. In the first two examples Λ is compact and we will see that therefore there exists a solution $\phi(\cdot,x)$ remaining in Λ for each $x \in \Lambda$ such that $D_\phi = [0,\infty)$.

(iii) If Λ is any set and $x \in \Lambda$ and $g(x)$ is not subtangential to Λ at x then there exists no right solution $\phi(\cdot,x)$ on any $[0,\varepsilon)$, $\varepsilon > 0$, for if $\phi(\cdot,x)$ were a solution, then $\phi(t,x) = x + \frac{d}{dt}\phi(0,x)t + o(t)$ and we would be able to choose t arbitrarily small such that

$$o(t) = d(x + t\,g(x),\ \phi(t,x)) < d(x + t\,g(x),\ \Lambda).$$

(iv) If $-g(x)$ is subtangential to Λ, locally compact, for all x in Λ $\partial\Lambda$, then for each $x_0 \in \Lambda$ there is a solution $\phi = \phi(\cdot,x_0)$ for the equation $\dot{x} = -g(x)$ on some interval $[0,\delta)$ for $\delta > 0$. But then ϕ^* given by $\phi^*(t) = \phi(-t)$ is a solution of (A) on $(-\delta,0]$. Therefore if $g(x)$ and $-g(x)$ are both subtangential to the locally compact set Λ for all $x \in \Lambda \cap \partial\Lambda$, then for each $x_0 \in \Lambda$ there is a solution $\phi(\cdot,x_0)$ defined on $(-\delta,\delta)$ for some $\delta(x_0) > 0$.

(v) Write $x = (x_1,x_2) \in E^2$ and choose $S \subset E^2$ such that $(x_1,x_2) \in S$ implies $x_1 = 0$. Let $\Lambda = E^2 - S$ and $g(x) = (0,1) \in E^2$ for all $x \in \Lambda$. Then $g(x)$ and $-g(x)$ are subtangential to Λ for all x but if Λ is <u>not</u> locally compact, there exists an $x_0 \in \Lambda$ with no solution $\phi(\cdot,x_0)$ in Λ defined on any interval $(-\delta,\delta)$ with $\delta > 0$.

We first state Lemma 3.4.13 which is essentially equivalent to Theorem 3.2 in Hartman [4], p. 14 . We need to consider ε-solutions which need not be everywhere differentiable, but Hartman's proof adapts easily. We will say the curves γ_i converge to γ, $\gamma_i \to \gamma$ if for each compact subset K of D_γ, K is a subset of D_{γ_i} for all but finitely many i and $\gamma_i(t) \to \gamma(t)$ uniformly for $t \in K$.

3.4.13 *LEMMA. Let* g *be continuous on the locally compact set* Λ *and assume the family* F *of solutions of* (A) *is a right* Λ*-family of curves. Choose* $\varepsilon_n \to 0^+$ *and* $x_n \to x$ *with* $x_n \in \Lambda$ *and* $x \in \Lambda$. *Let* $\phi_n = \phi_{\varepsilon_n}(\cdot,x_n)$ *be a maximal* ε_n*-solution. Then there exists a (right) maximal solution* $\phi = \phi(\cdot,x)$ *and a subsequence* $\{\phi_{n_i}\} \subset \{\phi_n\}$ *such that* $\phi_{n_i} \to \phi$. *Furthermore* $\sup D_\phi > 0$.

Although $d(x + sg(x),\Lambda)/s$ is assumed to tend to zero in theorem 3.4.11 for each x, it need not do so uniformly for x in a compact set. We, therefore, cannot use polygonal approximations with a "finite number of corners" as is done, for example, by

Coddington and Levinson [2,p.6].

Since Λ is locally compact, for each x there exists $b = b(x) > 0$ such that $K = B(2b(x),x) \cap \Lambda$ is compact. Let $U = \{y:$ for some $x \in \Lambda,\ d(x,y) < 2b(x)\}$. Then $\Lambda \subset U$ and Λ is relatively closed in U. By Tietze's extension theorem, (Lemma 3.4.9), g may be extended continuously to a function $g: U \to R^n$ without changing g on Λ. We shall assume for the duration of the proof that the domain of g is U for the equation (A).

3.4.14 *Proof of Theorem 3.4.11.*

We now define for each $x \in \Lambda$ and $\varepsilon > 0$ a particular ε-solution $\phi_\varepsilon(\cdot,x)$ (with range in U). We willlet $F_\varepsilon = \{\phi_\varepsilon(\cdot,x) : x \in \Lambda\}$. Fix $x \in \Lambda$ and ε.

Define $X(u,x) = x + ug(x)$ and $\rho(u) = d(X(u,x),\Lambda)$. Choose $s = s(x,\varepsilon) \in (0,\varepsilon$ sufficiently small that

$$\rho(s)/s \leq \varepsilon/2 ,$$

and if $\delta = \rho(s) + sG(B(b,x))$, then

$$|g(y) - g(x)| \leq \varepsilon/2 \quad \text{for all} \quad y \in B(\delta,x).$$

We assume s is sufficiently small that there exists an $x^* \in K$ such that

$$d(X(s),\Lambda) = d(X(s),x^*).$$

Let $[0,s] = D_{\phi_\varepsilon(\cdot,x)}$ and define

$$\phi_\varepsilon(t,x) = (s - t)s^{-1}x + ts^{-1}x^* \quad \text{for} \quad t \in [0,s].$$

We now show ϕ_ε is an ε-solution. For $t \in [0,s]$,

$$|\phi_\varepsilon(t) - \phi_\varepsilon(0)| \leq |x^* - x| \leq |x^* - X(s)| + |X(s) - \phi_\varepsilon(s)|$$

$$= \rho(s) + s|g(x)| \leq \delta$$

Hence for $t \in (0,s)$

$$|g(\phi_\varepsilon(t)) - \frac{d}{dt}\phi_\varepsilon(t)| \leq |g(\phi_\varepsilon(t)) - g(x)| + |g(x) - (x^* - x)s^{-1}|$$

$$\leq \frac{\varepsilon}{2} + |(x + sg(x)) - x^*|s^{-1}$$

$$\leq \frac{\varepsilon}{2} + \rho(s)s^{-1} \leq \frac{\varepsilon}{2} + \frac{\varepsilon}{2} \quad .$$

F_ε is, therefore, a Λ- family of ε-solutions. By Lemma 3.4.3 there exists an ε-solution $\sigma = \sigma(\cdot,x) \in P(F_\varepsilon)$ which is maximal in $P(F_\varepsilon)$. Let F_ε^U be the set of all right ε-solutions with range in U (domain of g is now U). Note $P(F_\varepsilon) \subset F_\varepsilon^U$. Suppose σ is not maximal in F_ε^U. Then if $T = \sup D_\sigma$, $\lim_{t\to T} \sigma(t)$ exists and equals some $y \in U$. We now show that y is in Λ. By definition of $P(\cdot)$, for each $0 \leq t \leq T$ there is $t_* \leq t$ and $t^* \in (t,T)$ such that for some $\phi_\varepsilon \in F_\varepsilon$

$$\sigma(t) = \phi_\varepsilon(t - t_*, \sigma(t_*)) \quad \text{for} \quad t \in [t_*, t^*]$$

and $[0, t^* - t_*] = D_{\phi_\varepsilon}$. By choice of F_ε, $\sigma(t^*)$ is some $x^* \in \Lambda$. By choosing $t_n \nearrow T$, there exists $t_n^* \in (t,T)$ such that $\sigma(t_n^*) \in \Lambda$. But $\sigma(t_n^*) \to y$ so y is in $\partial\Lambda$ and since Λ is closed relative to U, each y in $U \cap \overline{\Lambda}$ is also in $U \cap \Lambda$. Choose $\phi_\varepsilon(\cdot,y) \in F_\varepsilon$. We may extend σ by setting $\sigma(T + s) = \phi_\varepsilon(s,y)$ for $s \in D_{\phi_\varepsilon(\cdot,y)}$, contradicting the maximality of σ in $P(F_\varepsilon)$. Therefore each σ in F_ε is maximal in U.

Choose $\{\varepsilon_n\} \subset (0,\infty)$ with $\varepsilon_n \to 0^+$ and $\{\phi_n(\cdot,x_o)\}$, a sequence of maximal ε_n-solutions in F_{ε_n}. Applying Lemma 3.4.13 with the domain of g equal to U and $x_n \equiv x_o$ we find some subsequence $\{\sigma_i\} = \{\phi_{n_i}\}$ of $\{\phi_n(\cdot,x_o)\}$ converges to a <u>maximal solution</u> $\phi(\cdot,x_o) = \phi$ (with range in U) uniformly on compact subsets of D_ϕ. Choose $t \in D_\phi$. For each i, there are t_{i*} and t_i^*, $t_{i*} \leq t < t_i^*$, where $y_i = \sigma_i(t_{i*})$ and $x_i^* = \phi_i(t_i^*)$ are in Λ. But by choice of $s(\varepsilon_{n_i}, x_i) \leq \varepsilon_{n_i}$, $t_{i*} \to t$ and $\{\sigma_i(t_{i*})\}$ converges to $\phi(t)$ so $\phi(t) \in \Lambda$ for all $t \in D_\phi$, completing the proof of the theorem.

The usual global theorem for right-maximal solutions ϕ when Λ is open says if $T = \sup D_\phi < \infty$, then ϕ tends to the generalized boundary of Λ, that is, $\phi(t)$ leaves every compact subset of Λ as $t \to T$. The same result holds when Λ is locally compact but the result is stronger now since $\phi(t)$ must remain away from $\partial\Lambda \cap \Lambda$ as $t \to T$. We state the result as follows for Λ a locally compact set

3.4.15 *THEOREM*

Let ϕ *be a right-maximal solution and let* $T = \sup D_\phi$ *be less than* ∞. *Then for any sequence* $\{t_n\} \subset D_\phi, t_n \to T, \{\phi(t_n)\}$ *has no limit points in* Λ.

Proof: Assume the theorem is false and that $x_n = \phi(t_n)$ has a limit point x. Let $\phi_n(\cdot,x_n)$ be the right maximal solutions $\phi_n(t,x_n) = \phi(t + t_n)$. Theorem 3.4.13 (letting all $\varepsilon_n = 0$) implies some subsequence $\{\phi_{n_i}\}$ of $\{\phi_n\}$ converges to a right maximal solution ϕ^*. Choose $\varepsilon > 0$, $\varepsilon \in D_{\phi^*}$. Theorem 3.4.13 implies that for all but a finite number of n_i, ϕ_{n_i} is defined on $[0,\varepsilon]$, and so ϕ is defined at $t_n + \varepsilon$ which contradicts $T = \sup D_\phi$. Hence $\{\phi(t_n)\}$ has no limit points.

The rest of this section deals with invariant and weakly invariant sets though in some problems the connection with invariant sets becomes apparent only after some discussion. We assume that for the equation (A), g is continuous on the open set $U \subset R^n$.

3.4.16 *DEFINITION*

A set $W \subset U$ *is called* *positively (negatively) weakly invariant* *for* (A) *if for each* $x \in W$ *there exists a maximal solution* $\phi(\cdot,x) = \phi$ *such that* $\phi(t,x) \in W$ *for all* $t \in [0,\sup D_\phi)$ *(for all* $t \in (\inf D_\phi, 0]$). W *is* *weakly invariant* *if it is both positively and negatively invariant. A set* $S \subset U$ *is called* *positively (negatively) invariant* *if for each* $x \in S$ *and each* $\phi(\cdot,x)$, $\phi(t,x) \in S$ *for all* $t \in [0,\sup D_\phi)$ *(for all* $t \in (\inf D_\phi, 0]$).

The term "weak" invariance seems to have been first used by Roxin [5]. Yoshizawa [10] used the term semi-invariance.

Note that S is positively invariant or weakly positively invariant for (A) if and only if it is negatively invariant or weakly negatively invariant for

$$(-A) \qquad \dot{x} = -g(x),$$

since $\phi(t)$ is a solution if $\phi(-t)$ is a solution of (-A). Therefore when we state results for positive or negative (weak) invariance, "positive" and "negative" may be everywhere substituted for each other and the results will remain true.

We shall now give some simple propositions and non-trivial theorems. Results on "weak invariance" cannot be strengthened by substituting "invariance" nor can the word "positive" be inserted in these results. The "relative" closure of a set $S \subset U$, relative to U $(\bar{S} \cap U)$ will be denoted $\bar{S}^U$ and $\partial^U S = U \cap \partial S$.

3.4.17 *PROPOSITION*

The set W *is weakly invariant if and only if* W *is the union of trajectories .*

3.4.18 *PROPOSITION*

The set W *is positively invariant if and only if* $U \setminus W$ *is negatively invariant.*

3.4.19 *PROPOSITION*

If W *is compact, and* ϕ *is a maximal solution, such that* $\phi(t) \in W$, *for* $t \in D_\phi$, *then* $\sup D_\phi = + \infty$.

3.4.20 *PROPOSITION*

If W *is positively weakly invariant, then* $\bar{W}^U$ *is positively weakly invariant.*

Proof: Choose $x_n \in W$ such that $x_n \to x \in \partial^U W$ and choose $\phi_n = \phi(\cdot,x_n)$ right maximal solutions such that $\phi_n(t) \in W$ for $t \in D_{\phi_n}$. By Lemma 3.4.13 $\phi_n(t)$ converges (uniformly on compact sets) to a right solution $\phi(\cdot,x)$. Hence $\phi(t,x) \in \bar{W}^U$ for $t \in D_\phi$.

3.4.21 *THEOREM*

The relatively closed set W *is positively (or negatively) weakly invariant if and only if* $g(x)$ *(or* $-g(x)$*) is subtangential to* W *at* x *for all* $x \in W$.

Proof: Assume $g(x)$ is subtangential to W for all $x \in W$. Since W is relatively closed in an open set, it is locally compact. If we restrict the domain of g to W Theorem 3.4.11 says that for each x there is a solution $\phi = \phi(\cdot,x)$ in W for $t \geq$ which is maximal for (A) on W. Theorem 3.4.15 implies that if $T = \sup D_\phi < \infty$, then and $\phi(t_n) \to x_T$, then $x_T \in \partial W - W = \partial W \cap \partial U$. Hence ϕ is also maximal in U, letting the domain of g be U. Hence W is positively weakly invariant. If, however, there is an $x \in W$ such that $g(x)$ is not subtangential to W at x then by (3.4.12iii) there is no solution $\phi(\cdot,x)$ remaining in W and W is not weakly invariant. Similarly $-g(x)$ is subtangential for all $x \in W$ if and only if W is positively weakly invariant for the equation $\dot{x} = -g(x)$ which holds if and only if W is negatively weakly invariant for (A).

3.4.22 *COROLLARY*

If each solution $\phi(\cdot,x)$ *of* (A) *is uniquely determined by the initial condition* x *then* W *is invariant if and only if* $g(y)$ *and* $-g(y)$ *are subtangentia to* W *for all* $y \in W$.

3.4.23 *PROPOSITION (Roxin)*

If I *is invariant then* ∂I *and* $\bar{I} - I$ *are weakly invariant.*

Proof: Let $J = U - I$. By (3.4.20) $\bar{J}$ and $\bar{I}$ are weakly invariant so $x \in \partial I$ implies $g(x)$ and $-g(x)$ are subtangential to both $\bar{J}$ and $\bar{I}$ and so is subtangential to $\delta I = \bar{J} \cap \bar{I}$. Hence δI is weakly invariant. Since I is invariant $\bar{I} \setminus I = \partial I \setminus I$ is also weakly invariant.

The following proposition 3.4.24 is obvious, but Theorem 3.4.25 changes the conditions a little and becomes much tougher. We use all the machinery we have developed.

3.4.24 *PROPOSITION*

If W *is weakly positively invariant and* I *is positively invariant, then* $W \cap I$ *is weakly positively invariant.*

The following theorem is more significant that it first appears. We shall later show that the theorems of Kneser (3.2.14) and Fukuhara (3.2.21) are easy corollaries.

3.4.25 *THEOREM*

Let W_1 *and* W_2 *be closed (relative to* U*), positively weakly invariant sets; such that* $W_1 \cup W_2 = U$. *Then* $W = W_1 \cap W_2$ *is positively weakly invariant.*

Proof: For any $x \in W$ there exist a $\delta = \delta(x) > 0$ and solutions $\phi_1(\cdot,x)$ and $\phi_2(\cdot,x)$ defined for $t \in [0,\delta]$ such that $\phi_1(t,x) \in W_1$ and $\phi_2(t,x) \in W_2$ for $t \in [0,\delta]$. We may assume that δ is sufficiently small so that for $t \in [0,\delta]$, the straight line segment L_t between $\phi_1(t)$ and $\phi_2(t)$ is a subset of U. Choose $x_t \in L_t$ for $t \in [0,\delta]$ such that $x_t \in W_1 \cap W_2$. But x_t is between $\phi_1(t)$ and $\phi_2(t)$ so

$$d(W_1 \cap W_2, x + t\,g(x)) \leq d(x_t, x + tg(x)) \leq \sup_{i=1,2} d(\phi_i(t,x), x + tg(x)) \quad .$$

But the right-hand side is $o(t)$; therefore, $g(x)$ is subtangential to W at x for all $x \in W$, and W is weakly invariant.

3.4.26 *Counterexample*

To see that we need $W_1 \cup W_2 = U$ in theorem 3.4.25, let $\phi_1(\cdot,x)$ and $\phi_2(\cdot,x)$ be two distinct solutions through $x \in U$. Let $W_i = \{\phi_i(t,x) : t \in D_{\phi_i}\}$ for $i = 1,2,$. Assume further that x is chosen such that $g(x) \neq 0$ and $W_1 \cap W_2 = \{x\}$. Then $W = \{x\}$ is not weakly invariant since $g(x)$ is not subtangential to W at x. The proof actually uses only that for each $x \in W_1 \cup W_2$ there is a convex neighborhood of x in $W_1 \cup W_2$.

The previous results carry across to nonautonomous equations and time varying sets. If we let $n = m + 1$, $E^n = R \times E^m$, and let $x = (t,y)$ and $g(x) = (1, f(t,y)) \in E$, then $\phi(\cdot,x)$ is a maximal solution of (A) if and only if $\psi(\cdot,t,y)$ is a (maximal) solution of $\dot{y} = f(t,y)$ when we let $\phi(T,x) = (T + t, \psi(T + t, t,y))$. Hence the equations (A) and (B) are equivalent.

3.4.27 *DEFINITION*

Let S *be a subset of* $V \subset E^{m+1}$. *We will write* $S(t) = \{y : (t,y) \in S\} \subset E^m$. *We say* S *(or* $S(t)$*) is* <u>*invariant*</u> *for* (B) *if for each* $(t,y) \in S$ *and each* $\psi(\cdot;t\setminus y)$ *a* *each* $\tau \in D_{\psi(\cdot;t\ ,\ y)}$, $\psi(\tau;t,y) \in S(\tau)$. *We will use corresponding definitions for* S *and* $S(t)$ *being* <u>*weakly*</u> *and/or* <u>*positively*</u> *or* <u>*negatively invariant for*</u> (B).

3.4.28 *DEFINITION*

We say $v \in R^m$ *is subtangential to* $S(t)$ *at* (t,y) *if and only if*

$$d(y + sv, S(t + s)) = o(s) \qquad .$$

Note that if v is subtangential to $S(t)$ at (t , y), then $(1,v)$ is subtangential to S at $x = (t,y)$ (though the two are not equivalent) since

$$d((t,y) + (s,sv), S) \leq d(y + sv, S(t + s)) \quad .$$

Since (A) and (B) are equivalent, the following theorems are just forms of (3.4.11), (3.4.21) and (3.4.25).

3.4.29 *THEOREM*

Let Λ *be a locally compact subset of* E^{m+1} *and* $f:\Lambda \to E^m$ *be continuous. Assume* $f(t\ ,\ y)$ *is subtangential to* $\Lambda(t)$ *for all* $(t,y) \in \Lambda \cap \partial\Lambda$. *Then for each* (t_o,y_o) *there exists a* $\delta = \delta(t_o,x_o) > 0$ *and a solution* $\psi(\cdot\,;t_o,x_o)$ *for which* $[t_o,t_o + \delta(t_o,x_o)) \subset D_{\psi(\cdot\,;t_o,x_o)}$.

3.4.30 *THEOREM*

Let $f:V \to E^m$ *and assume* V *is open and* S *is closed relative to* V. *Assume* $f(t,y)$ *is subtangential to* $S(t)$ *for all* $(t,y) \in S$, *then* S *is positively (or negatively) weakly invariant for* (B).

3.4.31 *THEOREM*

If S_1 *and* S_2 *are closed relative to* V *and are positively weakly invariant and* V *is open in* E^{m+1}, *and if* $S_1 \cup S_2 = V$, *then* $S_1 \cap S_2$ *is positively weakly invariant.*

We shall also apply other results for autonomous systems to (B) when it suits us.

3.4.32 *DEFINITION*

U *will always denote an open subset of* V, *the domain of* f. *We shall say* $S \subset R^{m+1}$ *is* U *invariant if for each* $(t,y) \in U \cap S$ *and* $\psi(\cdot\,;t,y)$ *and each* T *such that* $\psi(\tau;t,y) \in U$ *when* τ *between* t *and* T, *we have* $\psi(T;t,y) \in S$. *This is equivalent to saying that* $S \cap U$ *is positively invariant for* (B) *when the domain of* f *is restricted to* U. *We will also use the corresponding terms with "negatively" or "positively" and/or "weakly."*

The positive solution funnel through (t,y) *is* $F_{t,y} = \{(s,\psi(s;t,y)) : s \geq t$ *and* ψ *is a solution of* $(B)\} \subset R^{m+1}$. *The* τ*-cross section is* $F_{t,y}(\tau) \subset R^m$.

3.4.33 *THEOREM (Fukuhara).*

Choose $(t,y) \in V$ *and* $t_1 > t$ *such that for any maximal solution* $\psi(\cdot,t,y)$, $\psi(t_1;t,y)$ *is defined. If* $y_1 \in \partial(F_{t,y}(t))$ *then there is a solution* ψ *such that*

$$\psi(\tau,t_1,y_1) \in \partial(F_{t,y}(\tau)) \qquad \tau \in [t, t_1].$$

We will prove the slightly more general result:

3.4.34 *THEOREM*

Choose $(t,y) \in V$. *Let* $U = \{(s,w) \in V : s > t\}$. *Then* $\partial F_{t,y}$ *is* U*-negatively weakly invariant.*

Proof: Note that by definition $F_{t,y}$ is positively invariant. Therefore $V \setminus F_{t,y}$ is negatively invariant. Therefore $V - F_{t,y}$ and $W_1 = \overline{V \setminus F_{t,y}}$ are (U-) negatively weakly invariant by (3.4.20). By definition $W_2 = \overline{F_{t,y}}$ is U-negatively invariant. Since $W_1 \cup W_2 = U$, (3.4.31) implies $W_1 \cap W_2 = \partial F_{t,y}$ is U-weakly invarian

To see that 3.4.33 implies 3.4.34 we prove a standard result.

3.4.35 *PROPOSITION*

If $t,y,$ *and* t_1 *are chosen as in* (3.4.33) *and* $F_1 = \{(t_o,y_o) \in F_{t,y} : t_o \in [t,t_1]$ *then* F_1 *is compact.*

Proof: If we choose any $\{t_n,y_n\} \subset F_1$ and $\psi_n = \psi_n(\cdot;t,y)$ such that $\psi_n(t_n) = y_n$, then (3.4.13) implies there is a subsequence $\{\psi_{n_i}\}$ converging to some solution

$\psi^*(\cdot,t,y)$ uniformly on $[t,t_1]$. We can assume the subsequence was chosen so that t_{n_i} converges to some t^*. Hence $(t_{n_i},x_{n_i}) \to (t^*,\psi^*(t^*,t,y)) \in F_1$. Therefore every sequence in F_1 has a convergent subsequence and F_1 is compact.

3.4.36 *COMMENTS*

(i) Theorem 3.4.34 says that for $(t_2,y_2) \in (\partial F_{t,y})(t_2)$ there is a solution $\psi(\cdot,t_2,y_2) = \psi$ such that $(\tau,\psi(\tau)) \in \partial F_{t,y}$ for $\tau \in D_\psi \cap [t_o,t_2]$. If t_1 is chosen as in (3.4.33) and $t_2 = t_1$, Proposition 3.4.35 implies that $[t_o,t_2] \subset D_\psi$.

(ii) Although $\partial(F_{t,y}(\tau)) \subset (\partial F_{t,y})(\tau)$ and $\psi(t_2)$ might not have been chosen in $\partial(F_{t,y}(t_2))$, in fact it can be shown that $\psi(\tau) \in \partial(F_{t,y}(\tau))$ for $\tau \in [t,t_2] \cap D_\psi$.

(iii) Let T be the sup $\{t_1 : t_1$ as in Theorem 3,4,33$\}$. Examples can be given with $V = R \times R$ such that $F_{t,y}(\tau)$ is compact for all $\tau \geq t$, $\tau \neq T$ yet for all $t_2 > T$ and for each ψ defined at t_2 and remaining in $\partial F_{t,y}$ on $D_\psi \cap [t,t_2]$ we have $D_\psi = (T,\infty]$; hence, the restriction on t_1 in (3.4.33) is necessary and (3.4.34) is in fact more general.

3.4.37 *THEOREM (Kneser).*

If $t,y,$ *and* t_1 *are chosen as in* (3.4.33), *then* $F_{t,y}(t_1)$ *is connected.*

Proof: Suppose $F_{t,y}(t_1)$ is not connected. Then $F_{t,y}(t_1) = C_1 \cup C_2$ where C_1 and C_2 are non-empty but $C_1 \cap \bar{C}_2$ and $\bar{C}_1 \cap C_2$ are empty. Let W_1 be the set of $(t_2,y_2) \in V$ such that there is a solution $\psi = \psi(\cdot,t_2,y_2)$ which is either not defined at t_1 or if defined satisfies

$$d(\psi(t_1),C_1) \leq d(\psi(t_1),C_2)$$

We call such a solution a W_1-defining solution for (t_2,y_2). Define W_2 similarly reversing the direction of the inequality. Clearly $W_1 \cup W_2 = V$. It is immediate using Lemma 3.4.13 that each convergent sequence lying, say, in W_1 converges to a point of W_1 so W_1 and W_2 are closed. Let $U = \{(t_2,y_2) \in V\colon t_2 < t_1\}$.

If (t_2,y_2) is in, say, W_1 and ψ is a W_1-defining solution for (t_2,y_2) and $\tau \in [t_2,t_1) \cap D_\psi$, then $(\tau,\psi(\tau)) \in W_1 \cap U$. Hence W_1 (and similarly W_2) are U-positively weakly invariant so $W_1 \cap W_2 = W$ is U-positively weakly invariant. But $(t,y) \in W_1 \cap W_2$ so there exists a solution $\psi(\cdot,t,y)=\psi$ such that $\psi(\tau) \in W$ for $\tau \in [t,t_1) \cap D_\psi$. By choice of t_1, $[t,t_1] \subset D_\psi$ and $\psi(t_1)$ is defined and $d(\psi(t_1),C_1) = d(\psi(t_1),C_2)$ which is a contradiction since $\psi(t_1) \in C_1 \cup C_2$. Therefore $F_{t,y}(t_1)$ is connected.

3.5 *Dynamical systems and nonautonomous differential equations*

In Section 3.1 we have investigated the relationships between the abstract theory of dynamical systems in the Euclidean n-space (Chapter 1) and the properties of flows defined by the solutions of an autonomous differential equation. Conditions have then been derived under which the flow defined by the solutions of such autonomous differential equations indeed defines a dynamical system. On the other hand, in Section 3.1 most of the theorems have been proved for the more general case of the time varying differential equation 2.1.5 for which the flow defined by its solution does not immediately define a dynamical system. In this section, without claim of completeness, we shall present the few general results available on time variable flows and in particular on the flows defined by the solutions of time-varying differential equations, having the property of uniqueness and existence in the large.

Given the time-varying differential system

3.5.1 $$\dot{x} = f(x,t)$$

we can introduce in the system a new independent variable instead of t, say τ, through the relation $\frac{dt}{d\tau} = 1$. Then the system 3.5.1 can be written in the following equivalent form (called parametric form)

3.5.2 $$\begin{cases} \dfrac{dx_i}{d\tau} = f_i(x,t) \\ \dfrac{dt}{d\tau} = 1 \end{cases} \qquad i = 1,\dots,n.$$

The next step is to introduce the $(n + 1)$-dimensional vector y through the relation:

3.5.3 $$\begin{aligned} y_i &= x_i \\ y_{n+1} &= t \end{aligned} \qquad i = 1,\dots,n$$

Then the system 3.5.2 takes the simpler form

3.5.4 $$\frac{dy}{d\tau} = g(y)$$

where $g(y)$ is an $n + 1$ dimensional vector defined through the relations

3.5.5 $$\begin{aligned} g_i(y) &= f_i(y) \qquad (i = 1,\dots,n) \\ g_{n+1}(y) &= 1 \end{aligned}$$

The differential system 3.5.4, which is formally of the same type as 3.1.1, has the property that, if its solutions have the uniqueness property and are extendable to $(-\infty,+\infty)$, then the flow induced by these solutions defines a dynamical system in the $(n + 1)$ dimensional Euclidean space. The dynamical system so defined has, however, very peculiar properties which follow from the very particular structure of the second equation 3.5.5, namely $g_{n+1}(y) = 1$. This dynamical system is parallelizable and in particular it does not have any bounded motions, thus no periodic orbits, no almost periodic motions and no equilibrium points. Because of this fact, the theory of dynamical systems presented in Chapter 1 has not been very helpful until now investigating the topological properties of flows defined by time-varying differential equations. As far as stability properties are concerned the situation is, on the other hand, not too bad. One can immediately rephrase the problems of stability of compact sets for the equation 3.5.1 as problems of stability of non-compact sets in the space E^{n+1}. For instance, the stability problem of the equilibrium point $x = 0$ of equation 3.5.1, i.e., of the point such that

3.5.6 $$f(0,t) \equiv 0 \qquad \text{for all } t,$$

is equivalent to the stability problem of the non-compact invariant set $y_i = x_i = 0$ $(i = 1,\dots,n)$, which is the axis y_{n+1}, i.e., a straight line in the space E^{n+1}. Then the theorems of the Liapunov second method for noncompact sets can and will be applied , obtaining in this way the classical stability theorems for equilibrium points of time-varying equations, with all their drawbacks and difficulty of application. In order to provide some tools for the study of the topological properties (recurrence, etc.) of the flow defined by the solutions of time-varying differential equations and with the hope of having in the future some new tools to investigate stability properties, we shall present some newly discovered alternative ways of studying the properties of the flow defined by such equations.

3.5.7 *DEFINITION*

Let $\Omega \subset E$ *be an open set.*

Let $C = C(\Omega \times R, E)$ *be the space of all continuous vector-valued functions* $f:\Omega \times R \to E$. *We shall say that a function* $f \in C$ *is* *__admissible__* *if the solutions of the differential equation* $\dot{x} = f(x,t)$ *are unique and are extendable in both directions.*

3.5.8 *THEOREM*

The mapping $\Pi^*: C \times R \to C$ *defined by*

3.5.9 $$\Pi^*(f,\tau) = f_\tau \quad , \quad \text{where} \quad f_\tau(x,t) = f(x,t+\tau),$$

defines a dynamical system on C *when* C *has the compact-open topology. The trajectory of* Π^* *is the set* $F = \{f_\tau : \tau \in R\}$. *The motion* $\Pi^*_f: R \to F$.

Proof: It must be shown that Π^* satisfies the axioms (1.1.2) of a dynamical system. Axioms (i) and (ii) are clearly satisfied. We want to show that also axiom (iii) (continuity) is satisfied. Let $\{\tau_n\}: \tau_n \in R: \tau_n \to \tau$. Then for each $(x,t) \in E \times R$

3.5.10 $$f_{\tau_n}(x,t) = f(x,t+\tau_n) \to f(x,\tau+\tau) = f_\tau(x,t)$$

since f is continuous. This proves the theorem.

It is interesting to study the properties of the motion Π_f^*.

3.5.11 *THEOREM*

The motion $\Pi_f^*: R \to F$ *of the dynamical system* $\Pi^*: C \times R \to C$ *is continuous in the compact-open topology on* F.

Proof: From the continuity of f on $E \times R$ its uniform continuity follows on every compact set in $E \times R$. Then the convergence of 3.5.10 is uniform on compact sets in $E \times R$.

It is easy to show from the theorems on existence, uniqueness, and global existence that if f is an admissible function, then all follows $f_\tau(x,t) = f(x, t + \tau)$ are also admissible. Thus

3.5.12 *THEOREM*

Consider the dynamical system $\Pi^*: C \times R \to C$. *Let* $f \in C$ *be an admissible function. Then for all* $t \in R$, *the image point* $\Pi^*(f,t) \in C$ *is also admissible.*

We are now ready for the presentation of the main theorem.

3.5.13 *THEOREM*

Assume that

i) $X = E \times F$ *is a metric space with metric*

3.5.14 $$d((x^1,f^1), (x^2,f^2)) = ||x^1 - x^2|| + \rho(f^1, f^2)$$

where $\rho(f^1,f^2)$ *is any metric which generates a compact-open topology of* C,

ii) $f \in C$ *is an admissible function,*

iii) $\phi(x,f,\tau)$ *denotes the solution of the differential equation* 3.5.1 *with* $\phi(x,f,0) = x$,

Then the mapping $\Pi: X \times R \to X$ *defined by*

5.15 $$\Pi((x,f),\tau) = (\phi(x,f,\tau), f_\tau)$$

a dynamical system.

oof: i) Notice that for each fixed τ the mapping $\Pi^\tau(x,f) = (\phi(x,f,\tau), f_\tau), \tau$ fixed, fines points in $X = E \times F$. Clearly from Theorem 3. 5.12, it follows that Π defined in all $X \times R$ and, in addition,

5.15 $$\Pi((x,f),0) = (\phi(x,f,0), f_0) = (x,f)$$

cause of the definition of $\phi(x,f,\tau)$ and the property 3.5.9. So the first property dynamical systems(1.1.2) is satisfied.

ii) Now let $\phi^1(t) = \phi(x,f,t)$ be the solution of 3.5.1 with $\phi^1(0) = x$ d $\phi^2(t) = \phi(\phi^1(\tau), f_\tau, t)$ be the solution of

.17 $$\dot{x} = f(x, t + \tau)$$

th $\phi^2(0) = \phi^1(\tau) = \phi(x,f,\tau)$. But $\phi^3(t) = \phi^1(t + \tau)$ is also a solution of 3.5.1 th $\phi^3(0) = \phi^1(\tau)$. Thus from the property of uniqueness of solutions of 3.5.17 we ve $\phi^2(t) = \phi^1(t + \tau)$ for all $t \in R$. Hence

.18 $$\begin{aligned} \Pi(\Pi((x,f),\tau);\sigma) &= \Pi((\phi^1(\tau), f_\tau);\sigma) = (\phi^2(\sigma), f_{\tau+\sigma}) = \\ &= (\phi^1(\tau+\sigma), f_{\tau+\sigma}) = \Pi((x,f);\tau+\sigma) \quad \text{for all } \tau,\sigma \in R. \end{aligned}$$

ich proves the second property of dynamical systems.

iii) Continuity of the mapping Π follows immediately from Theorem 3.5.8.

This proves the theorem.

.19 *Remark*

To fix the ideas a possible metric which generates a compact-open topology C may be given as follows:

3.5.20 $$\rho(f,g) = \sup_{\tau > 0} \{\inf[\sup(|f(x,t) - g(x,t)| : |x| + |t| \leq \tau), 1/\tau]\}.$$

3.5.21 *Notes and references*

Most of the material presented in this section is due to G. R. Sell [5].

3.6 *Classical results on the investigation of the stability properties of flows defined by the solutions of ordinary differential equations via the second method of Liapunov.*

The theorems that we shall prove in the sequel are given in the language and technique of differential equations. When not otherwise stated, these theorems will only apply to strong stability properties.

We shall present the Liapunov second method essentially for the case of the autonomous differential equation

3.6.1 $$\dot{x} = f(x),$$

where $f(x)$ is defined and continuous for all $x \in E$. From the material presented in Section 3.5 it must be obvious to the reader that also the case of the non-autonomous equation 3.5.1 can be included in this framework.

From the operational point of view in the second method of Liapunov, the stability properties of closed sets will be characterized by the relative properties of a pair of functions $v = \phi(x)$ and $w = \psi(x)$ connected to the differential system 3.6.1 through the relation

3.6.2 $$\psi(x) = \langle \operatorname{grad} \phi(x), f(x) \rangle = \sum_{i=1}^{n} \frac{\partial \phi}{\partial x_i} f_i(x)$$

For a given $\phi(x)$ the scalar function $\psi(x)$ is simply the total time derivative of the scalar function $v = \phi(x(t))$ along the solution curves of the differential system 3.6.1; thus

3.6.3 $$\frac{dv}{dt} = \psi(x)$$

For a <u>given $\psi(x)$</u> the relation 3.6.2 is a linear partial differential equation, which will have a solution $\phi(x)$ if integrability conditions are satisfied. These integrability conditions can be defined in the following way: given a real-valued function $\psi(x)$ and a vector $f(x) \neq 0$, a vector $b(x)$ may be chosen such that

3.6.4 $$\psi(x) = \langle b(x),\ f(x)\rangle .$$

Such a vector $b(x)$ is the gradient of a scalar function if the $\frac{n(n-1)}{2}$ integrality conditions:

3.6.5 $$\frac{\partial b_i(x)}{\partial x_j} = \frac{\partial b_j(x)}{\partial x_i} \qquad (i,j = 1,\dots,n)$$

are satisfied.

We shall now first prove a set of theorems which relates the stability properties of a given compact set M with the sign and uniform boundedness properties (see Chapter 0) of the real valued functions $v = \phi(x)$ and $w = \psi(x)$. The same theorem holds for the case of sets with a compact neighborhood.

3.6.6 *THEOREM*

Let $v = \phi(x)$ *and* $w = \psi(x)$ *be real-valued functions defined in an open neighborhood* $N(M) \subset E$ *of a* <u>*compact set*</u> M. *Assume that*

i) $v = \phi(x) \in C^1$,

ii) $v = \phi(x)$ *is definite for the set* M,

iii) $w = \psi(x)$ *is semidefinite for the set* M,

iv) for all $x \in N(M)$ *with* $\psi(x) \neq 0$, sign $\psi(x) \neq$ sign $\phi(x)$,

v) $\phi(x)$ *and* $\psi(x)$ *satisfy the relation* 3.6.2. *Then the compact set* M *is (uniformly)* <u>*stable*</u>.

Proof. Since the real-valued function $\phi(x)$ is definite for the set M, from Lemma 0.3.3, it follows that there exists a real number $\eta > 0$ and two strictly increasing function $\alpha(\nu)$ and $\beta(\nu)$, with $\alpha(0) = \beta(0) = 0$ such that it is

3.6.7 $$\alpha(\rho(x,M)) \leq \phi(x) \leq \beta(\rho(x,M)) \quad \text{for} \quad x \in S[M,\eta] \subset N(M)$$

Let $\varepsilon > 0 (\varepsilon \leq \eta)$ be given and choose $\delta > 0$ such that

3.6.8 $$\beta(\delta) < \alpha(\varepsilon)$$

that is, such that

3.6.9 $$0 < \delta < \beta^{-1}(\alpha(\varepsilon))$$

where β^{-1} denotes the inverse of the function $\beta(\nu)$. Obviously $\delta < \varepsilon$. We claim that $\rho(x^o,M) \leq \delta$ implies $\rho(x(x^o,t),M) < \varepsilon$, $t \in R^+$. In fact, in the set $S[M,\varepsilon]$

3.6.10 $$\psi(x) = \dot{\phi}(x(x^o,t)) \leq 0$$

which gives

3.6.11 $$\alpha(\rho(x(x^o,t),M) \leq \phi(x(x^o,t)) \leq \phi(x^o) \leq \beta(\rho(x^o,M)) \leq \beta(\delta).$$

If there would exist a $t_1 > t_0$ such that $\rho(x(x^o,t_1),M) = \varepsilon$, then we would have

3.6.12 $$\alpha(\varepsilon) \leq \beta(\delta)$$

which contradicts the choice of δ in 3.6.8 and proves the theorem.

For sake of completeness and for a better understanding of instability , we shall now state an obvious corollary regarding negative Liapunov stability of a compact set M.

3.6.13 *COROLLARY*

If a compact set M *satisfies Theorem 3.6.6 with the condition iv) replaced by iv')* sign $\psi(x)$ = sign $\phi(x)$ *for all* $x \in E$ *with* $\psi(x) \neq 0$, *then* M *is negatively stable.*

3.6.14 *Remark*

From the proof of Theorem 3.6.6, it is obvious (as already known for a dynamical system, as shown by Theorem 1.5.24) that a set M which satisfies

Theorem 3.6.6 is positively invariant.

3.6.15 *THEOREM*

Let $v = \phi(x)$ *and* $w = \psi(x)$ *be real-valued functions, defined in an open neighborhood* $N(M) \subset E$ *of a compact set* M. *Assume that*

i) $v = \phi(x) \in C^1$,

ii) $v = \phi(x)$ *is definite for the set* M,

iii) $w = \psi(x)$ *is definite for the set* M,

iv) $\operatorname{sign} \psi(x) \neq \operatorname{sign} \phi(x)$,

v) $\phi(x)$ *and* $\psi(x)$ *satisfy the condition* 3.6.2. *Then the compact set* M *is (uniformly) asymptotically stable for the system* 3.6.1.

Proof. In $S[M,\varepsilon_o]$, $\varepsilon_o > 0$, the inequalities 3.6.7 are again satisfied and, furthermore, there exist two additional strictly increasing functions $\omega(\nu)$ and $\gamma(\nu)$, $\omega(0) = \gamma(0) = 0$, such that

3.6.16 $$-\omega(\rho(x,M) \leq \psi(x) \leq -\gamma(\rho(x,M)).$$

From Theorem 3.6.6, it follows that M is uniformly stable. To prove the theorem we choose $\delta_o > 0$ such that $\beta(\delta_o) < \alpha(\varepsilon_o)$. Then $\rho(x^o,M) \leq \delta_o$ implies that $\rho(x(x^o,t),M) < \varepsilon_o$ for $t \in R^+$, since M is stable. We assert that $\rho(x^o,M) \leq \delta$ implies that:

$$\lim_{t \to +\infty} (\rho(x(x^o,t),M) = 0 .$$

For any $x(x^o,t)$ such that $\rho(x^o,M) < \delta$ we set $\phi(t) = \phi(x(x^o,t))$. We then hav

$$\psi(x(x^o,t)) = \dot{\phi}(t) \leq -\gamma(\rho(x(x^o,t),M)), \qquad t \geq t_o .$$

It follows then that

3.6.17 $$\phi(t) - \phi(t_o) \leq - \int_{t_o}^{t} \gamma(\rho(x(x^o,\tau),M))d\tau$$

Now let $\varepsilon > 0(\varepsilon < \delta_o)$ be any positive number. Choose $\delta > 0(\delta < \varepsilon)$ such that $\beta(\delta)<\alpha(\varepsilon)$, then if $\rho(x(x^o,t),M) = \delta$ then $\rho(x(x^o,t),M) < \varepsilon$ for $t \geq t_1$.

Now let $\rho(x^o,M) \leq \delta_o$. If $\rho(x^o,M) \leq \delta$, then $\rho(x(x^o,t),M) < \varepsilon$ for $t \geq t_o$. If $\delta < \rho(x^o,M) \leq \delta_o$, then as long as $\rho(x(x^o,t),M) > \delta$ we have

$$\phi(t) - \phi(t_o) \leq - \int_{t_o}^{t} \gamma(\delta)d\tau = -(t - t_o)\gamma(\delta)$$

or

3.6.18 $$t - t_o \leq \frac{\phi(t_o) - \phi(t)}{\gamma(\delta)} \leq \frac{\beta(\delta_o) - \alpha(\delta)}{\gamma(\delta)}$$

Let

3.6.19 $$T(\varepsilon) = \frac{\beta(\delta_o) - \alpha(\delta)}{\gamma(\delta)}$$

be the maximum time in which the solutions of the system 3.6.1 remain in the set $S[M,\delta_o] \setminus S(M,\delta)$. Since δ depends only upon ε, the inequality 3.6.16 and, therefore, 3.6.7 is violated if $t > t_o + T(\varepsilon)$. Hence there exists a t_1, with $t_o \leq t_1 < t_o + T(\varepsilon)$ such that $\rho(x(x^o,t_1),M) = \delta$. Thus $\rho(x(x^o,t),M) < \varepsilon$ for $t \geq t_o + T(\varepsilon)$ for all $t_o > 0$ and $\rho(x^o,M) \leq \delta$. This completes the proof.

3.6.20 *Remarks*

In the proof of the theorem no use has been made of the left hand part of the inequality 3.6.16. By proceeding as before, one can derive the analogue of inequality 3.6.18 as follows:

3.6.21 $$\tau(\varepsilon) = \frac{\alpha(\delta_o) - \beta(\delta)}{\omega(\delta)} \leq t - t_o$$

Now $\tau(\varepsilon)$ is the minimum time in which the solution of 3.6.1 can cross in the ring $S[M,\delta_o] \setminus S(M,\delta)$. By the same argument as in the above proof of Theorem 3.6.16, it follows that 3.6.21 does not hold for $t < t_o + \tau(\varepsilon)$. Thus $\rho(x(x^o,t),M) > \varepsilon$ for $t \leq t_o + \tau(\varepsilon)$ for all $t_o \geq 0$ and $\rho(x^o,M) \geq \delta$. Thus the solutions

$x(x^o,t)$ have a uniform rate of approach to M in $N(M)$.

From all theorems on asymptotic stability of compact sets it is possible to derive trivial corollaries on the complete instability of such sets by reversing the requirements of the relative sign of the independent variable t, and, therefore, inverting the direction of motion on each trajectory. For example, from theorem 3.6.15 it can be deduced that

3.6.22 *COROLLARY*

If a compact set M *satisfies Theorem 3.6.15 with condition iv) replaced by*

iv') $\operatorname{sign} \psi(x) = \operatorname{sign} \phi(x)$, *then the set* M *is completely unstable.*

We shall now prove the theorem which provides sufficient conditions for the instability of a compact set for the differential system 3.6.1.

3.6.23 *THEOREM*

Let $v = \phi(x)$ *and* $w = \psi(x)$ *be real-valued functions defined in an open non-empty set* $B \subset S(M,\eta) \subset E$, *where* $\eta > 0$ *and* M *is a compact set. Assume*

i) $\partial M \cap \partial B \neq \emptyset$,

ii) $\phi(x) = 0$ *for* $x \in [\partial B \cap S(M,\eta)]$, $\phi(x) \neq 0$ *for* $x \in [IB \cap S(M,\eta)]$,

iii) $v = \phi(x) \in C^1$,

iv) $\operatorname{sign} \phi(x) = \operatorname{sign} \psi(x)$, *for* $x \in [IB \cap S(M,\eta)]$,

v) for all $x \in B$, $|\phi(x)| \leq \beta(\rho(x,M))$ *and* $|\psi(x)| \geq \gamma(\rho(x,M))$,

vi) $\phi(x)$ *and* $\psi(x)$ *satisfy the condition* 3.6.2. *Then the compact set* M *is unstable for the system* 3.6.1.

Proof. Assume that $\phi(x) > 0$ in IB. For a sufficiently small $\delta > 0$ there exist $x^o \in IB$, $\rho(x^o,M) < \delta$, such that $\phi(x^o) > 0$. Consider the corresponding solution $x(t) = x(x^o,t)$ and the values of $\phi(x)$ along such solution $\phi(t) = \phi(x(x^o,t))$. Integrating $\psi(x) = \phi(x)$ along such solutions and taking into account the condition v) we obtain

$$\phi(t) - \phi(t_o) = \int_{t_o}^{t} \psi(\tau)\, d\tau \geq \int_{t_o}^{t} \gamma(\rho(x(\tau),M)d\tau$$

and

$$\phi(t) \geq \gamma(\rho(x(t_o),M)) \cdot (t - t_o) + \phi(t_o)$$

If for all $t \geq t_o$, $x(x^o,t) \in \mathcal{I}B$, then $\lim_{t \to +\infty} \phi(t) = +\infty$, which contradicts the hypothesis (v). Hence there exists $t = t_1 > t_o$ for which $x(t_1) \in \partial B \cap \partial S(M,\eta)$. Since, for all $t \geq t_o$ for which $x(t) \in \mathcal{I}B$ $\phi(x) \geq \alpha(\rho(x(x^o,t),M)) > 0$; we cannot have $\phi(t_1) = 0 \in \partial B$ thus $\rho(x(t_1),M) = \eta$ and the theorem is proved.

It must be pointed out that, from the hypothesis of Theorem 3.6.23, the set B cannot have any compact component which does not contain a component of M. In fact, if there would exist such compact component B_c then there would exist (3.8.26) at least one point $y \in \mathcal{I}B_c$ such that grad $\phi(y) = 0$. Hence $\psi(y) = 0$ which contradicts the hypothesis iv). On the other hand, B need not be a region, but it could be formed by a sequence of sets with non-compact closure which satisfy the conditions of the theorem.

From the theorems given it follows that

3.6.24 *THEOREM*

If there exists a pair of real-valued functions $\phi(x)$ *and* $\psi(x)$, *satisfying the condition* 3.6.2, *where* $\psi(x)$ *is definite for a compact set* M *in the neighborhood* $N(M) \subset E$ *and* $\phi(x) \in C^1$ *is such that* $\phi(x) = 0$ *for all* $x \in M$, *then the additional sign properties of the function* $\phi(x)$ *completely characterize the stability properties of the compact set* M.

Proof. i) If $\phi(x)$ is definite and sign $\phi(x) \neq$ sign $\psi(x)$, then from the theorem (3.6.15) it follows that M is asymptotically stable.

ii) If $\phi(x)$ is definite and sign $\phi(x) =$ sign $\psi(x)$, then from Corollary (3.6.22) it follows that M is completely unstable.

iii) If $\phi(x)$ is indefinite, then Theorem 3.6.23 insures that M is unstable. Finally

iv) If $\psi(x)$ is definite for M, $\phi(x)$ cannot be semi-definite for M.

In fact, if $\phi(x)$ is semi-definite the set $G \supset M$ such that if $y \in G$, $\phi(y) = 0$, is the absolute minimum of the $\phi(x)$, and since $\phi(x) \in C^1$, it follows that for all $y \in G \supset M$, grad $\phi(y) = 0$ and, thus, $\psi(y) = 0$ for some $y \notin M$ which contradicts the hypothesis and the theorem is proved.

Notice that Theorem 3.6.24 does not give necessary conditions for the stability of M. In fact, there do not always exist real-valued functions $\phi(x)$ and $\psi(x)$ satisfying 3.6.2 and such that $\psi(x)$ is <u>definite for a given</u> (positively) invariant set.

3.6.25 *DEFINITION*

A real-valued function $v = \phi(x)$ *which satisfies one of the stability theorems is called "Liapunov function".*

Theorem 3.6.15 and Corollary 3.6.22 define only local properties of the compact set M. That is, if Theorem 3.6.15 is satisfied, then there exists a sufficiently small $\delta > 0$, such that $S(M,\delta) \subset A(M)$ where $A(M)$ is the region of asymptotic stability of the set M. For the practical applications of the stability theorems, local properties are not very useful. It is, therefore, important to give theorems which provide sufficient conditions for global asymptotic stability or in the case in which the compact set M is not globally asymptotically stable, allow the exact identification of the region of asymptotic stability $A(M)$ or at least an approximate identification of the set $\partial(A(M))$.

Our first concern is to derive a theorem which will provide a sufficient condition for the global asymptotic stability of a compact set M.

3.6.26 *THEOREM*

If the conditions of Theorem 3.6.15 are satisfied in the whole space E *and, in addition,*

vi) $$\lim_{||x|| \to \infty} \phi(x) = \infty .$$

Then the compact set M *is globally asymptotically stable.*

Proof. Along the solutions of the system 3.6.1, let

$$\dot{\phi}(t) = \psi(x(t)) = -\chi(x(t))$$

Assume that $\chi(x)$ is positive definite for M. For all t with $t_0 \leq t \leq t_1$

3.6.27 $$0 \leq \phi(t) = \phi(t_0) - \int_{t_0}^{t} \chi(x(\tau))d\tau .$$

We claim that $\chi(x(\tau))$ is an integrable function in $[0, +\infty)$. In fact, from 3.6.27 and condition (vi), it follows that if $\chi(x(\tau))$ were not integrable, then $\lim_{t \to +\infty} \int_{t_0}^{t} \chi(x(\tau))d\tau \to -\infty$, which contradicts the hypothesis on the sign of $\chi(x)$. We shall now prove that

3.6.28 $$\lim_{t \to +\infty} x(x^o,t) = 0 \quad \text{for all} \quad x^o \in E.$$

In fact, if this were not true, then there would exist a $\varepsilon_1 > 0$ and a sequence of intervals $(t_n, t_n + \lambda)$ with $t_n \to +\infty$, $\lambda > 0$, such that

$$\rho(x(t), M) \geq \varepsilon_1 \quad \text{for} \quad t_n \leq t \leq t_n + \lambda; \quad n = 1,2,\ldots;\lambda > 0$$

But then condition vi) implies that for all $x \notin E$ we have $\chi(x(t)) \geq \varepsilon_2$ for $t_n \leq t \leq t_n + \lambda$; $n = 1,2,\ldots;\lambda > 0$ which contradicts the integrability of $\chi(x(t))$. Thus 3.6.28 follows. Since the hypothesis of the theorem obviously implies that M is stable, it follows from 3.6.28 that M is globally asymptotically stable. Q.E.D.

3.6.29 *Remark*

Theorem 3.6.26 would be also true if instead of condition vi), one simply required that condition 3.6.28 be satisfied for all $x \in E$. The fact that condition (vi) is not necessary will be shown by the following theorem which is a trivial corollary of Theorem 3.6.15. The Liapunov function commonly used in practice does, however, satisfy the condition vi).

3.6.30 *THEOREM*

Let $v = \phi(x)$ *and* $\theta = \theta(x)$ *be real-valued functions defined in the whole space* E. *Assume that*

i) $v = \phi(x) \in C^1$,

ii) $v = \phi(x)$ *is definite for a compact set* M,

iii) $\lim\limits_{||x|| \to +\infty} \phi(x) = \eta > 0$,

iv) $\theta = \theta(x)$ *be positive definite for the set* M,

v) $\psi(x) = \theta(x)(\phi(x) - \eta)$,

vi) $\phi(x)$ *and* $\psi(x)$ *satisfy the condition* 3.6.2. *Then the compact set* M *is globally asymptotically stable.*

By extending the definition of the function $\phi(x)$ and $\psi(x)$ to an open set B with noncompact closure one is able to show the existence in B of solutions which tend to infinity and have the so-called global (but not necessarily complete) instability.

3.6.31 *DEFINITION*

A compact set $M \subset E$ *will be called globally unstable (for the flow defined by the system of differential equations* 3.6.1) *if there is a sequence* $\{x^n\}$ *of points in* C(M), $x^n \to M$, *such that* $\|x(x^n,t)\| \to +\infty$ *as* $t \to +\infty$ *for each* n.

3.6.32 *THEOREM*

If in Theorem 3.6.23 *the set* $\bar{B}$ *is noncompact, then* M *is globally unstable*

.6.33 *Notes and References*

The idea of characterizing the stability properties of differential equations y means of the sign properties of a real-valued function is due to Liapunov [1]. similar idea in a much more geometrical context, quite near to our point of view s to be found in the work of Poincaré [1, Vol. 1, pg. 73 ff]. Here Poincaré evelops in E^2 a method which enables to prove the absence of limit cycles in a certain omains of the plane. This information is derived by analyzing the properties of the et $\psi(x) = \langle \operatorname{grad} \phi(x), f(x) \rangle = 0$ (contact curve) where $v = \phi(x)$ is a topological ystem, i.e., a real-valued function such that the curves $\phi(x) = \text{const.}$ are noninter- ecting, closed and differentiable.

Methods quite close to Liapunov's have also been suggested by J. Hadamard [1] ıd D. C. Lewis [2].

We want to emphasize again that Liapunov was originally interested only in ıe stability properties of a given motion. He formulated this problem as follows. et $\dot{y} = g(y,t)$. Let $y^1 = y^1(t)$ be a solution of such equation. In order to ıvestigate the stability of y^1, for all $\varepsilon > 0$, we shall consider solutions $= y(t)$ such that $||y^1(t_0) - y(t_0)|| < \eta$ and see if this implies that $y^1(t) - y(t)|| < \varepsilon$ for all $t \geq t_0$. This can be easily done by defining a new ıriable:

$$x = y - y^1(t) .$$

ıen from the differential equation $\dot{y} = g(y,t)$ one can obtain a new differential uation $\dot{x} = f(x,t) = g(x + y^1(t),t) - g(y^1(t),t)$. This equation is called equation of e perturbed motion. Notice that the stability problem for the motion $y^1 = y^1(t)$ r the equation $\dot{y} = g(y,t)$ is now reduced to the stability problem for the equilibrium int $x = 0$ of the equation $\dot{x} = f(x,t)$.

Theorem 3.6.6 and 3.6.15 are natural extensions of theorems of Liapunov [1].

Theorem 3.6.23 is the extension to compact sets of a theorem due to Chetaev [2].

Theorem 3.6.26 is due to E. A. Barbashin and N. N, Krasovskii [1].

Theorem 3.6.30 is an extension of a well known theorem due to Zubov [3,6].

Results for the stability of noncompact sets for differential equations are given in the works of G. P. Szegö [3], G. P. Szegö and G. R. Geiss [1], and Yoshizawa [7].

The problem of existence of Liapunov functions for differential equations (converse problem) has been discussed by many authors, notably J. L. Massera [5,6], N. N. Krasovskii [3,6,7,8,9], K. P. Persidski [2], Vrkoc [1] and J. Kurzweil [1,2] and Kurzweil and Vrkoc [1].

Stability problems for time-varying differential equations can be found in the excellent review paper by H. A. Antosiewicz [3] and in the books by W. Hahn [2] and T. Yoshizawa [10]. It has to be noted that most of the results for the stability of equilibrium points for time-varying differential equations presented in the classical literature can be derived as particular cases of stability theorems for noncompact sets.

3.7 *New results with relaxed conditions.*

Do we really need that a function $v = \phi(x)$ be (locally) positive definite for a compact set $M \subset E$ to be able to prove that M is stable? Even if this is necessary and sufficient, it may be simpler to use a function which is not definite even if there exists one which is. The answer is no. In fact, even indefinite functions may sometime be quite adequate to prove stability, as we shall show by an example. In what follows we shall restrict ourselves to the case of a continuum (a compact and connected set) $M \subset E$. This is not a restriction with respect to compact sets, since we know that if a compact set is stable all its components are stable. This stronger stability theorem for the differential equation

3.7.1 $$\dot{x} = f(x)$$

is based upon the following lemma whose proof is obvious.

3.7.2 *LEMMA*

Let $v = \phi(x)$ *and* $w = \psi(x)$ *be real-valued functions defined on an open neighborhood* $N(M) \subset E$ *of a continuum* M. *Assume that*

i) $v = \phi(x) \in C^1$,

ii) $\psi(x) \leq 0, \quad x \in N(M)$,

iii) $\psi(x) = \langle \text{grad}\, \phi(x), f(x) \rangle$,

iv) $Q_k = \{x: \phi(x) \leq k\} \qquad k$ *real*.

Then for every real k, *every compact component of* Q_k *which is contained in* $N(M)$ *is (strongly) positively invariant for the flow defined by the solution of the differential equation* 3.7.1.

Let now $\phi(x) = 0$ for $x \in M$ and let H_k $(k > 0)$ be the component of Q_k which contains M. Clearly then if H_k is compact then all solutions $x = x(t,x^0)$, $x^0 \in H_k$ of the differential equation 3.7.1 are bounded. The stability properties of M are then clearly related to the geometrical properties of H_k.

3.7.3 *THEOREM*

Let $v = \phi(x)$ *and* $w = \psi(x)$ *be real-valued functions defined on an open neighborhood* $N(M) \subseteq E$ *of a continuum* M. *Let*

3.7.4 $$||H_k|| = \sup \{\rho(x,M) : x \in H_k\} ,$$

i) $v = \phi(x) \in C^1$,

ii) $\phi(x) = 0, \; x \in M$,

iii) $\lim_{k \to 0^+} ||H_k|| \to 0$,

iv) $\psi(x) = \langle \operatorname{grad} \phi(x), f(x) \rangle \leq 0, \quad x \in \Gamma$.

Then the continuum M *is (positively strongly Liapunov) stable for the differential equation* 3.7.1.

Proof: The condition iii) of the theorem is equivalent to the following condition: given any $\eta > 0$, there exists a $k > 0$ such that $||H_k|| < \eta$. The proof is the usual one. From the continuity of $\phi(x)$ and condition ii) it follows that there exist $\varepsilon > 0$ such that $S(M,\varepsilon) \subset H_k$. From Condition (iii) we have the existence of $\eta(\varepsilon) > 0$ such that $H_k \subset S(M,\eta) \subseteq N(M)$. From Lemma 3.7.2 we have that $x^o \in S(M,\varepsilon) \subset H_k$ implies that all solutions $x = x(t,x^o)$ of the differential equation 3.7.1 have the property that for all $t \geq t_o$, $x(t, x^o) \subset H_k \subset S(M,\eta(\varepsilon)) \subseteq N(M)$, which is (positive strong Liapunov) stability and completes the proof.

3.7.5 *Remark*

If $v = \phi(x)$ is positive definite and continuous in $N(M)$, then condition (iii) of Theorem 3.7.3 is satisfied. However, there do exist semi-definite and even indefinite functions which satisfy condition (iii) in Theorem 3.7.3. Thus the above theorem seems stronger than the classical theorem of Liapunov on stability (3.6.6

3.7.6 *Example*

Consider the second order differential equation:

$$\ddot{x} + r^2 \left(r \sin \frac{1}{r} - \cos \frac{1}{r}\right) \dot{x} + x = 0, \qquad \text{where} \quad r = x^2 + y^2$$

or the equivalent system

$$\dot{x} = y, \quad \dot{y} = -x - r^2 \left(r \sin \frac{1}{r} - \cos \frac{1}{r}\right) y, \quad \text{where} \quad r = x^2 + y^2$$

We take

$$\phi(x,y) = (x^2 + y^2) \sin \left(\frac{1}{x^2 + y^2}\right) .$$

The function $\phi(x,y)$ is indefinite in any neighborhood of the origin, but satisfies the condition (ii) in Theorem 3.7.3. For the above system

$$\begin{aligned} \psi(x,y) &= -2\, y^2 r^2 \left(r \sin \frac{1}{r} - \cos \frac{1}{r}\right) \sin \frac{1}{r} \\ &\quad + r \cos \frac{1}{r} \left(- \frac{1}{r^2}\right) \left(-2y^2 r^2 \left(r \sin \frac{1}{r} - \cos \frac{1}{r}\right)\right) \\ &= -2y^2 \left(r \sin \frac{1}{r} - \cos \frac{1}{r}\right) \left(r^2 \sin \frac{1}{r} - r \cos \frac{1}{r}\right) \\ &= -2y^2\, r \left(r \sin \frac{1}{r} - \cos \frac{1}{r}\right)^2 . \end{aligned}$$

Notice that $\psi(x,y) \leq 0$, and all conditions of Theorem 3.7.3 are satisfied. The origin is thus stable for the given differential system, although the function $\phi(x,y)$ is not even semi-definite.

We shall now present a very general Theorem (3.7.11) which gives sufficient conditions for asymptotic stability and attraction of compact sets under much less stringent requirements then those of the classical theorems in Section 3.6. In particular, we shall relax condition iii) of Theorem 3.6.15.

Our main reason for relaxing condition (iii) of Theorem 3.6.15 is practical convenience. In fact, from the theoretical point of view, if a compact set is asymptotically stable, then there always exists a Liapunov function, i.e., a function which satisfies all the requirements of Theorem 3.6.15. This fact has been proved for a dynamical system in Section 1.7 and can be proved for the special case of a differential system. While the theory assures us of the existence of such a scalar function, in practical cases it may be rather difficult to find one which satisfies all requirements. The enlargement of the class of Liapunov functions may be extremely helpful for the solution of stability problems. The severity of condition (iii) of Theorem 3.6.15 can be quite well illustrated by the following example.

3.7.7 *Example*

Consider the second order differential system

3.7.8
$$\dot{\eta} = \chi$$
$$\dot{\chi} = \theta(\chi,\eta) \quad ,\theta(0,0) = 0$$

which is derived from the second order differential equation

$$\ddot{\eta} = \theta(\dot{\eta},\eta) \qquad .$$

We are interested in establishing the stability properties of the equilibrium point $\chi = \eta = 0$. For this consider the real-valued function

3.7.9
$$\phi(\chi,\eta) = a_{11}(\chi) + a_{22}(\eta)\eta^2$$

where the real-valued functions $a_{ii}(\chi,\eta)$ are defined in the whole plane χ,η. Consider then the total time derivative of 3.7.9 along the solutions of 3.7.8.

$$\dot{\phi} = \psi(\chi) = \eta[\frac{\partial a_{11}}{\partial \chi} + \frac{\partial a_{22}}{\partial \eta}\theta(\chi,\eta)\eta + 2a_{22}(\eta)\theta(\chi,\eta)]$$

This scalar function vanishes identically on the axis $\eta = 0$. Thus $\psi(x)$ is not definite for $M = \{0\}$ and the condition (iii) is never satisfied. This means that for all differential systems of the type 3.7.8 no scalar function of the class 3.7.9 can be used to prove either asymptotic stability or complete instability of the critical point $\chi = \eta = 0$.It is immediate that this is the case for all real-valued functions $\phi(x)$ whose level curves are orthogonal to the axis χ. In fact, all solutions of systems of the form 3.7.8 have have this property. Thus all systems of the type 3.7.8 have solution curves which are tangent to the level curves of the function 3.7.9 on the axis $\eta = 0$. Thus this particular property of the function $\phi(x)$ with respect to the solution curves of the differential system is, in most cases and in particular in the case of Example 3.7.8, not a property of the norm of the solutions and therefore is not a stability property. It seems obvious that, at least in some cases, it should still be possible to use such a real-valued function $v = \phi(x)$ for the characterization of the asymptotic stability properties of sets. This will be done in the next theorem.

The key of the whole problem is in the particular properties of the set $P = \{x \in E: \psi(x) = 0\}$. In this set we can distinguish 3 different components

i) $P_1 = \{x \in E: \text{grad } \phi(x) = 0\}$

ii) $P_2 = \{x \in E: f(x) = 0\}$

iii) $P_3 = \{x \in E: \text{grad } \phi(x) \text{ orthogonal to } f(x)\}$

or, which is the same, P_3 is the set of all points x^o, in which at least one of the corresponding solutions of the differential equation 3.7.1 is tangent to the level surface $\phi(x) = C$, defined by $\phi^{-1}(C) = x^o$. Along these lines the following theorem is of interest.

3.7.10 *THEOREM*

Let $v = \phi(x)$ *and* $w = \phi(x)$ *be real-valued functions defined in an open neighborhood* $N(M) \subset E$ *of a compact set* M. *Assume that:*

i) $v = \phi(x) \in C^1$,

ii) $\psi(x) = \langle \text{grad } \phi(x), f(x) \rangle \leq 0 \quad , \quad x \in N(M)$,

iii) There exists $x^0 \in N(M)$ *such that at least one solution* $x = x(t,x^0)$ *of* 3.7.1 *is bounded and such that* $\Lambda^+(x^0) \subset N(M)$. *Then* $\psi(x) \equiv 0$ *on* $\Lambda^+(x^0)$.

Proof: Let y_1 and y_2 be two points in $\Lambda^+(x^0)$. By definition, then, there exist sequences $\{t_n\} \in R^+, t_n \to +\infty$ and $\{\tau_n\} \in R^+, \tau_n \to +\infty$, such that $x(t_n,x^0) \to y_1$ and $x(\tau_n,x^0) \to y_2$. Since $x(t,x^0) \in N(M)$, by hypothesis (ii), $\phi(x(t,x^0))$ is a non-increasing function of t which is bounded from below because of continuity. Then from the hypothesis (iii) $\lim_{t \to +\infty} \phi(x(t,x^0))$ exists and is such that $\lim_{n \to \infty} \phi(x(t_n,x^0)) = \lim_{n \to \infty} \phi(x(\tau_n,x^0))$, which proves the theorem.

We shall now prove the main theorem on asymptotical stability and attraction. This will be done for the special case of a differential equation which defines a dynamical system.

3.7.11 *THEOREM*

Let Ω *be a compact, positively invariant set for the flow described by the differential equation* 3.7.1, *which defines a dynamical system. Let* $v = \phi(x)$ *and* $w = \psi(x)$ *be real-valued function defined on* Ω. *Assume that*

i) $v = \phi(x) \in C^1, \quad x \in \Omega$,

ii) $w = \psi(x) \leq 0, \quad x \in \Omega$,

iii) $\psi(x) = \langle \text{grad } \phi(x), f(x) \rangle$.

Consider the following sets

I) S = *largest invariant set in* Ω,

II) $Q = \{x \in \Omega$: $\phi(x)$ *is a minimum on* $\Omega\}$,

III) $P = \{x \in \Omega: \psi(x) = 0\}$,

IV) M = *largest invariant set in* P $(M \subseteq S)$,

V) U = *largest invariant set contained in* Q. *Then:*

a) All these sets are closed,

b) $Q \subseteq P$ *and* Q *is stable relative to* Ω *,*

c) M *is attracting relative to* Ω *,*

d) S *is asymptotic stable relative to* Ω *,*

e) $\partial M \cap \partial S \neq \emptyset$,

f) $S = D_\Omega^+(M) = \{y\colon \exists \{x_n\} \subset \Omega, \{t_n\} \subset R^+$ *such that* $x_n \to M,\ x_n t_n \to y\}$ *where* $D_\Omega^+(M)$ *is the first positive prolongation of* M *relative to* Ω*,*

g) S *is the smallest relatively asymptotically stable set containing* M*,*

h) if for all $x \in \partial M$, $\phi(x) = \text{const}$, M *is asymptotically stable relative to* Ω *and* $M = S$,

i) M *minimal implies* $M = S = U$,

j) $M = Q$ *implies* M *is asymptotically stable relative to* Ω *and* $M = S$,

k) if either Ω, S *or* M *are homeomorphic to the unit ball, then* M *contains a rest point,*

l) if either $M \subset I\Omega$ *or* $S \subset I\Omega$, *then the words "relative to* Ω*" may be deleted from the above statements.*

Proof: a) This is clear, for if a set is invariant, so is its closure.

b) For all $x \in Q$, grad $\phi(x) = 0$ implies $\psi(x) = 0$ which implies $Q \subseteq P$. Stability then follows from the usual theorem.

c) Since for each $y \in \Omega$, $\Lambda^+(y) \subset \Omega$ and $\psi(x) = 0$ for $x \in \Lambda^+(y)$, we get $\Lambda^+(y) \subset M$ as $\Lambda^+(y)$ is invariant.

d) Since $S \supset M$, S is attracting relative to Ω. Stability follows as S is the largest invariant set in Ω, $D_\Omega^+(S) \subset \Omega$, so that $D_\Omega^+(S) = S$.

e) For if $\partial M \cap \partial S = \emptyset$, then $\partial S \subset \Omega$ being invariant, we have $x \in \partial S$, $\Lambda^+(x) \neq \emptyset$ and $\Lambda^+(x) \cap M = \emptyset$, since $M \subset S$, we conclude that M is not attracting relative to Ω, contradicting (c).

f) Follows from proof of (d).

g) Obvious.

h) We need only show that M is stable relative to Ω. If not, then there is a sequence $\{x_n\}$ in Ω, $x_n \to x \in \partial M$, and a sequence $\{t_n\}$, $t_n \geq 0$, such that $x_n t_n \to y \notin M$. Indeed $y \in \Omega$ and $x_n[0,t_n] \subset \Omega$ as Ω is positively invariant and compact. Since $\phi(x_n) \geq \phi(x_n t_n)$, we get $\phi(x) \geq \phi(y)$ by continuity. However, if $z \in \Lambda^+(y)$, we get $\phi(z) \leq \phi(y)$, and since $z \in \partial M$, we have $\phi(x) \leq \phi(y) \leq \phi(z) = \phi(x)$, showing that $\phi(y) = \phi(x)$. This shows, however, that $yR \subset \Omega$ and $\phi(x)$ is constant on yR. Consequently $\psi(x) = 0$ on yR, showing that $yR \subset M$, a contradiction.

i) If M is minimal, then $\phi(x)$ is constant on M and the result follows from (h) and, the fact that $U \subset S$.

j) Follows from (h) and (c). Q.E.D.

We consider now the problem of the identification of a region Ω by means of real-valued functions $v = \phi(x)$, defined in a neighborhood of M. This ca be easily done with the help of Lemma 3.7.2 for pairs of real-valued functions $v = \phi(x)$ and $w = \psi(x)$ which satisfy the requirements of Theorem 3.7.11. In particular

3.7.12 *Remark*

Let $v = \phi(x)$ and $w = \psi(x)$ be as in Theorem 3.7.11. Assume that for some $k > 0$ we have $||H_k|| < \infty$, then we may take

3.7.13 $$\Omega = H_k.$$

3.7.14 *Remark*

Let $v = \phi(x)$, $w = \Psi(x)$ be continuous functions defined in a neighborh N(M) of M, where $\psi(x) \leq 0$. If for any $k > 0$, H_k is compact, then we may app Theorem 3.7.11 to H_k. If further $S \subset I(H_k)$, then notice that M, and S are respectively an attractor and an asymptotically stable set, and H_k is in their region of attraction. Moreover if for each $k > 0$ H_k is compact and for each k the same set S is the largest invariant set in H_k, then $\bigcup(H_k) \subset A(S)$ (region of attraction of S), and indeed in this case also $\bigcup(H_k) \subset A(M)$. Lastly if $\bigcup(H_k) = E$, then we can detect a globally asymptotically stable set by means of Theorem 3.7.11.

3.7.15 *Instability Theorems.*

Such theorems can easily be derived from Theorem 3.7.11. Note that if Ω is compact, negatively invariant, and $\psi(x) \geq 0$, then conclusion of Theorem 3.7.11 will be reversed in the sense that the set M will be negatively attracting, i.e., completely unstable with respect to Ω. This observation can be used to derive the classical instability theorem of Chetaev for example and many others.

3.7.16 Notes and References

The extension of the Liapunov Theory presented in Theorem 3.7.3 is due to A. Strauss [5]. The extension of the Liapunov theorem for asymptotic stability of the rest point $x=o$ allowing $\psi(x)$ to be semidefinite while the largest invariant set contained in the set $\{x \in E : \psi(x) = o\}$ is $\{o\}$, is due to E.A. Barbashin and N.N. Krasovskii [1]. An extension of Liapunov's theorem for attraction of compact sets leading to the approach stressed by the theorem 3.7.11 was originated by J.P. La Salle [3].

3.8 *The extension theorem.*

The "classical" theorems 3.6.15, 3.6.22 and 3.6.23 require a very well-defined behavior of both $\phi(x)$ and $\psi(x)$. Those two functions are, on the other hand, connected by the equation 3.6.2. We shall show that, if the given system 3.6.1 satisfies certain conditions, then from the global properties of $\psi(x)$ and the properties of $\phi(x)$ in an arbitrarily small neighborhood of the set M, one is able to deduce the global properties of $\phi(x)$. This fact will be of extreme help in many applications where the function $\psi(x)$ has known properties in the whole space, while the behavior of $\phi(x)$ is not known for large values of $\rho(x,M)$. The connection between $\psi(x)$ and $\phi(x)$ is given by equation 3.6.2 which contains information about the gradient of the real-valued function $\phi(x)$. This problem has, therefore, three steps: i) deduce from the global properties of $\psi(x)$ the global properties of grad $\phi(x)$; ii) deduce from the global properties of grad $\phi(x)$ and the local properties of $\phi(x)$ the global properties of $\phi(x)$; iii) deduce from the local stability properties of a compact set $M \subset E$ in the flow defined by the solutions of the ordinary differential equation 3.6.1 and the global properties of $\psi(x)$, the global stability properties of M. The theorems that we shall present are called <u>extension theorems</u> because they give conditions under which the local stability properties of compact sets can be extended to the whole space. This problem is essentially an investigation of the relationships between topological and analytical properties in E of the level lines of the real-valued, continuously differentiable function $v = \phi(x)$ and the stability properties of the ordinary differential equation

3.8.1 $$\dot{x} = \text{grad}\ \phi(x)$$

or

3.8.2 $$\dot{x} = \frac{\text{grad}\ \phi(x)}{1 + ||\text{grad}\ \phi(x)||}$$

In general, we will be interested in characterizing the stability properties of a compact set $M \subset E$ in the flow defined by the solutions of the ordinary differential equation

3.8.3 $$\dot{x} = f(x)$$

through the analytical and topological properties of the level lines of the real-valued, continuously differentiable function $v = \phi(x)$, which has the property that the real-valued function

3.8.4 $$\psi(x) = \langle \text{grad } \phi(x), f(x) \rangle$$

is definite in a suitable open set $N(M)$. The results that we shall obtain are related to various problems of differential geometry and topology. In the sequel we shall use the notion of critical points of a real-valued function.

3.8.5 *DEFINITION*

Let $v = \phi(x) \in C^1$ *be a real-valued function defined in* E. *A point* $x^c \in E$ *for which* $\text{grad } \phi(x^c) = 0$ *is called a* <u>*critical point*</u> *of* $\phi(x)$. *The real-valued function* $v = \phi(x)$ *is said to have an* <u>*infinite critical point*</u> *if there exists a sequence* $\{x^n\}: ||x^n|| \to \infty$, *such that* $\text{grad } \phi(x^n) \to 0$ *as* $||x^n|| \to \infty$. *By saying that* $\phi(x)$ *does not have any critical points in* $E \cup \{\infty\}$ *we mean that* $\phi(x)$ *has neither finite nor infinite critical points.*

We shall prove next the "extension theorems" for strong stability properties of compact sets. The first theorem is an extension theorem for asymptotic stability for the case in which the differential equation 3.8.3 defines a dynamical system. This theorem will then be used for the proof of a stronger result (Theorem 3.8.13) on the analytical properties of Liapunov functions.

3.8.6 *THEOREM*

Let $v = \phi(x)$ *and* $w = \psi(x)$ *be real-valued functions defined in the whole space* E. *Let* $M \subset E$ *be a compact set. Assume that*

i) $v = \phi(x) \in C^1$,

ii) $\phi(x) = 0$ *for* $x \in M$,

iii) there exists $\eta > 0$ *such that* $\phi(x) \neq 0$ *for* $x \in H(M,\eta)$,

iv) $\psi(x) = 0$ *for* $x \in M$, $\psi(x) \neq 0$ *for* $x \notin M$, $\psi(x)$ *does not have zeros at infinity,*

v) $\operatorname{sign} \phi(x) \neq \operatorname{sign} \psi(x)$ *for* $x \in S(M,\eta) \setminus M$,

vi) $\psi(x) = \langle \operatorname{grad} \phi(x), f(x) \rangle$,

vii) the differential system $\dot{x} = f(x)$ *defines a dynamical system.*

Then the compact set M *is globally asymptotically stable.*

Proof: The conditions of the theorem imply that $\phi(x) \neq 0$ for $x \in S(M,\eta) \setminus M$, i.e., the set $P = \{x : x \in S(M,\eta) \setminus M,\ \phi(x) = 0\}$ is empty. To see this notice first that P can have only isolated points, because at a point $y \in P$ which is a limit point of P one has $\operatorname{grad} \phi(y) = 0$, which implies by (vi) that $\psi(y) = 0$ and since $y \notin M$, we get a contradiction to (iv). If P has an isolated point z, then there is a neighborhood N of z, $N \subset S(M,\eta) \setminus M$, such that $\phi(x) \neq 0$ for $x \in N$, $x \neq z$. But then z is an extrenal point of $\phi(x)$ and therefore $\operatorname{grad} \phi(z) = 0$; also, by (vi), $\psi(z) = 0$, contradicting (iv). Thus P is empty. This shows that the set M is (locally) asymptotically stable.

Let then A(M) be the region of attraction of M; A(M) is an open invariant set. We will show that $A(M) = E$. This is equivalent to proving that $\partial A(M) = \emptyset$. In fact, if $A(M) \neq E$, then $\partial A(M) = \overline{A(M)} \cap \overline{C(A(M))}$, but if $\partial A(M) = \emptyset$, then $\emptyset = \overline{A(M)} \cap \overline{C(A(M))}$ and hence $E = \overline{A(M)} \cup \overline{C(A(M))}$ which implies that E is the union of two nonempty, disjoint closed sets and, therefore, not connected. This is a contradiction.

Now by asymptotic stability of M, there is a δ, $0 < \delta \leq \eta$, such that $S[M,\delta] \subseteq A(M)$. Assume for simplicity that $\phi(x) > 0$ for $x \in S[M,\eta] \setminus M$, and let

$$\nu = \min\{\phi(x) : x \in H(M,\delta)\} .$$

We claim that for all $x \in \partial A(M)$, $\phi(x) \geq \nu$. In fact, from the hypothesis made on $\phi(x)$ it follows that $\phi(x(x^o,t))$ is a strictly decreasing function of t for all $x \notin M$. Now, if for some $y \in \partial A(M)$, it were $\phi(y) < \nu$ it would be possible to find an $x^o \in A(M)$, $x^o \notin S[M,\delta]$ such that $\phi(x^o) < \nu$. As $x^o \in A(M) \setminus S[M,\delta]$, there is a $\tau > 0$ such that $x(x^o,\tau) \in H[M,\delta]$. Then $\nu \leq \phi(x(x^o,\tau)) < \phi(x^o) < \nu$ for $\phi(x(x^o,t))$ is strictly decreasing. This is absurd.

Since $\phi(x) \geq \nu$ for $x \in \partial A(M)$, we have by (v) that $\psi(x) < 0$ for $x \in \partial A(M)$. Let

$$-\mu = \sup\{\psi(x) : x \ \partial A(M)\} .$$

By (iv) $\mu > 0$ since $\partial A(M)$ is bounded away from M. Let now $x^o \in \partial A(M)$, then $x(x^o,t) \in \partial A(M)$ for $t \geq 0$, since $\partial A(M)$ is invariant. Then

$$\phi(x(x^o,t)) = \phi(x^o) + \int_{t_o}^{t} \psi(x(x^o,\tau))d\tau \leq \phi(x^o) - \int_{t_o}^{t} \mu \, d\tau = \psi(x^o) - \mu(t-t_o)$$

which shows that $\lim_{t \to +\infty} \phi(x(x^o,t)) = -\infty$, which is absurd, since we have proved that for all $x \in \partial A(M)$, $\phi(x) \geq \nu$. This contradiction shows that $\partial A(M) = \emptyset$ and proves the theorem.

3.8.7 *Remark*

Theorem 3.8.6 still holds if one replaces either condition (iii) with iii'): M is (locally) asymptotically stable, or if one replaces conditions (ii) and (iii) with

iii''): M is invariant and (locally) asymptotically stable.

We shall now prove an extension theorem for the case that equation 3.6.1 does not define a dynamical system. The proof of this extension theorem is based upon the following fundamental lemma on a property of real-valued functions.

3.8.8 *LEMMA*

Let $v = \phi(x)$, *be a real-valued function defined in* E. *Let* $M \subset E$ *be a compact set. Assume that*

i) $v = \phi(x) \in C^2$,

ii) $\phi(x) = 0$ *for* $x \in M$,

iii) $\phi(x) > 0$ *for* $x \in H(M,\delta)$, $\delta > 0$,

iv) for $\{x^n\} \subset E$, $\operatorname{grad} \phi(x^n) \to 0$ *implies* $x^n \to M$.

Then there exist two strictly increasing functions $\alpha(\mu)$ *and* $\beta(\mu)$, $\alpha(0) = \beta(0) = 0$, *such that*

3.8.9 $$\alpha(\rho(x,M)) \leq \phi(x) \leq \beta(\rho(x,M))$$

and, furthermore,

3.8.10 $$\lim_{\mu \to \infty} \alpha(\mu) = +\infty .$$

In addition to this, if

$$\nu = \min \{\phi(x) : x \in H(M,\delta)\},$$

then $\phi(x) > \nu$ *for* $x \in C(S[M,\delta])$.

Proof: Consider the differential system defined by

3.8.11 $$\dot{x} = f(x) = - \frac{\operatorname{grad} \phi(x)}{1 + ||\operatorname{grad} \phi(x)||} .$$

It is well known that the differential system 3.8.11 defines a dynamical system. Conditions (ii), (iii) and (iv) above imply that the function $\phi(x)$ and

3.8.12 $$\psi(x) = \langle \operatorname{grad} \phi(x), f(x) \rangle = - \frac{||\operatorname{grad} \phi(x)||^2}{1 + ||\operatorname{grad} \phi(x)||}$$

satisfy the conditions of the Theorem 3.8.6. Thus the set M is globally asymptotically stable for the system 3.8.11. Notice now that for any $x^o \in E \setminus S(M,\delta)$, there is a $\tau > 0$ such that $x(x^o,\tau) \in H(M,\delta)$. Thus $\phi(x(x^o,\tau)) \geq \nu$. Since $\phi(x(x^o,t))$, for any $x^o \in E \setminus M$, is a strictly decreasing function of t, we conclude that $\phi(x^o) > \phi(x(x^o,\tau)) \geq \nu$. Lastly as $\phi(x) = 0$, for $x \in M$, $\phi(x) > 0$ for $x \notin M$, and $\phi(x)$ is continuous, we can define two continuous increasing functions $\alpha(\eta)$ and $\beta(\eta)$ by

$$\alpha(\eta) = \min \{\phi(x) : x \in H(M,\eta)\}$$

and

$$\beta(\eta) = \max \{\phi(x) : x \in H(M,\eta)\} .$$

Notice that $\alpha(0) = \beta(0) = 0$ and $\beta(\eta) \geq \alpha(\eta) > 0$ for $\eta > 0$. There are thus strictly increasing continuous functions $\bar{\alpha}(\eta)$ and $\bar{\beta}(\eta)$ such that

$$\bar{\alpha}(\eta) \leq \alpha(\eta) \leq \beta(\eta) \leq \bar{\beta}(\eta) .$$

With these $\bar{\alpha}(\eta)$ and $\bar{\beta}(\eta)$ we have

$$\bar{\alpha}(\rho(x,M)) \leq \phi(x) \leq \bar{\beta}(\rho(x,M)) .$$

It remains to be proved that there exists one function $\alpha(\eta)$ which satisfies condition 3.8.10. We shall prove first that for all $x^o \in C(M)$, $\phi(x(x^o,t)) \to +\infty$ as $t \to -\infty$. In fact, notice that for any $x^o \in C(M)$

$$\phi(x(x^o,t)) = \phi(x^o) + \int_0^t \psi(x(x^o,\tau))d\tau .$$

For $\tau \leq 0$ we have $\rho(x(x^o,\tau),M) \geq \eta > 0$. Thus

$$\psi(x(x^o,\tau)) \leq \max\{\psi(x) : x \notin S(M,\delta)\} = -\chi < 0$$

where $\delta > 0$ is such that $x(x^o,\tau) \notin S(M,\delta)$ for $\tau \leq 0$. Hence for $t \leq 0$

$$\phi(x(x^o,t)) \geq \phi(x^o) + \int_t^o \chi d\tau = \phi(x^o) - \chi t .$$

Thus $\phi(x(x^o,t)) \to +\infty$ as $t \to -\infty$.

Now the existence of our $\alpha(\eta)$ with the property that $\alpha(\eta) \to +\infty$ as $\eta \to +\infty$ is equivalent to the property that $\phi(x) \to +\infty$ as $\rho(x,M) \to +\infty$. Assume that the last assertion is not true. Then there is a $h > 0$ such that the surfaces $\phi(x) = k$ are compact for $k < h$ and noncompact for $k \geq h$. Consider the open set P defined by

$$P = \{x: 0 < k < \phi(x) < h\}.$$

This open set is bounded away from the set M and is bounded by the surfaces $\phi(x) = k$ and $\phi(x) = h$. Notice that each point of the surface $\phi(x) = k$ is a strict egress point of P and the surface $\phi(x) = h$ consists of strict ingress points only. We claim that there is a point x^o, $\phi(x^o) = k$, such that $x(x^o,t) \in P$ for $t < 0$. For, otherwise, for every x^o, with $\phi(x^o) = k$ we will have a unique $t < 0$ such that $x(x^o,t) \in \{x: \phi(x) = h\}$, so that there will be continuous map of the compact set $\phi(x) = k$ onto the noncompact set $\phi(x) = h$ which is impossible. Notice that for such a x^o we have $\phi(x(x^o,t)) < h$ for $t \leq 0$ which contradicts the fact that $\phi(x(x^o,t)) \to +\infty$ as $t \to -\infty$ for all $x^o \notin M$. The theorem is completely proved.

By applying the fundamental lemma 3.8.8 we are now in the position of proving the extension theorem for differential systems which may not define dynamical systems.

3.8.13 *THEOREM*

Let $v = \phi(x)$ *and* $w = \psi(x)$ *be real-valued functions defined in the whole space* E. *Let* M *be a compact set. Assume that*

i) $v = \phi(x) \in C^2$,

ii) $\phi(x) = 0$ *for* $x \in M$,

iii) $\phi(x) \neq 0$ *for* $x \in H(M,\delta)$, $\delta > 0$,

iv) $\phi(x)$ *does not have (finite or infinite) critical points for* $x \notin M$, *v)* $\psi(x)$ *is semidefinite for* M *in* E,

vi) sign $\phi(x) \neq$ sign $\psi(x)$, $x \in S(M,\delta)$,

vii) $\psi(x) = \langle \text{grad } \phi(x), f(x) \rangle$.

Then the compact set M *is globally asymptotically stable, for the differential system* 3.6.1.

Proof: By Lemma 3.8.8 there are strictly increasing continuous functions $\alpha(r)$, $\beta(r)$ such that $\alpha(\rho(x,M)) \leq \phi(x) \leq \beta(\rho(x,M))$. Further $\alpha(\eta) \to +\infty$ as $\eta \to \infty$. We notice now that $\phi(x)$ satisfies conditions of usual theorem of global asymptotic stability (3.6.26). Hence the compact set M is globally asymptotically stable.

Similarly to what is done for Theorem 3.8.6 one has

3.8.14 *COROLLARY*

Theorem 3.8.13 *holds if condition (iii) is replaced by the condition:*

iii') the compact set M *is (locally) asymptotically stable.*

In what follows we shall prove a stronger version of Theorem 3.8.13 which is based upon the following Lemma 3.8.15. This lemma is an improvement of Lemma 3.8.8. Its proof, which was suggested by C. Olech in a private communication, is not based upon Theorem 3.8.6. An alternative possible proof of this theorem is based upon an improved version of Theorem 3.8.6 for flows without uniqueness.

3.8.15 *LEMMA*

Lemma 3.8.8 *still holds if condition i) is replaced by:*

i') $v = \phi(x) \in C^1$.

Proof: Assume that $\phi(x) > 0$ for $x \in H(M,\delta)$. Let

$$\nu = \min \{\phi(x) : x \in H(M,\delta)\} > 0 .$$

Consider the sets

$$N(\mu) = \{x \in E : \phi(x) < \mu\}$$

and

$$B(\mu) = \{x \in E : \phi(x) \leq \mu\}$$

The set $N(\nu/2)$ is obviously open, in addition every component $N_i(\nu/2)$ of $N(\nu/2)$ such that

3.8.16 $$N_i(\nu/2) \cap S(M,\delta) \neq \emptyset$$

is bounded. This is due to the fact that

3.8.17 $$\partial N(\tfrac{\nu}{2}) \cap H(M,\delta) = \emptyset .$$

Thus there exists at least one component $N_\ell(\nu/2)$ of $N(\nu/2)$ which is bounded ar has the properties 3.8.16 and 3.8.17. Let now $N_c(\beta)$, $\beta \geq \nu/2$ be that component of $N(\beta)$ with the property

3.8.19 $$N_\ell(\nu/2) \subset N_c(\beta) .$$

Notice that if $N_c(\beta)$ is bounded for $\beta = \beta_0$, then

a) $\overline{N_c(\beta_0)} = B_c(\beta_0)$, where $B_c(\beta_0)$ is the analogous component of $B(\beta_0)$

b) $\nu/2 \leq \phi(x) \leq \beta_0$ for $x \in B_c(\beta_0) \setminus N_c(\nu/2)$.

c) there exists $\varepsilon > 0$, such that $N_c(\beta)$ is bounded, for $\beta_0 \leq \beta \leq \beta_0 + \varepsilon$.

The statement above follows from (iv) and the implicit function theorem.

We shall next show that if $N_c(\beta)$ is bounded from $\beta < \alpha$, then also the set

3.8.20 $$A = \bigcup_{\beta < \alpha} N_c(\beta)$$

bounded and such that $A = N_c(\alpha)$.

Since A is open and connected and for $x \in \partial A$ $\phi(x) = \alpha$, $= N_c(\alpha)$. It must be shown now that the boundedness of $N_c(\beta)$ implies the undedness of A. Consider the set $A \setminus N_\ell(\nu/2)$. Clearly this set has a finite positive stance from M. Then from the hypothesis iv) it follows that there exists $k > 0$, ch that

8.21 $$||\text{grad } \phi(x)|| \geq k > 0 \qquad \text{for } x \in A \setminus N_\ell(\nu/2).$$

nsider next the differential equation

8.22 $$\dot{x} = \frac{-\text{grad } \phi(x)}{1 + ||\text{grad } \phi(x)||}$$

ich has global extendability (Theorem 3.1.62), but not necessarily uniqueness. t $x(x^o,t)$ be a solution of the differential equation 3.8.22 with $x^o \in A \setminus N_\ell(\nu/2)$; en the function $\phi(x(x^o,t))$ is a strictly decreasing function of t and, in dition

$$\frac{d}{dt} \phi(x(x^o,t)) \leq -L < 0$$

f $x(x^o,t) \in A \setminus N_\ell(\nu/2)$ or if $\phi(x(x^o,t)) \geq \nu/2$. Then for each solution $x^o,t)$ of the differential equation 3.8.22 with $x^o \in A \setminus N_\ell(\nu/2)$ there exists $= (\alpha - \nu/2) / L$ such that $x(x^o,\tau) \in N_\ell(\nu/2)$ for $\tau \geq T$ and $x^o \in A \setminus N_\ell(\nu/2)$. Hence ch point of A is at a finite distance from $N_\ell(\nu/2)$, which is bounded; also A bounded. Then for all $\beta > 0$ the set $N_c(\beta)$ is bounded and $\bigcup_{< +\infty} N_c(\beta) = E$. Thus

$$\lim_{||x|| \to \infty} \phi(x) \to + \infty$$

ich proves the most important part of the lemma. The remaining statements can be oved in exactly the same way as in Lemma 3.8.8 .

We can now apply Lemma 3.8.15 to the proof of the following result.

3.8.23 *THEOREM*

Theorem 3.8.13 *and Corollary* 3.8.14 *still hold if condition i) is replaced by*

i') $v = \phi(x) \in C^1$.

The proof of this theorem is exactly the same as the one of Theorem 3.8.13 when instead of Lemma 3.8.8, we use Lemma 3.8.15.

3.8.24 *Remark*

With obvious variations, theorems similar to 3.8.6, 3.8.7, 3.8.13, 3.8.14 and 3.8.23 can be proved also for the case of complete instability.

We shall now prove a theorem similar to 3.8.23 for the case of instability. This theorem is based upon two lemmas which have rather simple proofs.

3.8.25 *LEMMA*

Let $v = \phi(x)$ *be a real-valued function defined in* E. *Let* $M \subset E$ *be a compact set. Assume that*

i) $v = \phi(x) \in C^1$,

ii) $v = \phi(x)$ *is indefinite for* M *in* $S(M,\delta)$, $\delta > 0$,

iii) there exists $\eta > \delta$ *such that* $\phi(x) \neq 0$ *for* $x \in H(M,\eta)$.

Then there exists a point $x^c \in S(M,\eta) \setminus M$ *which is a critical point of* $\phi(x)$.

Proof: To fix the ideas assume that $\phi(x) > 0$ for $x \in H(M,\eta)$. Then there exists a set $Z = \{x \in S(M,\eta): \phi(x) = 0\}$. Furthermore, there exists an open set $\Gamma^- \subset S(M,\eta)$ such that $\phi(x) < 0$ for $x \in \Gamma^-$. By continuity $\partial\Gamma^- \subset Z$. The function $\phi(x)$ has then its least upper and greatest lower bounds in $\overline{\Gamma^-}$. Obviously the extremals cannot both bel[ong] to $\partial\Gamma^-$, since then it would follow that $\phi(x) \equiv 0$ for $x \in \overline{\Gamma^-}$, and then $\phi(x)$ would not be indefinite. Thus $\phi(x)$ has one extremal in $\partial\Gamma^-$ which is the critical poi[nt]

In the same fashion one may now prove the following lemma.

.8.26 *LEMMA*

Let $M \subset E$ *be a compact set, and let* $v = \phi(x)$ *be a* C^1 *function defined* *n* E *with the following properties:*

i) if for any sequence $\{x_n\}$, $\text{grad}\,\phi(x_n) \to 0$, *then* $x_n \to M$,

ii) there is an open connected set Γ *and an* $\eta > 0$ *such that* $\phi(x) = 0$ *for* $x \in \partial\Gamma \cap S(M,\eta)$, *and* $\phi(x) \neq 0$ *for* $x \in \Gamma \cap S(M,\eta)$, *and*

iii) $\partial\Gamma \cap M \neq \emptyset$.

en there exists an unbounded open connected set Γ^* *such that* $\Gamma^* \cap S(M,\eta) \equiv \Gamma \cap S(M,\eta)$, *ch that* $\phi(x) = 0$ *for* $x \in \partial\Gamma^*$, *and* $\phi(x) \neq 0$ *for* $x \in \Gamma^*$.

The above lemma is useful in deriving results on global instability ef. 3.6.30) of a compact set M.

8.27 *THEOREM*

Let $v = \phi(x)$ *and* $w = \psi(x)$ *be real-valued functions defined in* E, *and* *t* M *be compact set. If*

i) $v = \phi(x) \in C^1$,

ii) there is an open set Γ *such that* $\partial\Gamma \cap \partial M \neq \emptyset$,

iii) $\phi(x) = 0$ *for* $x \in \partial\Gamma \cap S(M,\eta)$, *and* $\phi(x) \neq 0$ *for* $x \in \Gamma \cap S(M,\eta)$,

iv) $\text{sign}\,\phi(x) = \text{sign}\,\psi(x)$ *for* $x \in \Gamma \cap S(M,\eta)$,

v) for any sequence $\{x_n\}$, $\psi(x_n) \to 0$ *implies* $x_n \to M$,

vi) $\psi(x) = \langle \text{grad}\,\phi(x), f(x) \rangle$.

en the compact set M *is globally unstable.*

The following theorem summarizes the results obtained above:

3.8.28 *Theorem (Extension Theorem)*

Let $v = \phi(x)$ *and* $w = \psi(x)$ *be real-valued functions defined on* E. *Let* $M \subset E$ *be compact. Assume that*

i) $v = \phi(x) \in C^1$,

ii) $\phi(x) = 0$ *for* $x \in M$,

iii) for any sequence $\{x_n\}$, $\psi(x_n) \to 0$ *implies* $x_n \to M$,

iv) $\psi(x) = \langle \text{grad}\, \phi(x), f(x) \rangle$.

Then whatever the local stability properties of M *for the system* 3.8.3, *these properties are global.*

3.8.29 *THEOREM*

Let $v = \phi(x)$ *and* $w = \psi(x)$ *be real-valued functions defined on* E. *Let* $M \subset E$ *be a compact set. Assume that*

i) $v = \phi(x) \in C^1$,

ii) $\phi(x) = 0$ *for* $x \in M$,

iii) for $\{x_n\} \subset E$, $\text{grad}\, \phi(x_n) \to 0$ *implies* $x_n \to M$,

iv) $\psi(x)$ *is semidefinite for* M *in* E,

v) $\psi(x) = \langle \text{grad}\, \phi(x), f(x) \rangle$,

vi) M *is the largest invariant set in the set* $\psi(x) = 0$. *Then, whatever the local stability properties of* M *maybe, they are global.*

3.8.30 *THEOREM*

Let $v = \phi(x)$ *and* $w = \psi(x)$ *be real-valued functions defined on* E. *Assume that for some* k $\|H_k\| < \infty$. *Then if* $\{x_n\} \subset E$, $\text{grad}\, \phi(x_n) \to 0$ *implies* $x_n \to M$, *the set* Ω *of Theorem* 3.7.11 *is the whole space and all the resul hold globally.*

Theorem 3.8.28 shows that if M is neither globally asymptotically stable nor globally unstable then there does not exist a real-valued function $v = \phi(x) \subset C^1$

such that $\psi(x)$ is a definite function for M. Practically then the problem of the construction of Liapunov functions for compact sets with global (strong) stability properties is reduced to a rather simple problem of searching a definite function $\psi(x)$ such that the usual equation :

3.8.31 $$\psi(x) = \langle b(x), f(x)\rangle$$

has a definite integrating factor.

On the other hand, the problem of extension theorems of sets with local strong stability properties is still not completely solved. The local version of the previously given extension theorems will be stated next. Its proof, which is not particularly difficult for the case of dynamical systems, requires rather involved machinery for the case of differential equations without uniqueness.

3.8.32 *THEOREM*

A necessary and sufficient condition for the invariant continuum $M \subset E$ *to be asymptotically stable and the open, invariant set* $A(M) \supset S(M,\varepsilon)$, $\varepsilon > 0$ *be its region of attraction is the existence of two real-valued function* $\phi(x)$ *and* $\psi(x)$ *such that*

i) $\phi(x) \in C^1$,

ii) $\phi(x) = 0$, $x \in M$,

iii) $\phi(x) \neq 0$, $x \in \partial S(M,\delta)$, *for some* δ, $0 < \delta < \varepsilon$,

iv) $\phi(x)$ *does not have finite or infinite critical points in* $\overline{A(M)} \setminus M$,

v) $\psi(x)$ *is semidefinite for* M *in* $A(M)$,

vi) sign $\phi(x) \neq$ sign $\psi(x)$ *for* $x \in \partial S(M,\delta) \cup \{x: \psi(x) \neq 0\}$,

vii) $\psi(x) = 0$ *for* $x \in \partial A(M)$,

viii) $\phi(x) = \text{const}$, $x \in \partial A(M)$,

ix) $\psi(x) = \langle \text{grad } \phi(x), f(x) \rangle$,

x) M *is the only invariant set contained in the set* $\{x \in E : \psi(x) = 0\}$.

Clearly the conditions iii), v) could be replaced by the usual conditions on the local stability properties of M.

3.1.33 *Notes and References*

Preliminary ideas leading to the extension theorems can be found in the wor of D. R. Ingwerson and W. Leighton. A complete preliminary statement was given by Szegö [4] with a complete proof of Lemma 3.8.25 and an incomplete proof of Theorem 3.8.6.

The complete proof of Theorem 3.8.6, Lemma 3.8.7 and Theorem 3.8.13 is due to G. P. Szegö and N. P. Bhatia [1]. The complete extension theorem 3.8.28 is due to G. P. Szegö [5].

The proof of Lemma 3.8.15 given in the text was suggested to us by C. Olech in a private communication.

3.9 *The use of higher derivatives of a Liapunov function.*

In the previous chapter the stability properties of sets with respect to the flow defined by the solutions of ordinary differential equations

3.9.1 $$\dot{x} = f(x), \qquad f(x) \in C^0$$

has been characterized by the properties of a real-valued function $v = \phi(x)$ and its total time derivative along the solutions of the differential equation 3.9.1.

3.9.2 $$\psi_1(x) = \langle \text{grad } \phi(x), f(x) \rangle \ .$$

In this section we shall briefly summarize some recent results obtained by various authors on the use of the total time derivative of order n of the real-valued function $v = \phi(x)$ along the solutions of 3.9.1, which is defined as follows

3.9.3 $$\psi_2(x) = \langle \text{grad } \psi_1(x), f(x) \rangle, \ldots, \psi_n(x) = \langle \text{grad } \psi_{n-1}(x), f(x) \rangle \quad .$$

where $f(x) \in C^{n-1}$ and $\phi(x) \in C^n$.

Most of the results obtained are not strictly stability results, but they lead to a more complete analysis of the qualitative behavior of the differential equation 3.9.1. This analysis is in accordance with the classification due to Nemytskii [13] of trajectories in the neighborhood of an isolated singular point into hyperbolic, parabolic and elliptic sectors.

The first use of $\psi_2(t)$ for the characterization of such qualitative properties seems to be due to N. P. Papush. The aim of his work is to identify the type of the Nemytskii classification of the solutions of 3.9.1 in a neighborhood of an equilibrium point by means of suitable sign combinations of ϕ, ψ_1 and ψ_2.

More recently M. B. Kudaev [1] has derived additional results on the behavior of the trajectories of the differential equation 3.9.1 in a neighborhood of an equilibrium point by suitable sign combinations of ϕ, ψ_1, ψ_2 and ψ_3.

Most of the results by Kudaev have been recently sharpened by J. Yorke [2] whose results are stated next. Notice that these results by Yorke have the extremely important and unique feature of having local conditions.

3.9.4 *THEOREM*

Let $v = \phi(x)$ *be a real-valued function defined in* E. *Let* H_k *be a bounded component of the set* $\{x \in E: \phi(x) < k\}$. *Assume*

i) $v = \phi(x) \in C^2$,

ii) for all $x \in \partial H_k$, $\psi_1(x) = 0$ *implies* $\psi_2(x) > 0$,

iii) there exists $y \in \partial H_k$ *such that* $\psi_1(y) \leq 0$,

iv) H_k *contains a compact invariant subset.*

Then there exists a point $z \in \partial H_k$ *such that*

$$x(t,z) \in H_k \qquad \text{for all } t > 0$$

where $x(t,z)$ *is the solution of the differential equation* 3.9.1 *with* $x(0,z) = z$.

3.9.5 *THEOREM*

If in Theorem 3.9.4 *conditions i) and ii) are satisfied and instead of (iii) and (iv) we assume that*

iii') the set $\{y \in \partial H_k: \psi_1(y) > 0\}$ *is nonempty and nonconnected.*

Then the set $\{z \in E: x(t,z) \in H_k$ *for all* $t > 0\}$ *has dimension at least* $n - 1$.

3.9.6 *THEOREM*

Let $M \subset E$ *be a compact invariant set and let* $v = \phi(x) \in C^2$ *be such that i)* $\phi(x) = 0$ *for all* $x \in M$,

$\phi(x) \geq 0$ *for all* $x \in E$,

ii) $\phi(x) \to \infty$ *as* $||x|| \to \infty$,

iii) $\psi_1(x) = 0$ *implies* $\psi_2(x) > 0$ *for all* $x \in E \setminus M$,

iv) $\psi_1(x) = 0$ *for some* $x \in E \setminus M$.

Then there exist points z_1 *and* z_2 *in* $E \setminus M$, *such that*

v) $x(t,z_1) \to M$ *as* $t \to +\infty$,

$|x(t,z_1)| \to \infty$ *as* $t \to -\infty$,

vi) $x(t,z_2) \to M$ *as* $t \to -\infty$,

$|x(t,z_2)| \to \infty$ *as* $t \to +\infty$,

vii) For all $z \in E \setminus M$ *either* $x(t,z)$ *behaves as in v) or vi) or*

$|x(t,z)| \to \infty$ *as* $|t| \to \infty$.

REFERENCES

AEPPLI, A.; MARKUS, L.,
[1] Integral equivalence of vector fields on manifolds and bifurcation of differential systems. Amer. J. Math., vol. 85 (1963), pp. 633-654.
AIZERMAN, Mark A.,
[1] On a problem concerning the stability 'in the large' of dynamical systems. Usp. mat Nauk 4, No. 4, (1949) pp. 187-188.
ALBRECHT, F.,
[1] Remarques sur un théorème de T. Ważewski relatif a l'allure asymptotique des integrales des équations différentielles, Bull. Acad. Polon. Sci. Cl. III, vol. 2, (1954) pp. 315-318.
[2] Un théorème de comportment asymptotique des solutions des équations des systèmes d'équations différentielles. Bull. Acad. Polon. Sci. Cl. III, vol. 4 (1956) pp. 737-739.
AMERIO, Luigi,
[1] Soluzioni quasi-periodiche, o limitate, di sistemi differenziali non-lineari quasi-periodici, o limitati, Ann. Mat. Pura Appl. vol. 39 (1955) pp. 97-119.
ANDREA, S. A.,
[1] On homeomorphisms of the plane, and their embedding in flows, Bull. Am. Math. Soc., vol. 71 (1965), pp. 381-383.
ANTOSIEWICZ, H. A.,
[1] An inequality for approximate solutions of ordinary differential equations. Math. Z., vol. 78 (1952) pp. 44-52.
[2] Stable systems of differential equations with integrable perturbation term. J. London Math. Soc., vol. 31, (1956) pp. 208-212.
[3] A survey of Liapunov's second method. Ann. Math. Studies 41 (Contr. theory nonlin. oscill. 4), (1958) pp. 141-166.
ANTOSIEWICZ, H. A.; DAVIS, P.,
[1] Some implications of Liapunov's conditions of stability. J. Rat. Mech. Analysis, vol. 3, (1954) pp. 447-457.
ANTOSIEWICZ, H. A.; DUGUNDJI, J.,
[1] Parallelizable flows and Liapunov's second method. Ann. Math., vol. 73 (1961), pp. 543-555.
ARZELÀ, C.,
[1] Funzioni di linee, Atti R. Accad. Lincei Rend. vol. 5 (1889) pp. 342-348.
[2] Sulle funzioni di linee, Mem. R. Accad. Bologna vol. 5 (1895) pp. 225-244.
ASCOLI, G.,
[1] Le curve limiti di una varietà data di curve, Mem. R. Accad. Lincei, vol. 18 (1883/4) pp. 521-586.
AUSLANDER, Joseph,
[1] Mean-L-stable systems, Ill. J. Math., vol. 3 (1959), pp. 566-579.
[2] On the proximal relation in topological dynamics. Proc. Am. Math. Soc., vol. 11 (1960), pp. 890-895.
[3] Generalized recurrence in dynamical systems. Contr. Diff. Equat., vol. 3 (1964), pp. 65-74.

AUSLANDER, J.; BHATIA, Nam P.; SEIBERT, Peter,
[1] Attractors in dynamical systems. Bol. Soc. Mat. Mexicana, vol. 9 (1964), pp. 55-66.
AUSLANDER, J.; HAHN, Frank,
[1] Point transitive flows, algebras of functions and the Bebutov system. (to appear).
AUSLANDER, J.; SEIBERT, P.,
[1] Prolongations and generalized Liapunov functions. Int. Symp. on Nonlinear Diff. Equations and Nonlinear Mech. Academic Press, New York and London, (1963) pp. 454-462.
[2] Prolongations and stability in dynamical systems. Annales de l'Inst. Fourier, vol. 14 (1964), pp. 237-267.
AUSLANDER, Louis; GREEN, L.; HAHN, F.,
[1] Flows on Homogeneous Spaces. Annals of Math. Studies, No. 53, Princeton, 1963.
AUSLANDER, L.; HAHN, F.,
[1] Real functions coming from flows on compact spaces and concepts of almost periodicity. Trans. Am. Math. Soc., vol. 106 (1963), pp. 415-426.
BAIDOSOV, V. A., (see also BARBASHIN, E. A.)
[1] Invariant functions of dynamical systems. Izvestiya Vysših Učebnyh Zavedenii, Matematika, 1959, No. 1 (8), pp. 9-15 (Russian).
[2] On homomorphisms of dynamical systems. Izvestiya Vysših Učebnyh Zavedenii, Matematika, 1960, No. 3 (16), pp. 21-29 (Russian).
BARBASHIN, E. A.,
[1] Sur certaines singularités qui surviennent dans us système dynamique quand l'unicité est en défaut. Comptes Rendus (Doklady) de l'Académie des Sciences de l'URSS, vol. 41 (1943), pp. 139-141.
[2] Les singularités locales des points ordinaires d'un système d'equations différentielles. Comptes Rendus (Doklady) de l'Académie des Sciences de l'URSS, vol. 41 (1943), pp. 183-186.
[3] Sur la conduite des points sous les transformations homéomorphes de l'espace. Comptes Rendus (Doklady) de l'Académie des Sciences de l'URSS, vol. 51 (1946), pp. 3-5.
[4] On homomorphisms of dynamical systems. Dokl. Akad. Nauk SSSR, vol. 61 (1948), pp. 429-432 (Russian).
[5] On the theory of generalized dynamical systems. Učenye Zapiski Moskov. Gos. Univ., No. 135, Matematika, vol. 2 (1948), pp. 110-133 (Russian).
[6] On homomorphisms of dynamical systems. Matemat. Sbornik, vol. 27 (1950), pp. 455-470 (Russian).
[7] Dispersive dynamical systems. Uspehi Matemat. Nauk, vol. 5 (1950), pp. 138-139 (Russian).
[8] On the theory of systems of multivalued transformations of a topological space. Učenye Zapiski Ural. Univ., No. 7 (1950) pp. 54-60 (Russian).

[9] The method of sections in the theory of dynamical systems. Mat. Sbornik, vol. 29 (1951), pp. 233-280 (Russian).

[10] On homomorphisms of dynamical systems. II, Mat. Sbornik, vol. 29 (1951), pp. 501-518 (Russian).

[11] On the behavior of points under homeomorphic transformations of a space. (Generalization of theorems of Birkhoff), Trudy Ural. politehn. in-ta, vol. 51 (1954), pp. 4-11 (Russian).

BARBASHIN, E. A.; BAIDOSOV, V. A.,

[1] On the question of topological definition of integral invariants. Izvestiya Vyssih Ucebnyh Zavedenii, Matematika, No. 3 (4), 1958, pp. 8-12 (Russian).

BARBASHIN, E. A.; KRASOVSKII, N. N.,

[1] On stability of motion in the large. Dokl. Akad. Nauk SSSR vol. 86, (1952), pp. 453-456.

[2] On the existence of Liapunov functions in the case of asymptotic stability in the large. Prikl. Mat. Mek., vol. 18, (1954) pp. 345-350.

BARBASHIN, E. A.; SOLOHOVICH, F. A.,

[1] The mapping of a dynamical system into a time-analytic dynamical system. Izvestiya Vysshih Uchebnyh Zavedenii, Matematika, No. 1 (14), 1960, pp. 11-15 (Russian).

BARBUTI, Ugo,

[1] Su alcuni teoremi di stabilità. Ann. Sc. Norm. Sup. Pisa, vol. 8 (1954) pp. 81-91.

BAROCIO, Samuel,

[1] On certain critical points of a differential system in the plane. Contr. Th. Nonlinear Osc. vol. 3 - (Annals Math. Studies vol. 36), Princeton Univ. Press, Princeton, N. J. (1956) pp. 127-137.

[2] On trajectories in the vicinity of a three-dimensional singularity. Boletin de la Sociedad Matematica Mexicana, vol. 1 (1956) pp. 57-58 (Spanish).

BASS, Robert W.,

[1] On the regular solutions at a point of singularity of a system of nonlinear differential equations. Amer. J. Math., vol 77, (1955) pp. 734-742.

[2] Zubov's stability criterion. Boletin de la Sociedad Matematica Mexicana, vol. 4 (1959), pp. 26-29.

BAUM, John D.,

[1] An equicontinuity condition for transformation groups. Proc. Am. Math. Soc., vol. 4 (1953), pp. 656-662.

[2] Asymptoticity in topological dynamics. Trans. Am. Math. Soc., vol. 77 (1954), pp. 506-519.

[3] P-recurrence in topological dynamics. Proc. Am. Math. Soc., vol. 7 (1956), pp. 1146-1154.

[4] An equicontinuity condition in topological dynamics. Proc. Am. Math. Soc., vol. 12 (1961), pp. 30-32.

[5] Instability and asymptoticity in topological dynamics. Pacific J. Math., vol. 12 (1962), pp. 25-34.

BEBUTOV, M. V.,
[1] Sur les systèmes dynamiques stables au sens de Liapounoff. Comptes Rendus (Doklady) de l'Académie des Sciences de l'URSS, vol. 18 (1938), pp. 155-158.
[2] Sur la représentation des trajectoires d'un système dynamique sur un système de droites parallèles. Bulletin Mathématique de l'Universite de Moscou, Série Internationale, vol. 2 (1939), Fasc. 3, pp. 1-22.
[3] Sur les systèmes dynamiques dans l'espace des fonctions continues. Bulletin Mathématique de l'Université de Moscou, Série Internationale, vol. 2 (1939), Fasc. 5.
[3] Sur les systèmes dynamiques dans l'espace des fonctions continues. Comptes Rendus (Doklady) de l'Académie des Sciences de l'URSS, vol. 27 (1940), pp. 904-906.
[4] On dynamical systems in the space of continuous functions. Bulletin Moskov. Gosuniversiteta, Matematika, vol. 2, no. 5 (1941), pp. 1-52 (Russian).
BEBUTOV, M. V.; STEPANOV, V. V.,
[1] Sur le changement du temps dans les systèmes dynamiques possédant une mesure invariante. Comptes Rendus (Doklady) de l'Académie des Sciences de l'URSS, vol. 24 (1939), pp. 217-219.
[2] Sur la mesure invariante dans les systèmes dynamiques qui ne diffèrent que par le temps. Recueil Mathématique (Mat. Sbornik), vol. 7 (1940), pp. 143-166.
BECK, Anatole,
[1] On invariant sets. Ann. Math., vol. 67 (1958), pp. 99-103.
[2] Continuous flows with closed orbits. Bull. Am. Math. Soc., vol. 66 (1960), pp. 305-307.
[3] Plane flows with few stagnation points. Bull. Am. Math. Soc., vol. 71 (1965), pp. 892-896.
[4] Plane flows with closed orbits. Trans. Am. Math. Soc., vol. 114 (1965), pp. 539-551.
BECKENBACK, Edwin F.; BELLMAN, Richard,
[1] Inequalities. Springer-Verlag, Berlin 1961.
BELLMAN, Richard, (see also BECKENBACK, E.F.)
[1] Stability Theory of Differential Equations, McGraw-Hill, New York-Toronto-London, 1953.
BELLMAN, R.; COOKE, Kenneth L.,
[1] Differential-Difference Equations. Academic Press, New York 1963.
BENDIXSON, Ivar,
[1] Sur les courbes définies par des équations différentielles. Acta Mathemat., vol. 24 (1901), pp. 1-88.
BESICOVITCH, A. S.,
[1] Almost Periodic Functions. Cambridge University Press, 1932, reprinted by Dover, New York, 1954.
BHATIA, Nam P., (see also AUSLANDER, J.)
[1] Stability and Lyapunov functions in dynamical systems. Cont. to Diff. Eqs., vol. III, No. 2 (1964), pp. 175-188.
[2] On exponential stability of linear differential systems. J. SIAM Control, Ser. A, vol. 2, No. 2 (1965), pp. 181-191.

[3] Weak attractors in dynamical systems. Bol. Soc. Mat. Mexicana, vol. 11, 1966, pp. 56-64.
[4] Criteria for dispersive flows. Mathematische Nachrichten, vol. 32, (1960), pp. 89-93.
[5] Lectures on ordinary differential equations: Stability Theory with Applications. Department of Mathematics, Western Reserve University, Spring, 1964.
[6] Asymptotic stability in dynamical systems. Tech. Note BN 462, Univ. Maryland, July 1966, to appear in Math. Systems Theory.

BHATIA, N. P.; LAKMIKANTHAM, V.,
[1] An extension of Liapunov's direct method. Mich. Math. J., vol. 12 (1965), pp. 183-191.

BHATIA, N. P.; LAZER, A.; LEIGHTON, W.,
[1] Applications of the Poincaré-Bendixon theory. Ann. Mat. Pura Applic. Ser. IV, vol. 73 (1966) pp. 27-32.

BHATIA, N. P.; LAZER, A. C.; SZEGÖ, G. P.,
[1] On global weak attractors in dynamical systems. To appear in J. Math. Anal. Appl. (the results in this paper were presented at an International Symposium on Differential Equations and Dynamical Systems, University of Puerto Rico, Mayaguez, Puerto Rico, held in December 1965 and a summary will appear in its Proceedings).

BHATIA, N. P.; SZEGÖ, G. P.,
[1] Weak attractors in R^n. Tech. Note BN 464, Univ. Maryland, July 1966, to appear in Math. Systems Theory.

BIELECKI, A.,
Une remarque sur la méthode de Banach-Cacciopoli-Tikhonov dans la théorie des équations differentielles ordinaires , Bull. Acad. Polon. Sci. Cl. III, vol. 4, (1956), pp. 261-264.

BIHARI, I.,
[1] A generalization of a lemma of Bellman and its applications to uniqueness problems of differential equations, Acta Math. Sci. Hungar. vol. 7 (1956) pp. 71-94.
[2] Researches on the boundedness and stability of the solutions of nonlinear differential equations. Acta Math. Ac. Sci. Hungar. vol. 8 (1957) pp. 261-278.

BIRKHOFF, George D.,
[1] Collected works. Vol. 1, 2, 3, Amer. Math. Soc., New York, 1950.
[2] Über gewisse Zentralbewegungen dynamischer Systeme, [Kgl. Ges. d. Wiss.,Nachrichten, Math.-phys. Klasse. 1926, Heft 1, pp. 81-92], Göttinger Nachrichten.
[3] On the periodic motions of dynamical systems. Acta Math., vol. 50 (1927), pp. 359-379.
[4] Stability and the equations of dynamics. Amer. J. Math., vol. 49 (1927), pp. 1-38.
[5] Dynamical Systems. Am. Math. Soc. Colloquium Publications, vol. 9, New York, 1927.
[6] Proof of a recurrence theorem for strongly transitive systems. Proc. Nat. Acad. Sciences of the United States of America, vol. 17 (1931), pp. 650-655.

BIRKHOFF, G. D.; LIFSHITZ, Jaime,
[1] Ciertas transformaciones en la dinamica sin elementos periodicos. Publicacione del Instituto de Mathematica, vol. 6 (1945), pp. 1-14.

BOCHNER, S.,
[1] Beiträge sur theorie der fastperiodischen Funktionen, I. Math. Annalen, vol. 96 (1926) pp. 119-147.

BOHR, H.,
[1] Zur theorie der fastperiodischen Funktionen I, II, III. Acta Math. vol. 45 (1924) pp. 29-127, vol. 46 (1925) pp. 101-214, vol. 47 (1926) pp. 237-281.

BORSUK, K.,
[1] Ueber die Abbildungen der metrischen Kompakten Raume auf die Kreislinie. Fund. Math., vol. 20, (1933) pp. 224.

BOUQUET, J. C.; BRIOT, C. A. A.,
[1] Recherches sur les fonctions définies par les équations différentielles. J. École Polytech. (Paris) vol. 21 cah. 36 (1856) pp. 133-198.

BRAUER, Fred,
[1] Some results on uniqueness and successive approximations. Canad. J. Math., vol. 11 (1959) pp. 527-533.
[2] Global behavior of solutions of ordinary differential equations. J. Math. Anal. Appl., vol. 2 (1961) pp. 145-158.
[3] Liapunov functions and comparison theorems. Proc. Int. Symp. Nonlinear Diff. Eqs. and Nonlinear Mech., Colorado Springs 1961, Academic Press 1963.
[4] Nonlinear differential equations with forcing terms. Proc. Amer. Math. Soc. vol. 15, (1964) pp. 758-765.
[5] The use of comparison theorems for ordinary differential equations. Proc. NATO Advanced Study Institute, Padua, Italy 1965, published by Oderisi, Gubbio, Italy, 1966.

BRAUER, F.; STERNBERG, Shlomo
[1] Local uniqueness, existence in the large, and the convergence of successive approximations. Amer. J. Math. vol. 80 (1958) p. 797.

BRIOT, C. A. A., (see BOUQUET, J. C.)

BRONSHTEIN, I. U., (see also SIBIRSKII, K. S.)
[1] Motions in partially ordered dynamical systems. Uchebnye Zapiski Kishinev. Univ., no. 39, 1959; pp. 249-251 (Russian).
[2] On dynamical systems without uniqueness, as semigroups of non-singlevalued mappings of a topological space. Dokl. Akad. Nauk SSSR, vol. 144 (1962), pp. 954-957 (Russian); Soviet Math. Doklady, vol. 3 (1962), pp. 824-827 (English translation).
[3] Recurrence, periodicity and transitivity in dynamical systems without uniqueness. Dokl. Akad. Nauk SSSR, vol. 151 (1963), pp. 15-18 (Russian); Soviet Math. - Doklady, vol. 4 (1963), pp. 889-892 (English translation).
[4] On dynamical systems without uniqueness, as semigroups of non-singlevalued mappings of a topological space. Izv. Akad. Nauk Moldavskoi SSR, Serija Estestien. Tehn. Nauk, no. 1 (1963), pp. 3-18 (Russian).
[5] Two examples of dynamical systems. Izv. Akad. Nauk Moldavskoi SSR, Serija Estestven. Tehn. Nauk, no. 1 (1963), pp. 73-74 (Russian).
[6] Recurrent points and minimal sets in dynamical systems without uniqueness. Izv. Akad. Nauk Moldavskoi SSR, Serija Estestven. Tehn. Nauk no. 7 (1965) pp. 14-21.
[7] On homogeneous minimal sets. Papers on Algebra and Analysis, Kishinev, 1965; pp. 115-118 (Russian).

BRONSHTEIN, I. U.; SHCHERBAKOV, B. A.,
[1] Certain properties of Lagrange stable funnels of generalized dynamical systems. Izv. Akad. Nauk Moldavskoi SSR, Serija Estestven. Tehn. Nauk, no. 5 (1962), pp. 99-102 (Russian).

BROUWER, L. E. J.,
[1] On continuous vector distributions. I, II, and III, Verh. Nederl. Akad. Wetersch. Afd. Natuurk. Sec. I. vol. 11 (1909) pp. 850-858; vol. 12 (1910) pp. 716-734; and vol. 13 (1910) pp. 171-186.

BROWDER, Felix E.,
[1] On the iteration of transformations in noncompact minimal dynamical systems. Proceedings of the American Mathematical Society, vol. 9 (1958), pp. 773-780.
[2] On a generalization of the Schauder fixed point theorem. Duke Math. J., vol. 26 (1959) pp. 291-303.
[3] On the continuity of fixed points under deformations of continous mappings. Summa Brasil. Math., vol. 4, (1960) pp. 183-191.
[4] On the fixed point index for continuous mappings of locally connected spaces. Summa Brasil. Math., vol. 4, (1960) pp. 253- 293.

BROWN, Morton,
[1] The Monotone Union of Open n-cells is an Open n-cell. Proc. Amer. Math. Soc., vol. 12 (1961), pp. 812-814.

BUDAK, B. M.,
[1] Dispersive dynamical systems. Vestnik Moskov. Univ., no. 8, 1947, pp. 135-137 (Russian).
[2] The concept of motion in a generalized dynamical system. Uchenye Zapiski Moskov. Gos. Univ., no. 155, Mathematika, vol. 5 (1952), pp. 174-194 (Russian).

BUSHAW, Donald,
[1] Dynamical polysystems and optimization. Contributions to Differential Equations, vol. 2 (1963), pp. 351-365.
[2] A stability criterion for general systems. Math. Systems Theory, vol. 1, (1967) pp 79-88.

CAIRNS, S. S.,
[1] Differential and Combinatorial Topology. Princeton University Press, Princeton, N. J., 1965.

CARATHEODORY, Costantin,
[1] Vorlesungen über reelle Funktionen. Teubner, Leipzig 1927 (or Chelsea, New York).
[2] Variationsrechnung und partielle Differentialgleichungen erster Ordnung, B. G. Teubner, Leipzig and Berlin (1935).

CARTWRIGHT, Mary L.,
[1] Topological aspect of forced oscillations. Research, vol. 1 (1948), pp. 601-606.
[2] Forced oscillations in nonlinear systems. Contributions to the Theory of Nonlinear Oscillations, vol. 1, Annals Math. Studies, Number 20, Princeton University Press, 1950, pp. 149-241.
[3] Some decomposition theorems for certain invariant continua and their minimal sets. Fundamenta Mathematicae, vol. 48 (1960) pp. 229-250.

[4] From non-linear oscillations to topological dynamics. J. London Math. Soc., vol. 39 (1964), pp. 193-201.
[5] Topological problems of nonlinear mechanics. Abhandlungen der Deutsch. Akad. der Wiss. Berlin, Klasse fur Math., Phys. Tech., (1965) Nummer 1; III. Konferenz uber nichtlineare Schwingungen, Berlin 1964, Teil I, Akademie-Verlag, Berlin, 1965; pp. 135-142.
[6] Equicontinuous mappings of plane minimal sets. Proc. London Math. Soc., vol. 14A (1965), pp. 51-54.

CARTWRIGHT, M. L.; LITTLEWOOD, J. E.,
[1] Some fixed point theorems. Ann. Math., vol. 54 (1951), pp. 1-37 (includes an appendix by H. D. Ursell, pp. 34-37).

CAUCHY, A. L.,
[1] Oeuvres completes vol. 1 Gauthier-Villars, Paris (1888).

CESARI, Lamberto,
[1] Asymptotic Behavior and Stability Problems in Ordinary Differential Equations. Ergebnisse der Mathematik und ihrer Grenzgebiete, Neue Folge, Heft 16, Springer-Verlag, 1959.

CHETAEV, N.,
[1] On stability in the sense of Poisson. Zap. Kazansk. Matem. Obsch., 1929.
[2] Un théorèm sur l'instabilité. C. R. (Doklady) Acad. Sci. URSS N.S. vol 2 (1934) pp. 529-531
[3] On instability of the equilibrium in certain cases where the force function does not have a maximum. Uch. Zapiski Kazansk. Univ. 1938.
[4] On unstable equilibrium in certain cases when the force function is not maximum. PMM vol. 16, (1952) pp. 89-93.
[5] The Stability of Motion. GITTL, Moscow 1946; 2nd edit. 1959 English translation, Pergamon Press, Ltd., London 1961

CHEN, K. T.,
[1] Equivalence and decomposition of vector fields about an elementary critical point. Amer. J. Math. vol. 85 (1963) pp. 693-722.

CHERRY, T. M.,
[1] Topological properties of solutions of ordinary differential equations. Amer. J. Math., vol. 59 (1937), pp. 957-982.
[2] Analytic quasi-periodic curves of discontinuous type on a torus. Proc. London Math. Soc., vol. 44 (1938), pp. 175-215.
[3] The pathology of differential equations. J. Australian Math. Soc., vol. 1 (1959), pp. 1-16.

CHU, Hsin,
[1] On totally minimal sets. Proc. Am. Math. Soc., vol. 13 (1962), pp. 457-458.
[2] Algebraic topology criteria for minimal sets. Proc. Am. Math. Soc., vol. 13 (1962), pp. 503-508.
[3] Fixed points in a transformation group. Pacific J. Math., vol. 15 (1965), pp. 1131-1135.
[4] A note on compact transformation groups with a fixed end point. Proc. Am. Math. Soc., vol. 16 (1965), pp. 581-583.

CLAY, Jesse Paul,
[1] Proximity relations in transformations groups, Trans. Am. Math. Soc., vol. 108 (1963) pp. 88-96.
[2] Invariant attractors in transformation groups, Ill. J. Math., vol. 8, (1964) pp. 473-479.
CODDINGTON, Earl A., and LEVINSON, Norman,
[1] Uniqueness and convergence of successive approximations, J. Indian Math. Soc., vol. 16 (1952) pp. 75-81.
[2] Theory of Ordinary Differential Equations, McGraw-Hill, New York (1955).
COFFMAN, C.V.,
[1] Linear differential equations on cones in Banach spaces, Pacific J. Math., vol. 12 (1962) pp. 69-75.
[2] Asymptotic behavior of solutions of ordinary difference equations, Trans. Amer. Math. Soc., vol. 110 (1964) pp. 22-51.
[3] Nonlinear differential equations on cones in Banach spaces. Pacific J. Math., vol. 14 (1964) pp. 9-16.
COLEMAN, Courtney;
[1] Equivalence of Planar Dynamical and Differential Systems, J. Diff. Eqs. vol. 1 (1965) pp. 222-233.
CONTI, Roberto,(see also REISSIG, R.)
[1] Sull' equivalenza asintotica dei sistemi di equazioni differenziali, Ann. di Mat. pura ed appl. vol. 41 (1965) pp. 95-104
[2] Limitazioni in ampiezza delle soluzioni di un sistema di equazioni differenziali ed applicazioni. Boll. Un. Mat. Ital. vol. 11 (1956) pp. 344-349.
[3] Sulla prolungabilitá delle soluzioni di un sistema di equazioni differenziali ordinarie Boll. Unione Mat. Ital. vol. 11 (1956) pp. 510-514.
CONTI, Roberto, and SANSONE, Giovanni,
[1] Equazioni differenziali non lineari, Cremonese, Rome (1956), English translation: Nonlinear Differential Equation, Pergamon, London, etc. 1965.
COPPEL, W.A.,
[1] Stability and Asymptotic Behavior of Differential Equations, Heath & Co., Boston, 1965.
CORDUNEANU, Constantin;
[1] Sur la stabilité asymptotique. An. Sti. Univ. Iasi, Sect. 1, vol. 5, pp. 37-39 (1959).
[2] Sur la stabilité asymptotique, II. Rev. math. pur. appl. vol. 5, pp. 573-576 (1960).
[3] Application des inégalités différentielles à la théorie de la stabilité. Abn. Sti. Univ. Iaşi, Sect., vol. 1, 6, pp. 46-58 (1960) (Russian, French summary).
[4] Sur certains systèmes différentielles non-linéaires, An. Sti. Univ. "Al.I. Cuza", Iaşi. Sec, 1 vol. 6 (1960) pp. 257-260

CRONIN, Jane,
[1] Fixed Points and Topological Degree in Nonlinear Analysis, Mathematical Surveys, Number 11, Am. Math. Soc., Providence, (1964)

DAVIS, P. see ANTOSIEWICZ, H.A.

DENJOY, Arnauld,
[1] Sur les caractéristiques à la surface du tore, Comptes Rendus Acad. des Sciences, Paris, vol. 194 (1932) pp. 830-833.
[2] Sur les caractéristiques du tore, Comptes Rendus, Acad. des Sciences, Paris, vol. 194 (1932), pp. 2014-2016.
[3] Sur les courbes définies par les equations différentielles à la surface du tore, J. Math. Pures et Appl., vol. 11 (1932), pp. 333-375.
[4] Les trajectoires à la surface du tore, Comptes Rendus, Acad. des Sciences, Paris, vol. 223 (1964) pp. 5-8.
[5] Sur les trajectoires du tore, Comptes Rendus, Acad. des Sciences, Paris, vol. 251 (1960), pp. 175-177.

DESBROW, D.,
[1] On connexion, invariance and stability in certain flows, Proc. Cambr. Phil. Soc., vol. 60 (1964), pp. 51-55.

DEYSACH, L.G. and SELL, G.R.,
[1] On the existence of almost periodic motions. Michigan Math. J. vol. 12 (1965), pp. 87-95.

DIGEL, E.,
[1] Zu einem Beispiel von Nagumo und Fakuhara, Math. Ziet., vol. 39 (1935) pp. 157-160.

DIRICHLET, G.L.,
[1] Ueber die Stäbilitat des Gleichgewichts, J. Reine Angew. Math., vol. 32 (1846) pp. 85-88.

DOWKER, Yael Naim,
[1] On minimal sets in dynamical systems, Quarterly J. Math., Oxford Second Series, vol. 7 (1956), pp. 5-16.

DOWKER, Yael Naim; FRIEDLANDER, F.G.,
[1] On limit sets in dynamical systems, Proc. London Math Soc., vol. 4 (1954), pp. 168-176.

DUBOSHIN, G.N.,
[1] On the problem of stability of a motion under constantly acting perturbations. Trudy gos. astron. Inst. Sternberg, vol. 14, No. 1 (1940).
[2] Some remarks on the theorems of Liapunov's second method. Vestnik Moscov. Univ.,.vol. 5 No. 10, pp. 27-31 (1950).
[3] A stability problem for constantly acting disturbances. Vesnik Moscov. Univ. vol. 7, No. 2, pp. 35-40 (1952).
[4] Foundations of the Theory of Stability of Motions, Moscow 1957.

DULAC, H.,
[1] Curves definidas por una ecuación diferencial de promer orden y de primer grade, Madrid (1933).

DUGUNDJI, J.(See also ANTOSIEWICZ, H.A.)
[1] Topology, Allyn and Bacon, Boston 1966.

EDREI, A.,
[1] On iteration of mappings of a metric space onto itself, Journal of the London Mathematical Society, vol. 26 (1951), pp. 96-103.

ELLIS, Robert,
[1] Continuity and homeomorphism groups, Proc. Am. Math. Soc., vol. 4 (1953), pp. 969-973.
[2] A note on the continuity of the inverse, Proc. Am. Math. Soc., vol. 8 (1957), pp. 372-373.
[3] Locally compact transformation groups, Duke Math. J., vol. 24 (1957), pp. 119-125.
[4] Distal transformation groups, Pacific J. Math., vol. 8 (1958), pp. 401-405.
[5] Universal minimal sets, Proc. Am. Math. Soc., vol. 11 (1960), pp. 540-543.
[6] A semigroup associated with a transformation group, Trans. Am. Math. Soc. vol. 94 (1960), pp. 272-281.
[7] Point transitive transformation groups, Trans. Am. Math. Soc., vol. 101 (1961), pp. 384-395.
[8] Locally coherent minimal sets, Michigan Math. J., vol. 10 (1963), pp. 97-104.
[9] Global sections of transformation groups, Ill. J. Math., vol. 8 (1964), pp. 380-394.
[10] The construction of minimal discrete flows, Am. J. Math., vol. 87, (1965), pp. 564-574.

ELLIS, R. and GOTTSCHALK, W.H.,
[1] Homomorphisms of transformation groups, Trans. Am. Math. Soc., vol. 94 (1960), pp. 258-271.

EL'SGOL'C, L.E.,
[1] An estimate for the number of singular points of a dynamical system defined on a manifold, Matemat. Sbornik, vol. 26 (1950), pp. 215-223 (Russian), English Translation: Am. Math. Soc. Translations No. 68, 14pp. (1952).

ENGELKING, R.,
[1] Quelques remarques concernant les operations sur les fonctions semi-continues dans les espaces topologiques, Bull. Acad. Polon. Sci. Ser. Sci. Math. Astronom. Phys., 11 (1963), pp. 719-726.

ENGLAND, James W.,
[1] A characterization of orbits, Proc. Am. Math. Soc., vol. 17 (1966), pp. 207-209.

ERUGIN, N.P.,
[1] On certain questions of stability of motion and the qualitative theory of differential equations. Prikl. Math. Mech. vol. 14, pp. 459-512 (1950).
[2] A qualitative investigation of integral curves of a system of differential equations. Prikl. Math. Mech., Vol. 14, pp. 659-664 (1950).
[3] Theorems on instability. Prikl. Math. Mech., vol. 16, pp. 355-361 (1952).
[4] The methods of A.M. Liapunov and questions of stability in the large. Prikl. Math. Mech., vol. 17 pp. 389-400 (1953)
[5] Qualitative methods in theory of stability. Prikl. Math. Mech. vol. 19, pp. 599-616 (1955).

FILIPPOV, A. F.,
[1] On certain questions in the theory of optimal control. Vestn. Moskov. Univ. Ser. Mat., Mekhan. Astron., Fiz., i Khim. No. 2 (1959), pp. 25-32, English Translation: SIAM J. Control. vol. 1, (1963).
FOLAND, N. E.; UTZ, W. R.,
[1] The embedding of discrete flows in continuous flows, Ergodic Theory (Proceedings of an International Symposium held at Tulane University, New Orleans, Louisiana, October 1961), pp. 121-134. Academic Press, New York, 1963.
FOMIN, S.,
[1] On dynamical systems in a space of functions. Ukrainskii Matematicheskii Zhurnal vol. 2 (1950), no. 2, pp. 25-47 (Russian).
FRIEDLANDER, F. G., (see also DOWKER, Y. N.)
[1] On the iteration of a continuous mapping of a compact space into itself. Proceedings of the Cambridge Philosophical Society, vol. 46 (1950), pp. 46-56.
FUKUHARA, Masuo, (see also NAGUMO, M.)
[1] Sur les systèmes des équations différentielles ordinaires. Jap. J. Mathematics, vol. 5 (1929) pp. 345-350.
[2] Sur les systèmes d'équations différentielles ordinaires. II, Jap. J. Math., vol. 6 (1930), pp. 269-299.
[3] Sur l'ensemble des courbes intégrales d'un système d'équations différentielles ordinaires. Proc. Imperial Acad. of Japan, vol. 6 (1930), pp. 360-362.
FURSTENBERG, H.,
[1] Disjointness in ergodic theory, minimal sets and diophantine approximations. Math. Systems Theory, vol. 1, no. 1.(1967) pp 1-50
GARCIA, Mariano; HEDLUND, G. A.,
[1] The structure of minimal sets. Bull. Am. Math. Soc., vol. 54 (1948), pp. 954-964.
GARY, John,
[1] The topological structure of trajectories. Mich. Math. J., vol. 7 (1960), pp. 225-227.
GEISS, Gunther R., (see SZEGÖ, George P.)
GERMAIDZE, V. E.; KRASOVSKII, N. N.,
[1] On stability under persistent disturbances. Prikl. Mat. Mek. vol. 21 (1957) pp. 133-135.
GHIZZETTI, Aldo,
[1] Sul comportamento asintotico degli integrali delle equazioni differenziali ordinarie, lineari ed omogenee. Giorn. Mat. Battaglini, vol. 1 (77) (1947) pp. 5-27.
[2] Un teorema sul comportamento asintotico degli integrali delle equazioni differenziali lineari omogenee. Rend. Mat. Univ. Roma, vol. 8 (1949) pp. 28-42.
[3] Stability problems of solutions of differential equations. Proc. NATO Advanced Study Institute, Padua, Italy 1965. Oderisi, Gubbio 1966.

GOTTSCHALK, Walter H.,

[1] A note on pointwise nonwandering transformations. Bulletin of the American Mathematical Society, vol. 52 (1946), pp. 488-489.

[2] Almost periodicity, equi-continuity and total boundedness. Bulletin of the American Mathematical Society, vol. 52, (1946) pp. 633-636.

[3] Recursive properties of transformation groups II. Bulletin of the American Mathematical Society, vol. 54 (1948), pp. 381-383.

[4] Transitivity and equicontinuity. Bulletin of the American Mathematical Society, vol. 54 (1948), pp. 982-984.

[5] Characterizations of almost periodic transformation groups. Proceedings of the American Mathematical Society, vol. 7 (1956), pp. 709-712.

[6] Minimal sets: and introduction to topological dynamics. Bulletin of the American Mathematical Society, vol. 64 (1958), pp. 336-351.

[7] The universal curve of Sierpinski is not a minimal set. Notices of the American Mathematical Society, vol. 6 (1959), p. 257.

[8] An irreversible minimal set, Ergodic Theory (Proceedings of an International Symposium held at Tulane University, New Orleans, Louisiana, October 1961), pp. 135-150. Academic Press, New York, 1963.

[9] Substitution minimal sets. Transactions of the American Mathematical Society, vol. 109 (1963), pp. 467-491.

[10] Minimal sets occur maximally. Transactions of the New York Academy of Sciences, vol. 26 (1964), pp. 348-353.

[11] A survey of minimal sets, Annales de l'Institut Fourier (Grenoble), vol. 14 (1964), pp. 53-60.

GOTTSCHALK, W. H.; HEDLUND, G. A.,

[1] Recursive properties of transformation groups. Bull. Am. Math. Soc., vol. 52 (1946), pp. 637-641.

[2] The dynamics of transformation groups. Trans. Am. Math. Soc., vol. 65 (1949), pp. 348-359.

[3] Asymptotic relations in topological groups. Duke Math. J., vol. 18 (1951), pp. 481-485.

[4] Topological Dynamics. Am. Math. Soc. Colloquium Publications, vol. 36, Providence, 1955.

[5] A characterization of the Morse minimal set. Proc. Am. Math. Soc., vol. 15 (1964), pp. 70-74.

GRABAR, M. I.,

[1] The representation of dynamical systems as systems of solutions of differential equations. Dokl. Akad. Nauk SSSR, vol. 61 (1948), pp. 433-436.

[2] Transformations of dynamical systems into systems of solutions of differential equations. Vestnik Moskov. Univer., 1952, no. 3, pp. 3-8 (Russian).

[3] On change of time in dynamical systems. Dokl. Akad. Nauk SSSR, vol. 109 (1956), pp. 250-252 (Russian).

[4] On a sufficient test for isomorphism of dynamical systems. Dokl. Akad. Nauk SSSR, vol. 109 (1956), pp. 431-433 (Russian).

[5] Isomorphism of dynamical systems differing only in time. Dokl. Akad. Nauk SSSR, vol. 126 (1959), pp. 931-934 (Russian).

GRAFFI, Dario,
[1] Sul periodo delle oscillazioni dei sistemi nonlineari a più gradi di libertà. Coll. Int. Vibr. non lin., Porquerolles 1951 pp. 189-193.
[2] Sul periodo delle oscillazioni nei sistemi nonlineari a due gradi di libertà. Mem. Accad. Sci. Bologna, vol. 9 (1952) pp. 17-22.

GREEN, L., (see AUSLANDER, L.)

GROBMAN, D. M.,
[1] Systems of differential equations analogous to linear ones. Dokl. Akad. Nauk SSSR, vol. 86 (1952) pp. 19-22.
[2] Homeomorphisms of systems of differential equations. Dokl. Akad. Nauk SSSR, vol. 128 (1959) pp. 880-881.
[3] Topological and asymptotic equivalence for systems of differential equation. Dokl. Akad. Nauk SSSR, vol. 140 (1961) pp. 746-747.
[4] Topological classification of the neighborhood of a singular point in n-dimensional space. Mat. Sb. (N.S.) vol. 56(98) (1962) pp. 77-94.

HAAS, Felix,
[1] A theorem about characteristics of differential equations on closed manifolds. Proc. Nat. Acad. of Sciences USA, vol. 38 (1952), pp. 1004-1047.
[2] On the global behavior of differential equations on two-dimensional manifolds. Proc. Am. Math. Soc., vol. 4 (1953), pp. 630-636.
[3] The global behavior of differential equations on n-dimensional manifolds. Proc. Nat. Acad. of Sciences U.S.A., vol. 39 (1953), pp. 1258-1260.
[4] Poincaré-Bendixson type theorems for two-dimensional manifolds different from the torus. Ann. Math., vol. 59 (1954), pp. 292-299.
[5] On the total number of singular points and limit cycles of a differential equation. Contributions to the Theory of Nonlinear Oscillations, vol. 3, pp. 137-172; Ann. Math. Studies, no. 36; Princeton, 1956.

HADAMARD, Jacques,
[1] Sur les trajectoires en dynamique. J. de Mathématiques, Ser. III, vol. 3 (1897) pp. 331-387.
[2] Sur les intégrales d'un system d'equations différentielles ordinaires, considérées comme fonctions des données initiales. Bull. Soc. Math. France, vol. 28 (1900) pp. 64-66.
[3] Sur l'itération et les solutions asymptotiques des équations différentielles, ibid. vol. 29 (1901) pp. 224-228.

HAHN, Frank J., (see also AUSLANDER J. and AUSLANDER, L.)
[1] Recursion of set trajectories in a transformation group. Proc. Am. Math. Soc., vol. 11 (1960), pp. 527-532.
[2] Nets and recurrence in transformation groups. Trans. Am. Math. Soc., vol. 99 (1961), pp. 193-200.
[3] On affine transformations of compact abelian groups. Amer. J. Math., vol. 85 (1963), pp. 428-446; errata, vol. 86 (1964), pp. 463-464.

[4] A fixed point theorem, Math. Systems Theory, vol. 1 (1967), pp 55-58

HAHN, Wolfgang,

[1] Uber die Anwendung der Methode von Ljapunov auf Differenzengleichungen. Math. Ann., vol. 136, pp. 430-441 (1958).

[2] Theorie und Anwendung der Direkten Methode von Ljapunov. Ergebnisse der Mathematik und ibrer Grenzgebiete, Neue Folge, Heft 22, Springer-Verlag, Berlin-Gottingen-Heidelberg, 1959, English translation: Theory and Application of Liapunov's Direct Method, Prentice-Hall, Englewood Cliffs, N. J., 1963.

[3] On the general concept of stability and Liapunov's direct method. MRC Tech. Report 485, Univ. Wisconsin, Madison, Wis., 1964.

HAJEK, Otomar,

[1] Critical points of abstract dynamical systems. Comm. Math. Universitatis Carolinae, vol. 5 (1964), pp. 121-124.

[2] Betti numbers of regions of attraction. Comm. Math. Univ. Carolinae, vol. 5 (1964), pp. 129-132.

[3] Structure of dynamical systems. Comm. Math. Univ. Carolinae, vol. 6 (1965), pp. 53-72. Correction of the above paper, Comm. Math. Univ. Carolinae, vol. 6 (1965) pp. 211-212.

[4] Flows and periodic motions. Comm. Math. Univ. Carolinae, vol. 6 (1965) pp. 165-178.

[5] Sections of dynamical systems in E^2, Czech. Math. J., vol. 15 (1965), pp. 205-211.

[6] Prolongations of sections in local dynamical systems. Czech. Math. J., vol. 16 (1966) pp. 41-45.

[7] Differentiable representation of Flows. Comm. Math. Univ. Carolinae, vol. 7 (1966) pp 219-225.

[8] Dynamical systems in the plane. Academic Press, New York 1967 (to be published).

HALANAY, Aristide

[1] Differential equations: stability theory, oscillations, time-lags. Academic Press, New York 1965.

HALE, Jack K.,

[1] Integral Manifolds of Perturbed Differential Systems. Ann. Math., vol. 73 (1961), pp. 496-531.

[2] Sufficient Conditions for stability and instability of autonomous functional differential equations. J. of Diff. Eq., vol. 1, (1965), pp. 452-482.

[3] Geometric Theory of Functional Differential Equations. Proc. of An International Symposium on Differential Equations and Dynamical Systems, University of Puerto Rico, Mayaguez, P.R., Dec. 1965, Academic Press, New York (to appear).

HALE, J. K.; STOKES, A. P.,

[1] Behavior of solutions near integral manifolds. Arch. Rat. Mech. Anal., vol. 6 (1960) pp. 133-170.

HALKIN, Hubert,

[1] Topological aspects of optimal control of dynamical polysystems. Contributions to Differential Equations, vol. 3 (1964), pp. 377-385.

[2] Finitely convex sets of nonlinear differential equations. To appear in Math. Systems Theory, vol. 1, (1967) pp 51-54.

HAMILTON, O. R.,
[1] A short proof of the Cartwright-Littlewood fixed point theorem. Canadian Journal of Mathematics, vol. 6 (1954), pp. 522-524.
HARTMAN, Philip,
[1] On local homeomorphisms of Euclidean spaces. Bol. Soc. Mat. Mexicana, vol. 5 (1960) pp. 220-241.
[2] On stability in the large for systems of ordinary differential equations. Canad. J. Math.,vol. 13 (1961) pp. 480-492.
[3] On uniqueness and differentiability of solutions of ordinary differential equations. Proceedings of a Symposium on Non linear Problems, Madison (Wis.) (1963) pp. 219-232.
[4] Ordinary Differential Equations. Wiley, New York, 1964.
[5] The Existence and Stability of Stationary Points. Duke Math. J., vol. 33,2 (1966), pp. 281-290.
HARTMAN, P.; OLECH, Czesław,
[1] On global asymptotic stability of solutions of ordinary differential equations. Trans. Amer. Math. Soc., vol. 104 (1962) pp. 154-178.
HARTMAN, P.; WINTNER, Aurel,
[1] Integrability in the large and dynamical stability. Amer. J. Math., vol. 65 (1943), pp. 273-278.
[2] On the asymptotic behavior of the solutions of a nonlinear differential equation. Amer. J. Math., vol. 68 (1946) pp. 301-308.
HEDLUND, G. A., (see also GOTTSCHALK, W. H.)
[1] Sturmian minimal sets. Amer. J. Math., vol. 66 (1944), pp. 605-620.
[2] A class of transformations of the plane. Proceedings of the Cambridge Philosophical Society, vol. 51 (1955), pp. 554-564.
[3] Mappings on sequence spaces (Part I). Communications Res. Div. Tech. Rep. No. 1, von Neumann Hall, Princeton, New Jersey, February 1961.
HILMY, Heinrich,
[1] Sur la structure d'ensemble des mouvements stables au sens de Poisson. Ann. Math., vol. 37 (1936), pp. 43-45.
[2] Sur les ensembles quasi-minimaux dans les systèmes dynamiques. Ann. Math., vol. 37 (1936), pp. 899-907.
[3] Sur les centres d'attraction minimaux des systèmes dynamiques. Compositio Math., vol. 3 (1936), pp. 227-238.
[4] Sur les mouvements des systèmes dynamiques qui admettent l'incompressibilité des domaines. Amer. J. Math., vol. 59 (1937), pp. 803-808.
[5] Sur une propriété des ensembles minima. Comptes Rendus (Doklady) de l'Academie des Sciences de l'URSS, vol. 14 (1937), pp. 261-262.
[6] Sur la théorie des ensembles quasi-minimaux. Comptes Rendus (Doklady) de l'Académie des Sciences de l'URSS, vol. 15 (1937), pp. 113-116.
[7] Sur les théorèmes de récurrence dans la dynamique générale. Amer. J. Math., vol. 61 (1939), pp. 149-160.
HIRASAWA, Yoshikazu (see URA, Taro)
HOCKING, J.; YOUNG, G.,
[1] Topology, Addison-Wesley, Reading, Mass., 1961.

HU, Sze-Tsen,
[1] Theory of Retracts , Wayne State Univ. Press, Detroit 1955.

INGWERSON, D.R.,
[1] A modified Liapunov method for nonlinear stability analysis. IRE Trans. Automatic Control vo. 6, pp. 199-210 (1961); Discussion, IRE Trans. Automatic Control, vol. 7, pp. 85-88 (1962).

JONES, G. Stephen,
[1] Asymptotic fixed point theorems and periodic systems of functional differential equations, Contributions to Differential Equations, vol. 2, pp. 385-405 (1963).
[2] Periodic motions in Banach space and applications to functional-differential equations, Contributions to Differential Equations vol. 3, pp. 75-106. (1964).
[3] Topologically convex sets and fixed point theory., RIAS Report 63-8. 1963.
[4] Fixed point theorems for convex sets with deleted boundary points, Not. AMS, vol. 10 No. 2, 191 (1963).
[5] Variable translation operators and periodic solutions of functional equations, Not. of AMS, vol. 10, No. 2, pp. 191-192, (1963).
[6] Stability and Asymptotic Fixed-Point Theory, Proc. Nat. Acad. Sc., U.S.A., 53, vol. 6 (1965), pp. 1262-1264.
[7] Mathematical aspects of control theory in G.S. Jones; A. Strauss and G.P. Szegö. Research report on "Mathematical aspects of control theory and the problem of satellite altitude control" Rept. NASA Cont. No. NAS. 5-9172
[8] Restrictive, retractable operators and fixed-point theorems. To appear in Math. Systems Theory.

KALMAN, R.E. and BERTRAM, J. E.,
[1] Control Systems analysis and design via the 'second method' of Liapunov. I. Continuous-time systems. II. Discrete-time systems. ASME J. of Basic Engineering, pp. 371-393, 394-400 (1960).

KAMENKOV, G.V.,
[1] On stability of motion over a finite interval of time, Prikl. Math. Mech. vol. 17, (1952), pp. 529-540.

KAMKE, E.
[1] Differentialgleichungen reeller Funktionen, Akademische Verlaggsgesellschaft, Leipzig (1930) [or Chelsea, New York (1947)]
[2] Zur Theorie der Systeme gewöhnlicher Differentialgleichungen, II, Acta Math. vol. 58 (1932) pp. 57-85.
[3] Differentialgleichungen, Losungsmethoden und Losungen, I (Gewohnliche Differentialgleichungen)(7th. ed.) Akademische Verlagsgesellschaft, Leipzig (1961)
[4] Differentialgleichungen. Losungsmethoden und Losungen, II (Partielle Differentialgleichungen erster Ordnung fur eine gesuchte Funktion) (4th ed.), Adademische Verlagsgesellshaft, Leipzig (1959)

KAMPEN van, E.R.,
[1] The topological transformations of a simple closed curve into itself, Amer. J. Math. vol. 57 (1935), pp. 142-152.
[2] Remarks on systems of ordinary differential equations, ibid. vol. 59 (1937) pp. 144-152.

[3] Notes on systems of ordinary differential equations, ibid. vol. 63 (1941) pp. 371-376.

KAPLAN, Wilfred,

[1] Regular curve-families filling the plane, I, Duke Math. J. vol. 7 (1940), pp. 154-185.

[2] Regular curve-families filling the plane, II, Duke Math. J. vol. 8 (1941), pp. 11-46.

[3] Differentiability of regular curve families on the sphere, Lectures in Topology, pp. 299-301, University of Michigan Press, Ann Arbor, 1941.

[4] The structure of a curve-family on a surface in the neighborhood of an isolated singularity, Am. J. Math. vol. 64 (1942), pp. 1-35.

[5] Dynamical systems with indeterminacy. Am. J. Math. vol. 72 (1950), pp. 573-594.

KATO, Junji,

[1] The asymptotic behaviour of the solutions of differential equations or the product space. Arch. Rat. Mech. Anal. vol. 6 (1960), pp. 133-170.

[2] The asymptotic relation of two systems of ordinary differential equations. Contr. Diff. Equa. vol. 3 (1964), pp. 141-161.

[3] Asymptotic Equivalences between systems of differential equations and their perturbed systems. Funkcialaj Ekvacioj, vol. 8 (1966) pp. 45-78.

KAYANDE, A. A.; LAKSHMIKANTHAM V.,

[1] Conditionally invariant sets and vector Liapunov functions. J. Math. Anal. App. vol. 14 (1966) pp. 285-293.

KELLEY, J. L.,

[1] General Topology. Van Nostrand, New York (1955).

KIMURA, Ikuo (see URA, T.),

KNESER, A.,

[1] Studien ueber die Bewegungsvorgänge in der Ungebung instabilen Gleichgevichtslagen, J. Reine u. Aug. Math. (I), vol. 115 (1895) pp. 308-327; II, vol. 118 (1897) pp. 186-223.

KNESER, Hellmuth,

[1] Über die Losungen eines Systemes gewohnlicher Differentialgleichungen das der Lipschitzschen Bedingung nicht genugt. Sitzungsberichte der Preussischen Akademie der Wissenschaften, Physische-Mathematische Klasse (1923), pp. 171-174.

[2] Regulare Kurvenscharen auf den Ringflachen. Math. Ann., vol. 91 (1924), pp. 135-154.

KOLMOGOROV, A. N.; FOMIN, S. V.,

[1] Elements of the Theory of Functions and Functional Analysis; Vol. 1, Metric and normed spaces. Graylock Press, Rochester, N. Y., 1957.

KRASOVSKII, Nikolai N.,

[1] On a problem of stability of motion in the large. Dokl. Akad. Nauk SSSR, vol. 88, (1953) pp. 401-404.

[2] On stability of motion in the large for constantly acting disturbances. Prikl. Mat. Mek., vol. 18, (1954), pp. 95-102.

[3] On the inversion of theorems of A. M. Liapunov and N. G. Chetaev on instability for stationary systems of differential equations. Prikl. Mat. Mek., vol. 18, (1954) pp. 513-532.

[4] On stability in the large of the solutions of a nonlinear system of differential equations. Prikl. Mat. Mek., vol. 18 (1954) pp. 735-737.

[5] Sufficient conditions for stability of solutions of a system of nonlinear differential equations. Dokl. Akad. Nauk SSSR, vol. 98 (1954), pp. 901-904.

[6] On the converse of K. P. Persidskii's theorem on uniform stability. Prikl. Mat. Mek., vol. 19 (1955), pp. 273-278.

[7] On conditions of inversion of A. M. Liapunov's theorems on instability for stationary systems of differential equations. Dokl. Akad. Nauk SSSR, vol. 101 (1955), pp. 17-20.

[8] Converse of theorems on Liapunov's second method and questions of stability of motion in the first approximation. Prikl. Mat. Mek., vol. 20 (1956), pp. 255-265.

[9] On the converse of theorems of the second method of A. M. Liapunov for investigation of stability of motion. Usp. Mat. Nauk, vol. 9, No. 3, (1956), pp. 159-164.

[10] On the theory of the second method of A. M. Liapunov for the investigation of stability. Mat. Sbornik, vol. 40, (1956), pp. 57-64.

[11] On stability with large initial perturbations. Prikl. Mat. Mek., vol. 21, (1957), pp. 309-319.

[12] Certain Problems of the Theory of Stability of Motion. Gos. Izd. Fiz.-mat. Lit. Moscow 1959, p. 211, English translation: Stability of motions. Stanford Univ. Press, Stanford (1963).

KRECU, V. I. (see SIBIRSKII, K. S.)

KREIL, K. A.,

[1] Das qualitative Verhalten der Integralkurven einer gewohnlichen Differentialgleichung erster Ordnung in der Umgebung eines singulären Punktes. Jahresber. Deutsch. Math. Verein, vol. 57 (1955) p. 111.

KREIN, M. G.,

[1] On some questions related to the ideas of Liapunov in the theory of stability. Uspehi Mat. Nauk (N.S.) vol. 3, no. 3, (1948), pp. 166-169.

KUDAEV, M. B.,

[1] The use of Liapunov functions for investigating the behavior of trajectories of systems of differential equations. Dokl. Akad. Nauk SSSR, vol. 147 (1962) pp. 1285-1287. English Trans: Soviet Math., pp. 1802-1804.

[2] Classification of higher-dimensional systems of ordinary differential equations by the method of Liapunov functions. Diff. Uravn., vol. 1, (1965) pp. 346-356, English Translation, Diff. Equations, pp. 263-269.

[3] Liapunov function for the region of influence of a single singular point of higher order. Vestnik Mosc. Univ. (1965) no. 1, pp. 3-13.

KURZWEIL, J.,

[1] On the reversibility of the first theorem of Liapunov concerning the stability of motion. Czechoslov. math. J., vol. 5, (1955) (English summary).

[2] The converse second Liapunov's theorem concerning the stability of motion. Czechoslov. math. J., vol. 6, (1956), pp. 217-259 pp. 455-473, (English summary).

KURZWEIL, J.; VRKOC , I.,

[1] The converse theorems of Liapunov and Persidskii concerning the stability of motion. Czechoslov. math. J., vol. 7, (1957) pp. 254-274, (English summary).

LAGRANGE, J. L.,

[1] Mecanique analytique, Desaint, Paris (1788).

[2] Oeuvres, I (1867) and IV (1869), Gauthier-Villars, Paris.

LASALLE, Joseph P.,

[1] A study of synchronous asymptotic stability. Annals of Mathematics, vol. 65 (1957), pp. 571-581.

[2] Asymptotic stability criteria. Proc. Symp. Appl. Math., vol. 13 (1962), pp. 299-307.

[3] Some extensions of Lyapunov's second method. IRE Trans. Circuit Theory, vol. CT-7, (1960) pp. 520-527.

[4] The Extent of Asymptotical Stability. Proc. Nat. Acad. Sc., vol. 46 (1960) pp. 363-365.

[5] Recent advances in Liapunov stability theory. SIAM Review, vol. 6 (1964), pp. 1-11.

[6] An invariance principle in the theory of stability. Brown University, Center for Dynamical Systems, Tech. Report 66-1.

LASALLE, Joseph P.; LEFSCHETZ, Solomon,

[1] Stability by Liapunov's Direct Method with Applications. Academic Press, New York, 1961.

LAZER, Alan C., (see BHATIA, N. P.)

LAKSHMIKANTHAM, V., (see also BHATIA, N. P. and KAYANDE, A. A.)

[1] On the boundedness of solutions of nonlinear differential equations. Proc. Am. Math. Soc., vol. 8 (1957) pp. 1044-1048.

[2] Upper and lower bounds of the norm of solutions of differential equations. Proc. Am. Math. Soc., vol. 13 (1962).

[3] Notes on a variety of problems of differential systems. Arch. Rat. Mech. Anal. (1962).

[4] Properties of solutions of abstract differential equations. MRC Tech. Rept. No. 334, Univ. Wisc., Madison, Wisc. (1962).

[5] Vector Liapunov functions and conditional stability. J. Math. Anal. Appl., vol. 10 (1965) pp. 368-377.

LEBEDEV, A. A.

[1] The problem of stability in a finite interval of time. Prikl. Mat. Mek., vol. 18 (1954), pp. 75-94.

[2] On stability of motion during a given interval of time. Prikl. Mat. Mek., vol. 18 (1954), pp. 139-148.

LEE, E. B.; MARKUS, L.,

[1] Optimal control for nonlinear processes. Arch. Rat. Mech. Anal., vol. 8 (1961), pp. 36-58.

LEFSCHETZ, Solomon,

[1] Differential equations: geometric theory, Pure and Applied Mathematics, vol. 6; Interscience, New York, first edition 1957; second edition, 1962.

[2] Liapunov and stability in dynamical systems. Bol. Soc. Mat. Mexicana, vol. 3 (1958), pp. 25-39.

LEIGHTON, Walter (see also BHATIA, N. P.)

[1] Morse theory and Liapunov functions. Rend. Circ. Mat. Palermo Ser. II. vol. 13 (1966) pp. 1-10.

LEVIN, J. J.,

[1] On the global asymptotic behavior of nonlinear systems of differential equations. Arch. Rat. Mech. Anal., vol. 6 (1960) pp. 65-74.

LEVINSON, Norman, (see CODDINGTON, E. A.)

[1] The asymptotic behavior of a system of linear differential equations. Am. J. Math., vol. 68 (1946) pp. 1-6.

LEWIS, Daniel C., Jr.,

[1] Metric properties of differential equations. Am. J. Math., vol. 71 (1949) pp. 294-312.

[2] Differential equations referred to a variable metric. Am. J. Math., vol. 73 (1951) pp. 48-58.

[3] Reversible transformations. Pacific Journal of Mathematics, vol. 11 (1961), pp. 1077-1087.

LIAPUNOV, A. M.,

[1] Problème géneral de la stabilité du mouvement. Annals of Mathematics Studies, no. 17, Princeton, 1947; reproduction of the French translation in Ann. de la Faculté des Sciences de Toulouse, vol. 9 (1907), pp. 203-474, of Russian memoir, Obshchaya Zadacha Ustoichivosti Dvizheniya, Kharkov, 1892, and a note, Comm. Soc. Math. Kharkov, vol. 3 (1893), pp. 265-272.

LIFSHITZ, Jaime, (see BIRKHOFF, G. D.)

LIPSCHITZ, R.,

[1] Sur la possibilité d'intégrer complétement un système donné d'equations différentielles. Bull. Sci. Math. Astro., vol. 10 (1876) pp. 149-159.

LOJASIEWICZ, S.,

[1] Sur l'allure asymptotique des intégrales du système d'équations différentielles au voisinage de point singulier. Ann. Polonici Math., vol. 1 (1954) pp. 34-72.

LITTLEWOOD, J. E., (see CARTWRIGHT, M. L.)

MALKIN, I. G.,

[1] Das Existenzproblem von Ljapunovschen Funktionen. Izv. fiz.-mat. Obsc. Kazan III, vol. 4, pp. 51-62 (German) and III, vol. 5, (1931) pp. 63-84.

[2] Certain Questions in the Theory of Stability of Motion in the Sense of Liapunov. American Math. Soc., No. 20, New York 1950 (English).

[3] Verallgemeinerung des Fundamentalsatzes von Liapounoff uber die Stabilität der Bewegungen. C. R. (Doklady) Acad. Sci. URSS vol. 18 (1938) pp. 162-164 (German).

[4] On the stability of motion in the sense of Liapunov. Mat. Sbornik vol. 3, (1938) pp. 47-100 (German summary).

[5] Sur un théorème d'éxistence de Poincaré-Liapounoff. C. R. (Doklady) Acad. Sci. URSS, vol. 27 (1940) pp. 307-310 (French).
[6] Basic theorems of the theory of stability of motion. Prikl. Mat. Mek., vol. 6 (1942), pp. 411-448 (English summary).
[7] Stability in the case of constantly acting disturbances. Prikl. Mat. Mek., vol. 8, (1944) pp. 241-245 (English summary).
[8] Theorie der Stabilität einer Bewegung. Verlag R. Oldenbourg, München 1959 (German translation of the Russian original published in 1952). (A very poor English translation of this book is "Theory of Stability of Motion," Atomic Energy Commission, Translation No. 3352, Dept. of Commerce, Washington, D. C. 1958.)

MANACORDA, T.,
[1] Sul comportamento asintotico di una classe di equazioni differenziali nonlineari. Boll. Un. Mat. Ital., vol. 7 (1952) pp. 137-142.

MARKOV, A. A.,
[1] Sur une propriété générale des ensembles minimaux de M. Birkhoff. Comptes Rendus Acad. Sciences, Paris, vol. 193 (1931), pp. 823-825.
[2] On a general property of minimal sets. Rusk. Astron. Zhurnal, 1932.
[3] Stabilität im Liapounoffschen Sinne und Fastperiodizitat. Mat. Zeit., vol. 36 (1933), pp. 708-738.
[4] Almost periodicity and harmonizability. L. Trudy vtorogo Vsesoyuzn. matem. s'ezda, vol. 2 (1936), pp. 227-231 (Russian).

MARKUS, Lawrence, (see also AEPPLI, A and LEE, E. B.)
[1] Escape times for ordinary differential equations. Rend. Sem. Mat. Politecnico Torino vol. 11 (1952) pp. 271-277.
[2] On completeness of invariant measures defined by differential equations. J. Math. Pure Appl. vol. 31 (1952) pp. 341-353.
[3] Invariant measures defined by differential equations. Proc. Ann. Math. Soc. vol. 4 (1953) pp. 89-91.
[4] A topological theory for ordinary differential equations in the plane. Colloque de Topol. et Geom. Diff., Strasbourg 1952.
[5] Global structure of ordinary differential equations in the plane. Trans. Am. Math. Soc. vol. 76 (1954) pp. 127-148.
[6] Asymptotically autonomous differential systems. Contrib. to Nonlinear Oscillations, vol. 3, (Annals Math. Studies vol. 36) Princeton Univ. Press, Princeton, N. J., 1956.
[7] Structurally stable differential systems. Ann. Math., vol. 73 (1961), pp. 1-19.
[8] Periodic solutions and invariant sets of structurally stable differential systems. Symp. Int. de Ecuac. Differenc. Ord., Mexico, 1961, pp. 190-194.
[9] The global theory of ordinary differential equations. Lecture Notes, Univ. of Minnesota, 1964-65.

MARKUS L.; YAMABE, H.,
[1] Global stability criteria for differential systems. Osaka Math. J., vol. 12 (1960) pp. 305-317.

MARCHAUD, M. A.,
[1] Sur les champs des demi-droites et les equations différentielles du premier ordre. Bull. Soc. Math. France, vol. 63 (1934), pp. 1-38.
[2] Sur les champs continus de demi-cones convexes et leur integrales. Compositio Math., vol. 3 (1936), pp. 89-127.
MARTIN, Monroe,
[1] A problem in arrangements. Bulletin of the American Mathematical Society, vol. 40 (1934), pp. 859-864.
MASSERA, Jose Luis,
[1] On Liapounoff's condition of stability. Ann. of Math., vol. 50 (1949) pp. 705-721.
[2] The existence of periodic solutions of systems of differential equations. Duke Math. J., vol. 17 (1950) pp. 457-475.
[3] Total stability and approximately periodic vibrations. Fac. Ing. Montevideo, Publ. Inst. Mat. Estad., vol. 2, (1954) pp. 135-145, (Spanish, English summary).
[4] Sobre la estabilidad en espacios de dimension infinta. Rev: Un. mat. Argentina, vol. 17, (1955) pp. 135-147.
[5] Contributions to stability theory. Ann. of Math., vol. 64, (1956) pp. 182-206. Correction, Ann. of Math., vol. 68, (1958) p. 202.
[6] On the existence of Liapunov functions. Publ. Inst. Mat. Estad. Montevideo, vol. 3, No. 4, (1960) pp. 111-124.
[7] Converse theorems of Liapunov's second method. Symposium intern. ecuac. dif. ord. Mexico, (1961) pp. 158-163.
MASSERA, J. L.; SCHÄFFER, J. J.,
[1] Linear differential equations and function spaces. Academic Press, New York, 1966.
MATROSOV, V. M.,
[1] On the theory of stability of motion. Prikl. Mat. Mek., vol. 26 (1962) pp. 992-1002.
MAYRHOFER, K.,
[1] Ueber die enden der Integralkurven bei gewohnlichen Differentialgleichungen. Monatsh. f. Math. v. Phys., vol. 41 (1934) pp. 183-187.
MENDELSON, Pinchas,
[1] On Lagrange stable motions in the neighborhood of critical points. Contr. Theory of Nonlinear Oscillations, vol. 5, (Annals of Mathematics Studies, No. 45), Princeton University Press, 1960, pp. 219-224.
[2] On unstable attractors. Bol. Soc. Mat. Mexicana, vol. 5 (1960), pp. 270-276.
MICHAEL, E.,
[1] Topologies on spaces of subsets. Trans. Am. Math. Soc., vol. 71, (1951) pp. 152-182.
MILLER, Richard K.,
[1] On almost periodic differential equations. Bul. Am. Math. Soc., vol. 70 (1964), pp. 792-795.

[2] Asymptotic behavior of solutions of nonlinear differential equations. Trans. Amer. Math. Soc., vol. 115 (1965), pp. 400-416.
[3] Almost periodic differential equations as dynamical systems with applications to the existence of a.p. solutions. J. Diff. Equat., vol. 1 (1965), pp. 337-345.

MILNOR, J.,
[1] Sommes de variétes differentiables et structures différentiables des spheres. Bull. Soc. Math. de France, vol. 87 (1959) pp. 439-444.
[2] Morse Theory, Ann. Math. Studies, Princeton Univ. Press, Princeton, N. J., 1963.
[3] Differential topology in lectures on modern mathematics, vol. II, edited by T. L. Saaty, Wiley 1964.

MINKEVICH, M. I.,
[1] The theory of integral funnels in generalized dynamical systems without a hypothesis of uniqueness. Dok. Akad. Nauk SSSR, vol. 59 (1948), pp. 1049-1052 (Russian).
[2] Closed integral funnels in generalized dynamical systems without a hypothesis of uniqueness. Dok. Akad. Nauk SSSR, vol. 60 (1948), pp. 341-343 (Russian).
[3] Theory of integral funnels in dynamical systems without uniqueness. Uchenye Zapiski Moskov. Gos. Univ., no. 135, Matematika, vol. 2 (1948), pp. 134-151 (Russian).
[4] Closed integral funnels in generalized dynamical systems without a hypothesis of uniqueness. Uchenye Zapiski Moskov. Gos. Univ., no. 6, Matematika, vol. 163 (1952), pp. 73-88 (Russian).

MIRANDA, C.,
[1] Un'osservazione su un teorema di Brouwer. Boll. Un. Mat. Ital. vol. 3 (1940) pp. 5-7.

MOISSEEV, N. D.,
[1] Summary of the History of Stability. Moscow 1949 (Russian).

MONTEL, Paul,
[1] Sur les suites infinies de fonctions. Ann. Sc. École. Nor. Sup., Ser. III, vol. 24 (1907) pp. 233-334.
[2] Sur l'intégral supérieure et l'intégrale inférieure d'une équation différentielle. Bull. Sc. Mathematiques, vol. 50 (1926) pp. 205-217.

MONTGOMERY, Deane; ZIPPIN, Leo,
[1] Topological Transformation Groups, Interscience, New York, 1955.

MORSE, Marston,
[1] A one-to-one representation of geodesics on a surface of negative curvature. Amer. J. Math., vol. 43 (1921), pp. 35-51.
[2] Recurrent geodesics on a surface of negative curvature. Trans. Am. Math. Soc., vol. 22 (1921), pp. 84-100.
[3] Relations between the critical points of a real function of n independent variables. Trans. Am. Math. Soc., vol. 27 (1925) pp. 345-396.

MOSTERT, Paul S.,
[1] One-parameter transformation groups in the plane. Proc. Am. Math. Soc., vol. 9 (1958), pp. 462-463.

MÜLLER, M.,
[1] Ueber das Fundamentaltheorem in der Theorie der gewöhnlichen Differentialgleichungen. Math. Zeit., vol. 26 (1927) pp. 619-645.
[2] Beweis eines Satzes des Herrn H. Kneser uber die Gesamtheit der Losungen, die ein System gewöhnlicher Differentialgleichungen durch einen Punkt schickt, ibid., vol. 28 (1928) pp. 349-355.
[3] Neuere Untersuchung über den Fundamentalsatz in der Theorie der gewöhnlichen Differentialgleichungen, Jber. Deutsch. Math. Verein., vol. 37 (1928) pp. 33-48.

MYSHKIS, A. D.,
[1] Generalizations of the theorem on a fixed point of a dynamical system inside of a closed trajectory. Mat. Sbornik, vol. 34 (1954), pp. 525-540 (Russian).

NAGUMO, M., (see also FUKUHARA, M.)
[1] Eine hinreichende Bedingung für die Unität der Lösung von Differentialgleichungen erster Ordnung, Jap. J. Math., vol. 3 (1926) pp. 107-112.
[2] Un théorème relatif a l'ensemble des courbes intégrales d'un système d'équations différentielles ordinaires. Proc. Phys. Math. Soc., Japan (III), vol. 12 (1930), pp. 233-239.

NEMYTSKII, V. V.,
[1] Sur les systèmes dynamiques instables. Comptes Rendus de l'Académie des Sciences, Paris, vol. 199 (1934), pp. 19-20.
[2] Über vollständig unstabile dynamische Systeme. Annali di Matematica Pura ed Applicata, vol. 14 (1936), pp. 275-286.
[3] Sur les systèmes de courbes remplissant un espace métrique. Comptes Rendus (Doklady) de l'Académie des Sciences de l'URSS, vol. 21 (1938), pp. 99-102.
[4] Sur les systèmes de courbes remplissant un espace métrique. (Généralisation des théorèmes de Birkhoff.), Recueil Mathématique (Mat. Sbornik), vol. 6 (1939), pp. 283-292.
[5] Systèmes dynamiques sur une multiplicité intégrale limité. Comptes Rendus (Doklady) de l'Académie des Sciences de l'URSS, vol. 47 (1945), pp. 535-538.
[6] Sur les familles de courbes du type de Bendixon. Comptes Rendus (Doklady) de l'Académie des Sciences de l'URSS, vol. 21 (1938), pp. 103-105.
[7] Les systèmes dynamiques généraux. Comptes Rendus (Doklady) de l'Academie des Sciences de l'URSS, vol. 53 (1946), pp. 491-494.
[8] On the theory of orbits of general dynamical systems. Mat. Sbornik, vol. 23 (1948), pp. 161-186 (Russian).
[9] The structure of one-dimensional limiting integral manifolds in the plane and three-dimensional space. Vestnik Moskov. Univ., no. 10, 1948, pp. 49-61 (Russian).
[10] Topological problems of the theory of dynamical systems. Uspehi Mat. Nauk, vol. 4 (1949), pp. 91-153 (Russian), English translation: Am. Math. Soc. Translation no. 103, (1954) p. 85.

[11] Generalizations of the theory of dynamical systems. Uspeki Mat. Nauk, vol. 5 (1950), pp. 47-59 (Russian).
[12] Some general theorems on the distribution of integral curves in the plane. Vestnik Moskov. Univ. Ser. I Mat. Meh.(1960), no. 6, pp. 3-10.
[13] Sur une classe importante des systémes caracteristiques sur le plan. Annali di Matematica Pura ed Applicata, vol. 49 (1960), pp. 11-24.
[14] Some modern problems in the qualitative theory of ordinary differential equations. Russian Math. Survey, 1965, pp. 1-34.

NEMYTSKII, V.V.; STEPANOV, V. V.,
[1] Qualitative Theory of Differential Equations. Moscow-Leningrad; first edition 1947; second edition 1949 (Russian), English translation: Princeton University Press, 1960, Princeton Mathematical Series, no. 22.

OLECH, Czesław, (see also HARTMAN, P.)
[1] On the asymptotic behavior of the solutions of a system of ordinary nonlinear differential equations. Bull. Acad. Polon. Sci. Cl. III, vol. 4 (1956) pp. 555-561.
[2] Remarks concerning criteria for uniquesness of solutions of ordinary differential equations, ibid., vol. 8 (1960) pp. 661-666.
[3] On the global stability of autonomous systems in the plane. Contributions to Differential Equations, vol. 1 (1963) pp. 389-400.

ONUCHIC, Nelson, (see also HARTMAN, P.)
[1] Applications of the topological method of Ważewski to certain problems of asymptotic behavior in ordinary differential equations. Pacific Journal of Mathematics, vol. 11 (1961), pp. 1511-1527.
[2] Relationships among the solutions of two systems of ordinary differential equations. Mich. J. Math., vol. 10 (1963) pp. 129-139.

OPIAL, Z.,
[1] Sur l'allure asymptotique des solutions de certaines équations différentielles de la mecanique nonlinéaire. Ann. Polon. Math. vol. 8 (1960) pp. 105-124.
[2] Sur la dépendence des solutions d'un système d'équations differentielles de leurs seconds membres. Ann. Polonici Math., vol. 8 (1960), pp. 75-89.

OSGOOD, W.,
[1] Beweis der Existenz einer Lösung der Differentialgleichung dy/dx = f(x,y) ohne Hinzunahme der Cauchy-Lipschitzschen Bedingung, Monatsh. Math. Phys., vol. 9, (1898) pp. 331-345.

OXTOBY, J. C.,
[1] Stepanoff flows on the torus. Proceedings of the American Mathematical Society, vol. 4 (1953), pp. 982-987.

PAPUSH, N. P.,
[1] A study of the distribution of integral curves occupying a domain containing one singular point. Mat. Sbornik, vol. 38 (1956) p. 337.

PEANO, G.,
[1] Sull' integrabilità delle equazioni differenziali di primo ordine. Att. R. Accad. Torino, vol. 21 (1885/1886) pp. 677-685.
[2] Démonstration de l'integrabilité des équations différentielles ordinaires. Math. Ann., vol. 37 (1890) pp. 182-228.

PEIXOTO, Mauricio M.,
[1] On structural stability. Ann. Math., vol. 69 (1959), pp. 199-222.
[2] Some examples on n-dimensional structural stability. Proc. Nat. Acad. Sciences U.S.A., vol. 45 (1959), pp. 633-636.
[3] Structural stability on two-dimensional manifolds. Symposium Internacional de Ecuaciones Diferenciales Ordinarias, Mexico, 1961, pp. 188-189.
[4] Structural stability on two-dimensional manifolds. Topology, vol. 1 (1962), pp. 101-120.

PEIXOTO, Marilia C.; PEIXOTO, M. M.,
[1] Structural stability in the plane with enlarged boundary conditions. Anais da Academia Brasileira de Ciencias, vol. 31 (1959), pp. 135-160.

PERRON, O.,
[1] Ueber Ein- und Mehrdeutigkeit des Integrales eines Systems von Differentialgleichungen. Math. Ann., vol. 95 (1926) pp. 98-101.
[2] Ueber Existenz und Nichtexistenz von Integralen partieller Differentialgleichungssysteme im reellen Gebiet. Math. Zeit., vol. 27 (1928) pp. 549-564.
[3] Eine hinreichende Bedingung fur die Unitat der Losung von Differentialgleichungen erster Ordnung, ibid., vol. 28 (1928) pp. 216-219.
[4] Uber Stabilität und asymptotisches Verhalten der Integrale von Differentialgleichungssystemen. Math. Z., vol. 29 (1928) pp. 129-160.
[5] Uber Stabilität und asymptotisches Verhalten der Losungen eines Systems endlicher Differenzengleichungen. J. reine angew. Math., vol. 161 (1929) pp. 41-61.
[6] Die Stabilitätsfrage bei Differentialgleichungen. Math. Z., vol. 32 (1930) pp. 703-728.

PERSIDSKII, K. P.,
[1] Au sujet du problème de stabilité. Bull. Soc. phys.-math. Kazan III vol. 5, No. 3, (1931) pp. 56-62 (French).
[2] Un theorème sur la stabilité du mouvement. Bull. Soc. phys.-math. Kazan III vol. 6, (1934) pp. 76-79.
[3] On the stability theory of the solutions of systems of differential equations. Bull. Soc. phys.-math. Kazan III vol. 8, (1936).
[4] On a theorem of Liapunov. C. R. (Doklady) Acad. Sci. URSS vol. 14, (1937) pp. 541-544.
[5] On the theory of stability of solutions of differential equations. Thesis, Moscow 1946, summary: Usp. Mat. Nauk vol. 1, No. 1, 5-6, (1946) pp. 250-255.
[6] On the stability of the solutions of an infinite system of equations. Prikl. Mat. Mek., vol. 12, (1948) pp. 597-612.
[7] On stability of solutions of a system of countably many differential equations. Izv. Akad. Nauk Kazach. SSr vol. 56, Ser. Mat. Mekh., No. 2, (1948) pp. 3-35.

[8] Countable systems of differential equations and the stability of their solutions. Uch. Zapiski Kazach. Gos. Univ. Mat. Fiz: No. 2 (1949)
[9] On stability of solutions of differential equations. Izv. Akad. Nauk Kazach. SSr vol. 60, Ser. Mat. Mekh., No. 4, (1950) pp. 3-18.
[10] On Liapunov's second method in linear normed spaces. Vestnik Akad. Nauk Kazach. SSR, No. 7, (1958) pp. 89-97.
[11] Inversion of Liapunov's second theorem on instability in linear normed spaces. Vestnik Akad. Nauk Kazach. SSR, No. 10 (1959) pp. 31-35.

PERSIDSKII, S. K.,
[1] On the second method of Liapunov. Izv. Akad. Nauk Kazach. SSR No. 4 (1956) pp. 43-47.
[2] On stability in a finite interval. Vestnik Akad. Nauk Kazach. SSR, No. 9 (1959) pp. 75-80.
[3] Some theorems on the second method of Liapunov. Vestnik Akad. Nauk Kazach. SSR, No. 2 (1960) pp. 70-76.
[4] On Liapunov's second method. Prikl. Mat. Mek., vol. 25 (1961) pp. 17-23.

PETROVSKII, I. G.,
[1] Lectures on the theory of ordinary differential equations. Fifth augmented edition. Izdah. "Nauka", Moscow, 1964, p.272 (Russian). English translation, Prentice Hall, Englewood Cliffs, N. J., 1966.

PICARD, E.,
[1] Mémoire sur la théorie des équations aux derivées partielles et la méthode des approximations successives. J. Math. Pures Appl., vol. 6 (1890) pp. 423-441.
[2] Leçons sur quelques équations fonctionelles. Gauthier-Villars, Paris (1928).
[3] Lecons sur quelques problèmes aux limites de la théorie des équations différentielles. Gauthier-Villars, Paris (1930).

PLIŚ, A.,
[1] On a topological method for studying the behavior of the integrals of ordinary differential equations. Bull. Acad. Polon. Sci. Cl., vol. 2 (1954) pp. 415-418.
[2] Characteristics of nonlinear partial differential equations, ibid., vol. 2 (1954) pp. 419-422.
[3] On sets filled by asymptotic solutions of differential equations. Am. Inst. Fourier, vol. 14 (1966) pp. 191-194.

PLISS, V. A.,
[1] A qualitative picture of the integral curves in the large and the construction with arbitrary accuracy of the region of stability of a certain system of two differential equations. Prikl. Mat. Mek., vol. 17 (1954) pp. 541-554.
[2] Certain Problems of the Theory of Stability of Motion in the Whole. Izd. Leningradsk. Univ. (1958).

POINCARÈ, Henri,
[1] Oeuvres, Gauthier-Villars, Paris (1929).
[2] Les méthodes nouvelles de la mecaniques celeste, Gauthier-Villars, Paris (1892-1899) reprint, Dover, New York.

PUGH, Charles C.,
[1] Cross-sections of solution funnels, Bull. Am. Math. Soc., vol. 70 (1964) pp. 580-583.
[2] The closing lemma and structural stability, Bull. Am. Math. Soc., vol. 70 (1964), pp. 584-587.
PUTNAM, C.R.,
[1] Unilateral stability and almost periodicity. J. Math. Mech. vol. 9 (1960), pp.915-917.
REEB, Georges,
[1] Sur certaines propriétés topologiques des trajectoires des systèmes dynamiques, Acad. Royale Belgique, Classe des Sciences, Memoires, vol. 27, no. 9, Brussels, 1952.
[2] Sur la théorie générale des systèmes dynamiques, Annales de l'Institut Fourier (Grenoble), vol. 6 (1955-56), pp. 89-115.
[3] Sur certains problèmes de topologie algébrique et de topologie générale en dynamique, Bol. Soc. Mat. Mexicana, vol. 5 (1960), pp. 199-202.
REISSIG, Rolf,
[1] Kriterien für die Zugehorigkeit dynamischer Systeme zur Klasse D. Math, Nachr. vol. 20, pp. 67-72 (1959).
[2] Stabilitätsprobleme in der qualitativen Theorie der Differentialgleichungen. J. Ber. Deutsche Math.-Verein. vol. 63, pp. 97-116 (1960).
[3] Neue Probleme und Methoden aus der qualitativen Theorie der Differentialgleichungen. M. Ber. Deutsche Akad. Wiss. vol. 2, pp. 1-8 (1960).
REISSIG, Rolf, SANSONE, G., CONTI, R.,
[1] Qualitative Theorie nichtlinearer Differentialgleichungen, Publicazioni dell'Instituto Nazionale di Alta Matematica, Edizioni Cremonese, Roma, 1963.
REMAGE' Russell, Jr.,
[1] On minimal sets in the plane, Proceedings of the American Mathematical Society, vol. 13 (1962), pp. 41-47.
ROSEAU, Maurice,
[1] Vibrations nonlinéaires et théorie de la stabilité. Springer Tracts in Nat. Phyl. vol. 8. Springer-Verlag, Berlin, New York, 1966.
ROOS, B.W.,
[1] Note on generalized dynamical systems, SIAM Review, vol. 6 (1964), pp. 269-274.
ROXIN, Emilio,
[1] Reachable zones in autonomous differential systems. Bol. Soc. Mat. Mexicana vol. 5 (1960), pp. 125-135.
[2] The existence of optimal controls. Mich. Math J. vol. 9 (1962), pp. 109-119.
[3] Axiomatic foundation of the theory of control systems, Second Int. Conf. IFAC, Basel, 1963.

[4] Stabilität in allgemeinen Regelungssystemen, Third Conference on Nonlinear Oscillations, Berlin, 1964.
[5] Stability in general control systems, J. Diff. Equa. vol. 1 (1965)pp. 115-150.
[6] On generalized dynamical systems difined by contingent equations, J. Diff. Equa. vol. 1 (1965) pp. 188-205.
[7] On Stability in Control Systems, J. SIAM Control vol. 3 (1966), pp. 357-372.
[8] Local definition of generalized control systems, Mich. Math. J. vol. 13 (1966), pp. 91-96.

ROXIN, E.O., and SPINADEL, V.W.,
[1] Reachable zones in autonomous differential systems, In "Contributions to Differential Equations," vol. 1, pp. 275-315. Interscience, New York, 1963.

SAITO, Tosiya,
[1] On the measure-proserving flow on the torus, J. Math. Soc. Jap. vol. 3 (1951), pp. 279-286.
[2] On dynamical systems in n-dimensional torus, Funkcialaj Ekvacioj, vol. 7 (1965), pp. 91-102.

SANDOR, St., WEXLER, D.,
[1] Sur la stabilité dans les systèmes dynamiques, Revue de Mathématiques Pures et Appliquées, vol. 3 (1958), pp. 325-328.

SANSONE, Giovanni, (see also CONTI, R. and REISSIG, G.)
[1] Equazioni differenziali nel campo reale, Zanichelli, Bologna (1948).

SANTORO, Paolo.,
[1] Sulle Stabilitá intermedie tra quella uniforme e quella asintotica uniforme per sistemi lineari, Le Matematiche, vol. 20(1965) pp. 41-45.

SARD, A.,
[1] The measure of the critical values of differentiable maps, Bull. Amer. Math. Soc. vol. 48 (1942), pp. 883-896.

SCHÄFFER, Juan Jorge, (see MASSERA, J.L.)

SCHAUDER, J.
[1] Der Fixpunktsatz in Funktionalraumen, Studia Math. vol. 2 (1930) pp. 171-180.

SCHWARTZ, Arthur J.,
[1] A generalization of a Poincaré-Bendixson theorem to closed two-dimensional manifolds, Amer. J. Math. vol. 85 (1963), pp. 453-458; errata, p. 753.

SCHWARTZMAN, Sol.,
[1] Asymptotic cycles, Ann. Math. vol. 66 (1957), pp. 270-284.
[2] On the existence of strongly recurrent and periodic orbits, Bol. Soc. Mat. Mexicana, vol. 5 (1960), pp. 181-183.
[3] Global cross sections of compact dynamical systems, Proc. Nat. Acad. Sciences, U.S.A., vol. 48 (1962), pp. 786-791.

SCHWEIGERT, G.E.,
[1] A note on the limits of orbits, Bull. Am. Math. Soc. vol. 46., (1940), pp. 963-969.
SEIBERT, Peter, (see also J. AUSLANDER and BHATIA, N.P.)
[1] Stability under perturbations in generalized dynamical systems, International Symposium on Nonlinear Differential Equations and Nonlinear Mechanics, Academic Press, New York and London, 1963, pp. 463-473.
[2] Zum Problem der Stabilität unter standig wirkenden Storungen bei dynamischen Systemen, Archiv der Math. vol. 15 (1964), pp. 108-114.
SEIFERT, George,
[1] Stability conditions for the existence of almost-periodic solutions of almost-periodic systems, J. Math. Anal. Appl., vol. 10 (1965), pp. 409-418.
SELL, George R. (see also DEYSACH, L.G.)
[1] Stability theory and Liapunov's second method. Arch. Rat. Mech. Anal. vol. 14 (1963), pp. 108-126.
[2] A note on the fundamental theory of ordinary differential equations, Bull. Am. Math. Soc., vol. 70 (1964), pp. 529-535.
[3] On the fundamental theory of ordinary differential equations, J. Diff. Equa. vol. 1 (1965), pp. 370-392.
[4] Periodic solutions and asymptotic stability, J. Diff. Equa., vol. 2 (1966), pp. 143-157.
[5] Nonantonomous differential equations and topological dynamics I and II. To appear.
SHCHERBAKOV, B.A. (see also BRONSTEIN, I.V.)
[1] Classifications of motions stable in the sense of Poisson. Pseudorecurrent motions, Dokl. Akad. Nauk. SSSR, vol. 146 (1962), pp. 322-324. (English Translation) Soviet Math. Dokl. pp, 1320-1322.
[2] On classes of motions stable in the sense of Poisson. Pseudorecurrent motions, Izv. Akad. Nauk Moldavskoi SSR, Serija Estestvennyh u Tehniceskih Nauk, no. 1 (1963), pp. 58-72 (Russian).
[3] Decomposition of a set of Poisson-stable motions 152 (1963), pp. 71-74 (Russian).
[4] Constituent classes of Poisson-stable motions, Sibirskii Mat. Zhurnal, vol. 5 (1964), pp. 1397-1417 (Russian).
[5] Dynamical systems. Review of papers given at the Kishinev seminar on the qualitative theory of differential equations, Differencial'nye Uravnenija, vol. 1 (1965), pp. 260-266.

[6] Minimal motions and the structure of minimal sets, Papers on Algebra and Analysis, Kishinev, 1965; pp. 99-110 (Russian).
[7] On a class of motions stable in Poisson's sense, Papers on Algebra and Analysis, Kishinev, 1965; pp. 155-160 (Russian).

SHUBIN, M.A.,
[1] On some properties of generalized ω-limit sets in dynamical systems, Vestnik Mosc. Univ. (1966) No. 3, 58-60.

SIBIRSKII, K.S.,
[1] Uniform approximation of points of dynamical limit sets and motions in them, Dokl. Akad. Nauk SSSR, vol. 146 (1962), pp. 307-309 (Russian).
[2] Uniform approximation of points and properties of motions in dynamical limiting sets, Izv. Akad. Nauk Moldavskoi SSR, Serija Estestv. Tehn. Nauk, no. 1 (1963), pp. 38-48.

SIBIRSKII, K.S.; BRONSTEIN, I.U.,
[1] Partially ordered group dynamical systems, Učebnye Zapiski Kishinev. Univ. No. 54, 1960; pp. 33-36 (Russian).

SIBIRSKII, K.S.; KRECU, V.I.; BRONSTEIN, I.U.,
[1] Liapunov stability in partially ordered dynamical systems, Učebnye Zapiski Kishinev. Univ., No. 54, 1960, pp. 29-32 (Russian).

SIBIRSKII, K.S.; STAKHI, A.M.,
[1] Limit properties of partially ordered dispersive dynamical systems, Izv. Akad. Nauk Moldavskoi SSR, Serija Estestv. Tehn. Nauk, no. 11 (1963), pp. 42-49 (Russian).
[2] On partial ordering of groups, Papers on Algebra and Analysis, Kishinev, 1964. (Russian).

SIEGEL, Carl Ludwig,
[1] Note on differential equations on the torus, Ann. Math., vol. 46 (1945), pp. 423-428.

SMALE, Stephen,
[1] Morse inequalities for a dynamical system, Bul. Am. Math. Soc., vol. 66 (1960), pp. 43-49.
[2] On gradient dynamical systems, Ann. Math. vol. 74 (1961), pp. 199-206.
[3] On dynamical systems, Symposium Internacional de Ecuaciones, Diferenciales Ordinarias, Mexico, 1961, pp. 195-198.
[4] Generalized Poincare' conjecture in dimensions greater than four, Am. Math. vol. 74 (1961), pp. 361-406.

SOLNCEV, Yu.K.,
[1] Two examples of dynamical systems determined by infinite systems of differential equations,

Učenye Zapiski Moskov. Gos. Univ., vol. 155 (1952). Matematika, No. 5, pp. 156-167 (Russian).

SHOLOHOVICH, F. A., (see also BARBASHIN, E.A.)

[1] The relationship between a linear dynamical system and a certain differential equation in Banach space. Dokl. Akad. Nauk SSSR, vol. 120 (1958), pp. 43-46, (Russian).

[2] Linear Dynamic systems, Izvestiya Vysših Učebnyk Zavedenii Matematika, 1957, No. 1, pp. 249-257 (Russian).

SPINADEL, V.W., (see ROXIN, E.O.)

STAKHI, A.M. (see SIBIRSKII, K.S.)

STALLINGS, J.,

[1] Polyhedral homotopy sphere, Bull. Am. Math. Sec. vol. 66 (1960), pp. 485-488.

[2] The piecewise linear structure of Euclidean space, Proc. Cambr. Phil. Soc. vol. 58 (1962), pp. 481-488.

STEPANOV, V.V. (see also BEBUTOV, M.V., and NEMYTSKII, V.V.)

[1] Lehrbuch der Differentialgleichungen, VEB Deutscher Verlag der Wiss. Berlin 1956, German translation of the Russian edition. Gos. Iz. Tekh. Teor. Lit., Moskow 1953.

STERNBERG, Shlomo, (see also BRAUER, F.)

[1] On differential equations on the torus, Am. J. Math. vol. 79 (1957), pp. 397-402.

[2] On Poincare's last geometrical theorem, Proc. Am. Math. Soc., vol. 8 (1957) pp, 787-789.

[3] On the structure of local homeomorphisms of Euclidean n-space. Am. J. Math., vol. 80 (1958), pp. 623-631.

STRAUSS, Aaron, (see also JONES, G.S.)

[1] Continuous dependence of solutions of ordinary differential equations, Amer. Math Monthly June 1964.

[2] Liapunov functions and L^p solutions of differential equations, Trans. AMS, vol 119 (1965) pp.37-50.

[3] Liapunov functions and global existence, Bull. AMS, vol. 71, pp. 519-520 (1965).

[4] On the stability of a perturbed nonlinear equation, Proc. AMS, vol 17 (1966), pp 803-807.

[5] A Geometric Introduction to Liapunov's Second Method, Proc. NATO Advanced Institute, Padova, 1965.

STRAUSS, A. and YORKE, J.A.

[1] Perturbation theorems for ordinary differential equations, J. Diff. Eq., vol 3 (1967), pp 15-30.

[2] On asymptotically antonomous differential equations. Math. Systems Theory, to appear.

SZARSKI, J.,

[1] Remarque sur un critère d'unicité des intégrales d'une équation différentielles ordinaire, Ann. Polon. Math. vol. 12 (1962), pp. 203-205.

SZEGÖ, George, P., (see also BHATIA, N.P. and JONES, G.S.)
[1] Contributions to Liapunov's Second Method: Nonlinear Autonomous Systems , Proc. Int. Symp. Nonlinear Diff. Eqs., Academic Press, 1963, pp. 421-430.
[2] Contributions to Liapunov's Second Method: Nonlinear Autonomous Systems , Jour. Basic. Eng. Trans. ASME(D) 84 (1962) pp. 571-578.
[3] On a New Partial Differential Equation for the Stability Analysis of Time-invariant Control Systems SIAM J. Control 1 (1962), pp. 63-75.
[4] On Global Stability Properties of Nonlinear Control Systems , Rept. Contract NONR-1228 (23) Sept., 1964.
[5] New Theorems on Stability and Attraction , Proc. NATO Adv. Study Insti., Padova, 1965, Oderisi, Gubbio, 1966.

SZEGÖ, G.P.; BHATIA, N.P.,
[1] An extension theorem for asymptotic stability and its application to a control problem. Math. Symp. Diff. Eqs. and Dyn.,Sy. Mayaguez, Puerto Rico, 1965. To be published in the proceedings.

SZEGÖ, G.P., GEISS,Gunther, R.,
[1] A Remark on 'A New Partial Differential Equation for the Stability Analysis on Time-invariant Control Systems' , SIAM J. Control 1 (1963) pp. 369-376.

SZMYDT, Z.,
[1] Sur la structure de l'ensemble engendré par les intégrales tendant vers le point singulier du système d'équations différentielles. Bull. Acad. Polon. Sci. Cl. III, vol. 1 (1953) pp. 223.
[2] Sur les systèmes d'équation différentielles dont toutes les solutions sont bornées. Am. Pol. Math. vol. 2 (1955) pp. 234-236.

SZMYDTOWNA, Z.,
[1] Sur l'allure asymptotique des intégrales des équations différentielles ordinaires, Ann. Soc. Polon. Math. vol. 24 (1951), pp. 17-34.

TAAM, C.T.,
[1] Asymptotic relations between systems of differential equations, Pacific J. Math. vol. 6 (1956), pp. 373-388.

TA, Li.,
[1] Die Stabilitätsfrage bei Differenzengleichungen. Acta math. vol. 63, pp. 99-141 (1934).

TONELLI, Leonida,
[1] Sull' unicità della soluzione di un' equazione differenziale ordinaria. Rend. R. Acad. Naz. Lincei, Ser. 6, vol. 1, (1925) pp. 272-277.
[2] Opera scelte (Selected Works), Published by UMI, Edited by S. Cinquini, vol. 3 Cremonese, Roma (1960-1962).

TRJITZINSKY, W.J.,
[1] Problèmes dans la théorie des systèmes dynamiques, Acta Math., vol. 95 (1956), pp. 191-289.
[2] Aspects topologiques de la théorie des fonctions réeles et quelques conséquences dynamiques, Annali di Matematica Pura ed Applicata, vol. 42 (1956), pp. 51-117.

TROICKII, S.,
[1] On dynamical systems defined by an everywhere dense set of recurrent motions, Učenye Žapiski Moskov. Gos. Univ., vol. 15 (1939), Matematika, (Russian).

TURAN, P.,
[1] On the instability of systems of differential equations Acta, Math. Acad. Sci. Hung., vol. 6 (1955), pp. 257-270.
[2] On the property of the stable or conditionally stable solutions of systems of nonlinear differential equations, Ann. di. Mat. pura ed appl. vol. 48 (1959), pp. 333-340.

TUROWICZ, A.,
[1] Sur les trajectoires et les quasitrajectoires des systemes de commande nonlinéaires. Bull. Acad. Polon. Sci. vol. 10 (1962), pp. 529-531.
[2] Sur les zones d'émission des trajectoires et des quasitrajectoires des systèmes de commande nonlinéaires. Bull. Acad. Polon. Sci. vol. 11 (1963) pp. 47-50.

URA, Taro,
[1] Sur les courbes définies à la surface du tore par des équations admettant un invariant intégral, Annales Scient. École Normale Supérieure, vol. 69 (1952), pp. 259-275.
[2] Sur les courbes définies par les équations différentielles dans l'espace à m dimensions, Annales Scient. École Normale Superieure, vol. 70 (1953) pp. 287-360.
[3] Sur les periodes fundamentales de solutions periodiques. Comment. Math. Univ. Sancti Pauli, vol. 4 (1955) pp. 113-130.
[4] Sur le courant exterieur a une region invariante; prolongements d'une caracteristique et l'ordre de stabilité, Funkcialaj Ekvacioj, vol.2 (1959)pp. 143-200.
[5] Stability and limit sets, RIAS Technical Report No. 62-4, February 1962, 24 pp.
[6] Stability of a closed invariant set and its nearby lying invariant sets, RIAS Technical Report No. 62-14, July 1962, 61 pp.
[7] On the flow outside a closed invariant set; stability, relative stability and saddle sets, Contr. Diff. Equations, vol. 3 (1964), pp. 249-294.

URA, Taro; HIRASAWA, Yoshikazu,
[1] Sur les points singuliers des equations differentielles admettant in invariant integral, Proc. Jap. Acad. vol. 30 (1954), pp. 726-730.

URA, Taro; KIMURA, Ikuo,
[1] Sur le courant exterieur a une region invariante; Théorème de Bendixson, Comment. Math. Univers. Sancti Pauli, vol. 8 (1960), pp. 23-39.
[2] Stability in topological synamics, Proc. Jap. Acad., vol. 40 (1964), pp. 703-706.

VEJVODA, O.,
[1] The stability of a system of differential equation in the complex domain. Czechoslov. Math.J. vol. 7 (82), pp. 137-159 (1957).

VINOGRAD, R.E.,
[1] On the limiting behavior of unbounded integral curves, Dokl. Akad. Nauk SSSR, vol. 66 (1949), pp. 5-8 (Russian).
[2] On the limit behavior of an unbounded integral curve, Učenye Zapiski Moskov. Gos. Univ., vol. 155 (1952), Matematika, No. 5, pp. 94-136 (Russian).

VINOKUROV, V.R.,
[1] On the definition of a dynamic limit point in general dynamical systems, Izvestija Vysših Učebnyh Zavedenii, Matematika, 1964 No. 3 (40) pp. 36-38 (Russian).

VRKOC, J., (see also KURZWEIL, J.L.,)
[1] On the inverse theorem of Chetaev. Czechoslov. Math. J. vol. 5 (80), pp. 451-461.
[2] Integral stability. Czechoslov. Math. J. vol. 9 (84), pp. 71-129 (1959) (Russian).

VRUBLEVSKAYA, I.N.,
[1] On trajectories and limiting sets of dynamical systems, Dokl. Akad. Nauk SSSR, vol. 97 (1954) pp. 9-12 (Russian).
[2] Some criteria of equivalence of trajectories and semitrajectories of dynamical systems, Dokl. Akad. Nauk SSSR, vol. 97 (1954), pp. 197-200 (Russian).
[3] On geometric equivalence of the trajectories of dynamical systems, Mat. Sbornik, vol. 42 (1957), pp. 361-424 (Russian).

WALTER, W.,
[1] Bemerkung zu verschiedenen Eindentigkeits-kriterien für gewöhnliche Differentialgleichungen. Math. Z. vol. 84 (1964), pp. 222-227.

WAŻEWSKI, T.,
[1] Sur l'unicité et la limitation des integrales des équations aux dérivées partielles du premier ordre, Atti R. Accad. Naz. Lincei Rend. Cl. Sci. Fis. Mat. Nat. (6) vol. 18 (1933), pp. 372-376.

[2] Sur l'evalutation du domaine d'existence des fonctions implicites réelles ou complexes, Ann. Soc. Polon. Math. vol. 20 (1947), pp. 81-125.
[3] Sur un principe topologique de l'examen de l'allure asymptotique des intégrales des équations différentielles ordinaires, ibid. vol. 20 (1947), pp. 279-313.
[4] Sur les intégrales d'un systeme d'équations différentielles ordinaires, ibid vol. 21 (1948), pp. 277-297.
[5] Sur quelques definitions équivalentes des quasitrajectoires des systèmes de commande. Bull. Acad. Polon. Sci. vol. 40 (1963), pp. 460-474.

WAŻEWSKI, T., ZAREMBA, S.,
[1] Sur les ensembles de condensation des caracteristiques d'un système d'équations différentielles ordinaires, Ann. Soc. Polonaise de Math. vol. 15 (1936), pp. 24-33.

WEXLER, D. (see also SANDOR, St.),
[1] Stability theorems for a system of stationary differential equations. Rev. Math. pur. appl. vol. 3, pp. 131-138 (1958).

WINTNER, A., (see also HARTMAN, P.),
[1] The nonlocal existence problem of ordinary differential equations, Amer. J. Math. vol. 67 (1945) pp. 277-284.
[2] On the convergence of successive approximations, ibid. vol. 68 (1946) pp. 13-19.
[3] Asymptotic equilibria, ibid. vol. 68 (1946) pp. 125-132.
[4] The infinities in the nonlocal existence problem of ordinary differential equations, ibid. vol. 68 (1946) pp. 173-178.
[5] On the local uniqueness of the initial value problem of the differential equation $d^n x/dt^n = f(t,x)$, Boll. Un. Mat. Ital. (3) vol. 11 (1956) pp. 496-498.

WHITNEY, Hassler,
[1] Regular families of curves, Proc. Nat. Acad. Sciences U.S.A. vol. 18 (1932), I, pp. 275-278; II. pp. 340-342.
[2] Regular families of curves, Ann. Math. vol. 34 (1933), pp. 244-270.
[3] On regular families of curves, Bull. Am. Math. Soc. vol. 47 (1941), pp. 145-147.

WHYBURN, G.T.,
[1] Analytic Topology, Am. Math. Soc. Colloquium Publications, vol. 28, New York, 1942.

WU, Ta-Sun,
[1] Continuous flows with closed orbits, Duke Math. J. vol. 31 (1964), pp. 463-469.
[2] Proximal relations in topological dynamics, Proc. Am. Math. Soc. vol. 16 (1965), pp. 513-514.
[3] Left almost periodicity does not imply right almost periodicity, Bull. Am. Math. Soc., vol. 72 (1966) pp. 314-316.

YAMABE, H., (see MARKUS, L.),
YORKE, James, A., (see also STRAUSS, A.,)
[1] Asymptotic preperties of solutions using the second derivative of a Liapunov function. Ph.D. Thesis, University of Maryland, College Park, M.D. June 1966.
[2] Liapunov functions and the existence of solutions tending to zero, Math. Systems Theory (to appear).
YOSHIZAWA, Taro,
[1] On the stability of solutions of a system of differential equations. Mem. Coll. Sci. Univ. Kyoto A. vol. 29, pp. 27-33 (1955).
[2] Note on the solutions of a system of differential equations. Mem. Coll. Sci. Kyoto A, vol. 29, pp. 249-273 (1955).
[3] On the equiasymptotic stability in the large. Mem. Coll. Sci. A. vol. 32, pp. 171-180 (1959).
[4] "Liapunov's functions and boundedness of solutions." Funkcialaj Ekvacioj, vol 2, pp. 95-142 (1959).
[5] Stability and boundedness of systems. Arch. rat. Mech. Anal. vol. 6, pp. 409-421(1960).
[6] Asymptotic behavior of solutions of nonautonomous systems near sets, J. Math. Kyoto University, vol. 1 (1962), pp. 303-323.
[7] Stability of sets and perturbed systems, Funkcialaj Ekvacioj, vol. 5 (1963), pp. 31-69.
[8] Asymptotic stability of solution of an almost periodic system of functional-differential equations. Rend. Circ. Mat. Palermo Ser. II, vol. 13 (1964) pp. 1-13.
[9] Eventual properties and quasi-asymptotic stability of a non-compact set. Funkcialaj Ekvacioj vol. 8 (1966), pp. 79-90.
[10] The Stability Theory by Liapunov's Second Method (book) Mathematical Society of Japan, Tokyo 1966.
ZAREMBA, M.S.C. (see also Ważewski, T.)
[1] Sur les équations au paratingent. Bull. Sci. Math. I. vol. 60 (1936), pp. 139-160.
ZEEMAN C.,
[1] The generalized Poincarè conjecture, Bull. Am. Math. Soc. vol. 67 (1961), pp. 270.
ZHIDKOV, N.P.,
[1] Certain properties of discrete dynamical systems, Uchenye Zapiski Moskov. Gos. Univ. no. 163, Matematika, vol. 6 (1952), pp. 31-59 (Russian).
ZIPPIN, Leo,
[1] Transformation groups, Lectures in Topology, Ann Arbor, 1941, pp. 191-221.

ZUBOV, V.I.,
[1] Some sufficient criteria for stability of a nonlinear system of differential equations. Prikl. Math. Mech. vol. 17, pp. 506-508 (1953).
[2] On the theory of A.M. Liapunov's second method. Dokl. Akad. Nauk. SSSR. vol. 99, pp. 341-344. (1954).
[3] Questions of the theory of Liapunov's second method, construction of a general solution in the region of asymptotic stability. Prikl. Math. Mech., vol. 19, pp. 179-210, (1955).
[4] On the theory of A.M. Liapunov's second method. Dokl. Akad. Nauk. SSSR., vol. 100, pp. 857-859 (1955).
[5] An investigation of the stability problem of systems of equations with homogeneous right hand members. Dokl. Akad. Nauk. SSSR., vol. 114, pp. 942-944. (1957).
[6] The Methods of Liapunov and their Applications Leningrad 1957, English Translation: Noordhoff, Groningen, The Netherlands, 1964.
[7] Conditions for asymptotic stability in the case of nonstationary motions and estimate of the rate of decrease of the general solution. Vestnik Leningradsk. Univ. Ser. Mat. Mekh. Astron, vol. 12, No. 1, pp. 110-129 (1957)
[8] On a method of investigating the stability of a null-solution in doubtful cases. Prikl. Math. Mech., vol. 22, pp. 46-49 (1958).
[9] On stability conditions in a finite time interval and on the computation of the length of that interval. Bull. Inst. Politechn. Iaşi, N.S. vol. 4 (8), pp. 69-74, (1958) (German summary).
[10] Mathematical Methods for the Investigation of Systems of Automatic Control. Gos. Sojus. Izd. Sudostroit. Promysl., Leningrad 1959, p. 324 English Translation, Pergamon, Oxford, 1964.
[11] Some problems in stability of motion. Mat. Sbornik vol 48 (90), pp. 149-190 (1959).
[12] On the theory of recurrent functions, Sibirskii Mat. Zh. vol. 3 (1962), pp. 532-560 (Russian).

INDEX